LIPOLYTIC ENZYMES

Lipolytic Enzymes

Hans Brockerhoff

Department of Neurochemistry
New York State Institute for Basic Research
in Mental Retardation
Staten Island, New York

Robert G. Jensen

Department of Nutritional Sciences
University of Connecticut
Storrs, Connecticut

Academic Press   *New York San Francisco London   1974*

A Subsidiary of Harcourt Brace Jovanovich, Publishers

ACADEMIC PRESS, INC.
111 Fifth Avenue, New York, New York 10003

United Kingdom Edition published by
ACADEMIC PRESS, INC. (LONDON) LTD.
24/28 Oval Road. London NW1

Library of Congress Cataloging in Publication Data

Brockerhoff, H
     Lipolytic enzymes.   I.   Jensen, Robert Gordon,
Date   joint author.   II.   Title.   [DNLM: 1. Enzy-
mes.   2.   Lipase.   3.   Phospholipase.   QU135 B864L 1974]
QP609.L55B76   574.1'9253   73-18992
ISBN 0−12−134550−5

Contents

Preface

Lipids constitute a large part of the biomass of the earth, and lipolytic enzymes play an essential role in the turnover of this material. Apart from their general biological significance, these enzymes are especially important in the fields of nutrition, food technology, clinical medicine, analytical and preparative lipid chemistry, and biochemistry, and, above all, in research on disturbances of lipid metabolism and the diseases of the circulatory and the nervous system.

Recent advances in protein chemistry, preparative biochemistry, and enzymology have benefited research on lipolytic enzymes. Some can now be isolated in a pure state. Biochemical research on these enzymes is now concentrated on their structure and catalytic mechanism, and especially on the property that distinguishes them from most other enzymes: the ability to act at interfaces on substrates that are insoluble in water. An understanding of how lipases and phospholipases function under this condition may lead to a better understanding of general intracellular metabolism, which also takes place, to a large degree, at interfaces, namely at biological membranes.

We have attempted to summarize the present knowledge of lipolytic enzymes in this book. We have discussed only those publications which seemed to offer some information on the biochemistry of the enzymes. Two enzymes have been given particularly lengthy treatment, pancreatic lipase and phospholipase 2, not only because they are the most thoroughly investigated lipolytic enzymes, but also because they may serve as models for all others.

Thanks are due to these colleagues who have obliged us by critically reviewing parts of the manuscript: J. A. Alford, H. Blair, G. Assmann, B. Borgström, G. H. de Haas, D. J. Hanahan, R. L. Ory, H. van den Bosch, R. Verger, and M. A. Wells. We are especially grateful to Drs. G. Assmann, G. H. de Haas, L. Sarda, H. van den Bosch, and R. Verger for letting us read several pertinent manuscripts before their publication.

Hans Brockerhoff
Robert G. Jensen

Abbreviations and Definitions

DFP	diisopropyl fluorophosphate
DG	diglyceride
DNP	diethyl-p-nitrophenyl phosphate
DPG	diphosphatidylglycerol
EDTA	ethylenediaminetetraacetic acid (sodium salt)
FFA	free fatty acids
GDG	galactosyl diglyceride
GGDG	digalactosyl diglyceride
GPC	sn-glycerol-3-phosphorylcholine
HDL	high density lipoproteins
"K_m"	apparent Michaelis constant of lipolytic reactions
LDL	low density lipoproteins
LPL	lipoprotein lipase
MG	monoglyceride
NEM	N-ethylmaleimide
PA	phosphatidic acid
PC	phosphatidylcholine
PCMB	p-chloromercuribenzoate
PE	phosphatidylethanolamine
PG	phosphatidylglycerol
PI	phosphatidylinositol
PS	phosphatidylserine
specific activity (s.a.)	μmoles fatty acids produced/min/mg of protein (unless otherwise defined)
SDS	sodium dodecyl sulfate
TC	taurocholate

TEAE-cellulose	triethylaminoethyl cellulose
TG	triglyceride
Units	units of enzymatic activity $= \mu$moles substrates transformed/min or μmoles fatty acid produced/min (unless otherwise defined)
V	maximal rate of enzymatic hydrolysis
VLDL	very low density lipoproteins

LIPOLYTIC
ENZYMES

I

Introduction

Lipolytic enzymes are indispensible for the biological turnover of lipids. They are required as digestive enzymes in the transfer of lipid from one organism to another, that is, from plant to animal and from animal to animal. Within the organisms, they are instrumental in the deposition and mobilization of the fat that is used as an energy reservoir, and they are also involved in the metabolism of intracellular lipids and therefore in the functioning of biological membranes. Often the enzymes are found where their importance is less obvious, as in the venoms of reptiles and invertebrates and among extracellular microbial enzymes.

Lipolytic enzymes may be defined as long-chain fatty acid ester hydrolases, with "long-chain fatty acids" meaning aliphatic acids, saturated or unsaturated, with twelve or more carbon atoms. The most "average" fatty acid chemically, physically, and biologically, being oleic acid, it can be said that any esterase capable of hydrolyzing esters of oleic acid is a lipolytic enzyme. The alcohol moiety of the ester may be glycerol (lipases) or a glycerol derivative (phospholipases 1 and 2, lysophospholipase, galactolipase) or a sterol (cholesterol esterase). While these esters are the natural, or physiological, substrates, most lipolytic enzymes will also hydrolyze artificial esters with often quite dissimilar structures.

In this book we have also included as "lipolytic enzymes" several phosphoric ester hydrolases and an amide hydrolase, partly because these enzymes have historically been included in the group and partly because they are poor relatives (phospholipases 3 and 4, sphingomyelinase) and an orphan (ceramidase) that have nowhere else to go. However, we have excluded the glycosidases which split the various carbohydrates from glycolipids and conjugated sterols, or enzymes that act as remote from the

1

lipid part of a complex lipid as, for instance, a triphosphoinositide phosphatase.

IMPORTANCE OF LIPOLYTIC ENZYMES

Apart from their obvious biological importance, there are other, more specific, reasons that make the study of lipolytic enzymes interesting and rewarding. The medical and therapeutic applications come to mind first. The manipulation of lipolytic activities will probably play a part in future methods for treating malfunctions of fat metabolism and thus control cardiovascular diseases. It may be mentioned here that pancreatic lipase is necessary for the absorption of fat, cholesterol esterases for the absorption of cholesterol esters and perhaps of cholesterol, hormone-sensitive lipase for the mobilization of fat from adipose tissue, and pancreatic and intracellular phospholipases for the absorption and metabolism of polyunsaturated fatty acids. The assay of serum lipase is a clinical tool of some importance. Congenital hyperlipemia (Type I hyperlipoproteinemia) may arise from a deficiency in lipoprotein lipase, and lipid-storage diseases have been connected with other enzymes described here, cholesterol esterase and sphingomyelinase. Even more apt to provoke speculations and hopes is the fact that cholesterol fatty acid esters are prominent components of atherosclerotic plaques and that the myelination of the developing brain correlates with a decrease in the cholesterol ester content.

Of similar interest are the practical aspects of lipases in food and other industries. Uncontrolled lipolysis can cause a flavor defect in milk and cause an accumulation of free fatty acids of (FFA) in, for example, cottonseed oil, requiring an extra centrifuge operation for removal. On the other hand, cheese flavor is due in part to short-chain fatty acids hydrolyzed by both microbial and milk lipases. The pungent and characteristic flavor of Italian cheeses is produced by the deliberate addition to the milk of pregastric esterase, a lipase that preferentially releases short-chain acids. Microbial lipases are used in combination with other enzymes to help degrade sewage. A recently marketed test kit for triglyceride glycerol employs a lipase to provide the glycerol. An acid lipase has been added to bread dough for the uniform production of monoglycerides; the monoglycerides greatly improve the resistance of the bread to staling. These few examples emphasize the industrial importance of controlled and selective lipolysis and suggest additional promising developments.

For the chemist interested in lipids, whether working in industrial or biochemical research, lipolytic enzymes have for years been important as analytical and synthetic tools. Pancreatic lipase and snake venom phospholipase 2 in particular are used extensively to determine the "positional distribution" of fatty acids in the glyceride molecule; such methods are now indispensible routines in many investigations of fats and fat metabolism. Because of their positional specificity, the enzymes are also used to prepare intermediates in the chemical synthesis of lipids. The search for enzymes with different specificities has been partly successful and is still going on.

Phospholipases, especially types 2 and 3, can attack and degrade biological membranes in a controllable fashion by virtue of their selective and precise action on phospholipids, and these enzymes are therefore widely used in experimental biochemistry, for instance, in studies on red cell or mitochondrial membranes.

For the biochemist, perhaps the most important and fascinating aspect of lipolytic enzymes is the unique physicochemical character of the reactions they catalyze. Since the enzymes are water soluble but the substrates are not, these reactions have to take place at an oil–water or micelle–water interface. Conventional enzyme kinetics is of little relevance under these conditions. On the other hand, a satisfactory system of interface kinetics is not yet available; but the uniqueness of the situation has now been generally recognized, and tentative molecular and kinetic models of enzymatic action at the interface have been advanced. There is hope that the next years will bring not only further developments of such models, but especially new experimental methods to put enzymology at the interface on a secure quantitative basis. Such an "interfacial enzymology" would be of universal importance for biological science because in nature, as opposed to the laboratory, enzymatic reactions more often take place at membranes—that is, interfaces—than in aqueous solutions. As a damper for overenthusiasm, it must be conceded that the analogy between the interfacial lipolysis of an oil and intracellular-interfacial enzymatic processes is not altogether perfect; the chapter on kinetics (Chapter III) will enlarge on this question. The expectation can still be maintained, however, that the development of the kinetics of lipolysis and an understanding of the orientation and action of lipolytic enzymes in interfaces will illuminate much more general and extended areas of biochemistry.

II

Nomenclature

All lipolytic enzymes are hydrolases and as such belong to enzyme class 3 according to the classification recommended by the Enzyme Commission of the International Union of Biochemistry (Florkin and Stotz, 1965). Apart from the glycosidases, which split the carbohydrate residues from glycolipids and which are not discussed in this book, all lipolytic enzymes are ester hydrolases, enzyme group 3.1, with the single exception of the ceramidase, a C-N hydrolyzing enzyme (EC 3.5). "Lipolytic enzymes" in the narrowest sense hydrolyze esters of fatty acids and are therefore carboxyl ester hydrolases (EC 3.1.1). The phospholipases 3 and 4, included here, are phosphoric diester hydrolases (EC 3.1.4); phosphatidate phosphohydrolase is a phosphoric monoester hydrolase (EC 3.1.3).

I. SUBSTRATES

After their classification according to the type of bond they hydrolyze, the enzymes are further classified according to the substrates on which they act. Lipolytic enzymes (with the exception of phospholipase 4, sphingomyelinase, cholesterol esterase, and ceramidase) all act on esters of glycerol, which is the molecular backbone of most of the lipid material occurring in nature. Glycerol is a trihydroxy compound with two primary and one secondary hydroxyl groups. The two primary groups are sterically distinct, although the molecule has a plane of symmetry (Hirschmann, 1960); substitution with two different substituents will lead to optically active derivatives. In a nomenclature that has been generally adopted (IUPAC–IUB Commission on Biochemical Nomenclature, 1967), glycerol is written in a Fischer projection with the secondary hydroxyl to the left, and the carbon

4

Position

$$
\begin{array}{ccc}
\text{H} & & \\
\text{HCOH} & \text{H}_2\text{COH} & 1 \\
\text{HO} \blacktriangleright \text{C} \blacktriangleleft \text{H} \quad \text{or} & \text{HOCH} & 2 \\
\text{HCOH} & \text{H}_2\text{COH} & 3 \\
\text{H} & &
\end{array}
$$

sn-Glycerol

atoms are numbered 1, 2, and 3 from top to bottom. Glycerol so numbered is named *sn*-glycerol (i.e., stereospecifically numbered). The unambiguous description of isomeric glycerides becomes now possible. For instance, one of the six possible isomers of a triglyceride containing one residue each of palmitic, oleic, and stearic acid may be described as 1-palmitoyl-2-oleoyl-3-stearoyl-*sn*-glycerol, or *sn*-glycerol-1-palmitate-2-oleate-3-stearate.

$$
\text{C}_{17}\text{H}_{33}\overset{\overset{\text{O}}{\|}}{\text{C}}\text{O} - \left[\begin{array}{l} -\overset{\overset{\text{O}}{\|}}{\text{O}}\text{CC}_{15}\text{H}_{31} \\ \\ -\overset{\overset{\text{O}}{\|}}{\text{O}}\text{CC}_{17}\text{H}_{35} \end{array} \right. \quad \text{or} \quad \text{Oleoyl} - \left[\begin{array}{l} \text{Palmitoyl} \\ \\ \text{Stearoyl} \end{array} \right. \quad \text{or} \quad \text{O} - \left[\begin{array}{l} \text{P} \\ \\ \text{S} \end{array} \right.
$$

1-Palmitoyl-2-oleoyl-3-stearoyl-*sn*-glycerol

In this book, we shall use an abbreviated nomenclature for fatty acids (Holman, 1966). They are described by configuration and location of the double bond—length of the chain: number of double bonds. For example, *cis* 9, *cis* 12-18:2 is *cis*-Δ9, *cis*-Δ12-octadecadienoic (linoleic) acid; 16:0 is hexadecanoic (palmitic) acid. Since most natural unsaturated acids have cis double bonds, this sign can usually be omitted, and the location of the double bonds is also usually understood without being spelled out; e.g., 18:2 is linoleic, 18:3 linolenic acid. The double bonds of natural fatty acid, it should be remembered, are usually separated by methylene (CH_2) groups. A triglyceride can now be written in an abbreviated and sterically defined formula.

$$
18:1 - \left[\begin{array}{l} 16:0 \\ \\ 18:0 \end{array} \right. \quad \text{or} \quad sn\text{-}16:0\text{-}18:1\text{-}18:0
$$

If the triglyceride is a racemate of *sn*-16:0-18:1-18:0 and *sn*-18:0-18:1-16:0, it is written *rac*-16:0-18:1-18:0 or *rac*-18:0-18:1-16:0. This racemate and its two enantiomers represent the triglycerides that are of most value for the characterization of lipolytic enzymes. A more thorough treatment of triglyceride nomenclature is given by Litchfield (1972).

Diglycerides are diacylglycerols. A monoacid (e.g., palmitic acid) diglyceride can occur in three isomeric forms:

$$
16{:}0 -\left[\begin{array}{l}16{:}0 \\ OH\end{array}\right. \qquad
16{:}0 -\left[\begin{array}{l}OH \\ 16{:}0\end{array}\right. \qquad
HO -\left[\begin{array}{l}16{:}0 \\ 16{:}0\end{array}\right.
$$

$$
1,2\text{-} \qquad\qquad\qquad 2,3\text{-} \qquad\qquad\qquad 1,3\text{-}
$$

Diglyceride

The first two diglycerides form a racemate, *rac*-1,2 or 1,2(2,3)-diglyceride; this racemate is often the first produce of the hydrolysis of a triglyceride by a lipase.

Monoglycerides (monacylglycerol) can occur as the 1-, 2-, or 3-isomer; the mixture of the 1- and 3-isomers is often loosely named 1-monoglyceride. Alternative names are α-monoglyceride for the mixture of 1 and 3; and β-monoglyceride for the 2-isomer.

Natural phosphoglycerides (with one exception that is of no interest here) are all of the *sn*-glycerol-3-phosphate type.

$$
\begin{array}{cl}
\text{Position} & \\
1 & \\
2 & \overset{\overset{\textstyle O}{\|}}{RCO} -\left[\begin{array}{l} \overset{\overset{\textstyle O}{\|}}{OCR} \\ \\ \underset{\underset{\textstyle O^-}{|}}{\overset{\overset{\textstyle O}{\|}}{OPO}} -X \end{array}\right. \\
3 &
\end{array}
$$

The terms "phosphoglyceride" and "phospholipid" are usually interchangeable, although "phospholipid" also includes the phosphate-containing sphingolipids, in particular, sphingomyelin. If there is the possibility of confusion, the narrower terms should be used.

Phosphoglycerides contain four ester bonds: two carboxyl esters, positionally different, in *sn* 1 and 2; and two phosphate ester bonds, one toward the hydrophobic diglyceride moiety, the other toward the hydrophilic substituent X which is either a polyhydroxy or an amino alcohol. If X is hydrogen, the phosphoglyceride is the monophosphate ester, phosphatidic acid. Table II-1 lists the more common substituents X and the names of the corresponding phospholipids.

Lysophosphoglycerides contain only one fatty acid, either in position 1 or 2. The 1-acyl lysophospholipids are better known in the laboratory; "lysolecithin," for example, usually refers to the 1-acyl compound (lysoPC); but both isomers are important in intermediary metabolism.

"Galactolipids" are 1,2-diacyl glycerols that carry galactose or galactosylgalactose in a glycosidic linkage in position 3 of the glycerol. A related

TABLE II-1

Nomenclature of Phosphoglycerides According to their Hydrophilic Substituent X

X	Phosphoglyceride	Abbre-viation	Trivial name
H^+	Phosphatidic acid	PA	
$CH_2CH_2N^+(CH_3)_3$	Phosphatidylcholine	PC	Lecithin
$CH_2CH_2NH_2$	Phosphatidylethanolamine	PE	(Cephalin)
$CH_2CH(NH_3)^+COO^-$	Phosphatidylserine	PS	(Cephalin)
$CH_2CHOHCH_2OH$	Phosphatidylglycerol	PG	
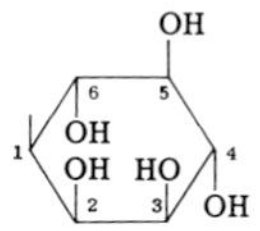	Phosphatidylinositol	PI	Monophosphoinositide
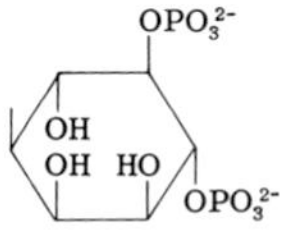	Phosphatidylinositol 4,5-diphosphate	PIP_2	Triphosphoinositide
$-CH_2CHOHCH_2$ / O / Phosphatidic acid	Diphosphatidylglycerol	DPG	Cardiolipin

compound is the "sulfolipid." Formulas are given in the chapter on glycosyldiglyceride lipases (Chapter IV, Section XI).

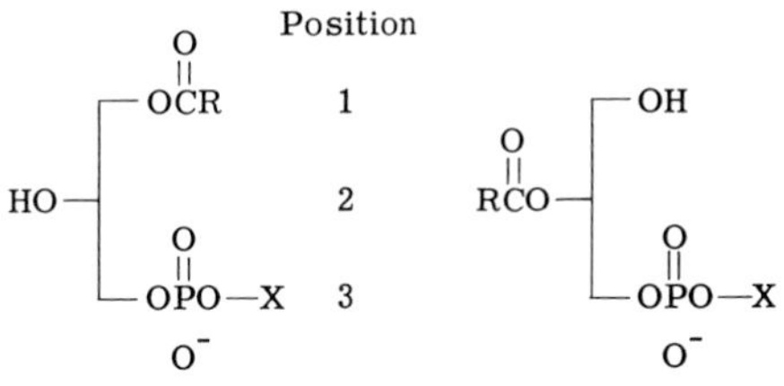

1-Acyl- 2-Acyl-
-glycerylphosphoryl-X
(lysophospholipid)

The general formula of sphingomyelins is shown in the introduction to phospholipases (Chapter VI). The molecule has two phosphate ester bonds corresponding to those of the phosphoglycerides, and in addition a fatty acid amide bond.

TABLE II-2

Lipolytic Enzymes

EC	Systematic name	Trivial name
3.1.1.3	Glycerol-ester hydrolase	Lipase
3.1.1.13	Sterol-ester hydrolase	Cholesterol esterase
3.1.1.4	Phosphoglyceride 2-acyl-hydrolase	Phospholipase 2
	Phosphoglyceride 1-acyl-hydrolase	Phospholipase 1
3.1.1.5	Lysophosphoglyceride acyl-hydrolase	Lysophospholipase
3.1.3.4.	Phosphatidate phosphohydrolase	Phosphatidic acid phosphatase
3.1.4.3	Phosphoglyceride diglyceride-hydrolase	Phospholipase 3
3.1.4.–	Sphingomyelin *N*-acylsphingosine-hydrolase	Sphingomyelinase
3.1.4.4	Phosphoglyceride phosphatidate-hydrolase	Phospholipase 4
3.1.5.–	*N*-Acylsphingosine acyl-hydrolase	Ceramidase

II. TABLE OF LIPOLYTIC ENZYMES

The numbering recommended by the Enzyme Commission is used in Table II-2. The resulting classification is in many respects unsatisfactory and uncertain; this will appear clearly throughout this book. Unfortunately, a more logical classification is at present not available or even possible, mainly because the substrate specificity of most lipolytic enzymes is only incompletely known. The classification of lipolytic enzymes refers to their "natural" or "physiological" substrates. This is precise enough for those enzymes with well-defined structural substrate requirements, such as phospholipase 2, or for those enzymes whose physiological purpose is obvious. Pancreatic lipase, for example, although it will hydrolyze many other esters almost as fast as triglycerides, should yet, obviously, be described as a triglyceride lipase, just as the digestive enzyme chymotrypsin, although it can hydrolyze esters, is obviously a proteinase. However, many lipolytic enzymes that are described in the literature, and especially most of the intracellular enzymes, have been identified by their action on one substrate only, and it is often impossible to ascertain if this substrate has been the true physiological substrate. In recent years, assay procedures have been much refined; it is now possible, for instance, to discriminate between "lipase" and phospholipase 1. On the other hand, intracellular "monoglyceride lipase," "lysophospholipase," and "cholesterol esterase" activities have not yet been clearly distinguished, and in many cases they may quite possibly be properties of the same enzyme. The same may be true for the "galactolipase," "phospholipase," and "lysophospholipase" of plants. Since these activities appear at present under different systematic numbers, the entire nomenclature of lipolytic enzymes may have to be changed. A tentative outline for a future classification is given in Chapter VIII.

The list of enzymes in Table II-2 requires some additional discussion for each item.

Glycerol-ester hydrolases, EC 3.1.1.3, are the "lipases" in the narrowest sense, i.e., long-chain triglyceride acyl-hydrolases. They may or may not have positional specificity for the primary ester bonds. Usually the enzymes also hydrolyze di- and mono-glycerides. The most prominent member of the group is pancreatic lipase, the lipolytic enzyme *par excellence*. In the same group we must include "monoglyceride lipase" and "galactolipase."

Cholesterol esterase, EC 3.1.1.13, may also be a "monoglyceride lipase," but this question is not yet settled.

For the phospholipases, this book introduces new trivial names, for reasons that are discussed in the introduction to phospholipases (Chapter VI).

The phospholipases 2 (EC 3.1.1.4), which used to be named lecithinase or phospholipase A and are now commonly called phospholipase A_2, are the lipolytic enzymes with the strictest substrate specificity; they hydrolyze only the fatty acid ester in position 2 of *sn*-3 phosphoglycerides. Phospholipases 1 (Phospholipase A_1), with the same systematic number EC 3.1.1.4, hydrolyze the ester in position 1; in one case it has been shown that only the *sn*-3-phosphoglyceride can serve as a substrate.

Lysophospholipases (EC 3.1.1.5) might conceivably be specific for either position 1 or 2 of lysophospholipids, but no such specificity has yet been found; in fact, there is reason to believe that "lysophospholipases" are enzymes of very broad specificity which may be members of a group of amphiphilic carboxyl-ester hydrolases.

The following enzymes are phosphohydrolases: phosphatidate phosphohydrolase, EC 3.1.3.4; phospholipase 3, EC 3.1.4.3, which hydrolyzes the phosphate ester bond in position 3 of phosphoglycerides; sphingomyelinase, EC 3.1.4.–, which attacks the corresponding bond in sphingomyelin. It is possible that enzymes exist that have both phospholipase 3 and sphingomyelinase activity. Phospholipases 4, EC 3.1.4.4, split the hydrophilic alcohol from both phosphoglycerides and sphingomyelins.

Phospholipase 3 has been called phospholipase D or C in the literature; since 1960, "C" has been generally used. Phospholipase 4, formerly phospholipase C or D, has been called phospholipase D in recent years. *N*-Acylsphingosine acyl-hydrolase (ceramidase) EC 3.1.5.–, splits the acyl–amide bond of its substrate.

III

Kinetics of Lipolysis

I. SUBSTRATE AND SUPERSUBSTRATE

According to the fundamental tenet of enzymology, an enzyme and its substrate form a complex in which the reaction takes place and which then dissociates into enzyme and products; these reactions are reversible.

$$E + S \underset{k_2}{\overset{k_1}{\rightleftharpoons}} ES \underset{k_4}{\overset{k_3}{\rightleftharpoons}} E + P \tag{III-1}$$

The kinetic equations that can be derived from this concept can be expected to apply also to the reactions of lipolytic enzymes with their substrates. It has, in fact, been demonstrated that the hydrolysis of triglycerides by pancreatic lipase obeys the fundamental Michaelis–Menten equation (Eq. III-5) (Sarda and Desnuelle, 1958). Nevertheless, this agreement is largely formal and does not shed much light on the nature of lipolytic reactions; the kinetics appropriate for lipolytic enzymes has yet to be developed.

Conventional enzyme kinetics has been developed for reactions in aqueous solutions. Typically, an enzyme with a molecular weight of perhaps 20,000 to 100,000 reacts with substrates with molecular weights of 50 to 500, and the concentrations of all reactants enter into the rate equations. The enzymatic hydrolysis of a lipid is different in one essential aspect: it is a heterogeneous reaction because the enzyme is water soluble but the substrate is not. Therefore, the enzyme–substrate interaction must take place at the interface of the aggregated substrate and water. The kinetic treatment of such enzymatic reactions is still in the embryonic stage.

It has often been remarked that conventional enzymology is an artificial science because enzymatic reactions do not usually take place in aqueous solutions but much more often at interfaces, insofar as the cell with its membranes and organelles can be considered as a conglomerate of interfaces. It appears, then, that the *in vitro* study of lipolytic reactions can provide a lifelike model for a "natural" enzymology, and thus be of inestimable value for the understanding of cellular metabolism. It must be admitted that this conclusion is partly based on an upside down premise. In intracellular metabolism, it is usually the enzymes that are immobile and the substrates that have to diffuse to the enzymes; the *in vitro* studies that most closely mimic this situation are the studies with the matrix-bound enzymes that have recently become available. Among the lipolytic enzymes, many intracellular phospholipases are also membrane bound and do not differ, in this respect, from many other enzymes. The typical lipolytic enzymes, however—the lipases and phospholipases of the digestive glands, of venoms, of adipose tissue, and of the blood—operate under opposite conditions: it is the substrate which is immobile relative to the enzyme. This is the consequence of substrate aggregation. For example, the lamellar "liposomes" formed by lecithins in water may be as large as 10^9 daltons, and the fat globules that are the substrates for the digestive lipases are much larger still. Under physiological conditions, the substrate aggregates may admittedly be smaller. Sonicated lecithin aggregates may have a particle weight of 2×10^6 (Attwood and Saunders, 1965), phospholipid–cholate micelles a particle weight of 5×10^5 to 1×10^6; but this is still 35–70 times the weight of (porcine) pancreatic phospholipase. In lipolysis, obviously, the enzyme has to seek out the substrate, not the substrate the enzyme—"substrate" being understood, in this case, not as a single molecule but as the nonaqueous phase of aggregated lipid. Nevertheless, lipolysis and the enzymology of membranes have in common that their kinetics are restricted to two dimensions instead of three and that at least one of the reactants, enzyme or substrate, is partly immobilized in a larger matrix.

The term "supersubstrate" has been proposed (Brockerhoff, 1974) to describe the matrix in which a substrate molecule is embedded. In lipolytic reactions, this matrix may be the surface of a triglyceride droplet or a phospholipid micelle, and thus consists of an aggregate of many substrate molecules; or it may be such a surface modified by inclusion of nonsubstrates, such as amphiphilic cations or anions; or it may be a biological membrane in which the substrate molecules are few and far between. It is necessary that the enzymes not only approach the supersubstrate and bind to it, but that they do so in the right orientation to bring their reactive sites into the vicinity of the substrate molecules in the supersubstrate phase.

It is clear, then, that the kinetics of lipolysis will have to differ from conventional enzyme kinetics in many respects. Since the methods of kinetic analysis that take the peculiarities of lipolysis into account are as yet undeveloped, this chapter will offer mainly negative, cautionary illustrations concerning the application of conventional kinetics to lipolytic reactions.

No attempt is made in the following sections to explain the derivation of the usual kinetic equations. In treatment and notation, a well-known textbook of biochemistry (Mahler and Cordes, 1971) is followed, except that K_m rather than K_s is used for the Michaelis constant.

II. MAXIMAL VELOCITY

For an enzymatic one-substrate reaction that has reached the steady state but has not yet accumulated enough product, P, to make the reverse reaction, K_4, noticeable, it follows from Eq. (III-1) that the reaction rate $v = dp/dt$ equals

$$v = \frac{k_3 e_0 s}{(k_2 + k_3)/k_1 + s} \tag{III-2}$$

with e_0 and s the concentrations of enzyme and substrate. Setting

$$V = k_3 e_0 \tag{III-3}$$

and

$$K_m = (k_2 + k_3)/k_1 \tag{III-4}$$

we obtain

$$v = Vs/(K_m + s) \tag{III-5}$$

the Michaelis–Menten equation.

At infinite substrate concentrations, the velocity v becomes V, the maximal velocity of the reaction. In lipolytic reactions, which are not usually subject to substrate inhibition, V is easy enough to determine. The most common procedure is to measure the initial velocity v at several substrate concentrations and to plot $1/v$ against $1/s$ according to Lineweaver and Burk (1934) (Fig. III-1). According to the reciprocal Michaelis–Menten equation

$$\frac{1}{v} = \left(\frac{K_m}{v}\frac{1}{s}\right) + \frac{1}{V} \tag{III-6}$$

the intercept on the $1/v$ axis equals $1/V$. The catalytic rate constant $k_3 = v/e_0$ (often written k_{cat}) of an enzymatic reaction is a more interesting constant because it is independent of the enzyme concentration and equals the turnover number, i.e., the number of substrate molecules converted by one enzyme molecule in the unit of time. However, k_{cat} can only be calculated if molar enzyme concentration—and therefore molecular weight—is known; more often than not this is not the case. Usually, V is expressed as specific activity, e.g., as micromoles of substrate converted by 1 mg of enzyme in 1 minute. The rate constant k_{cat} can be obtained from Eq. III-3. For instance, porcine pancreatic lipase, with a molecular weight of 48,000 and the specific activity of 10,000 against tributyrin at 25°C will give a k_{cat} of 4.8×10^5 min^{-1} or 8×10^3 sec^{-1}. Specific activities are, in this book, always expressed in micromoles fatty acid released per minute per milligram of protein, unless otherwise stated.

Straightforward as the determination of V and k_{cat} for lipolytic reactions may appear, the difference from reactions in a homogeneous phase must not be overlooked. A maximal rate is reached asymptotically with increasing substrate concentration; this means, in an aqueous solution, increasing numbers of substrate molecules surrounding the enzyme. In a lipolytic reaction, V is approached with increasing concentrations of the super-substrate, i.e., lipid particles or oil–water interface, and it is reached when all of the enzyme is adsorbed to the lipid–water interface. However, the molecular concentration of the substrate in the vicinity of the reactive site of a lipase cannot be manipulated by changing the supersubstrate concentration, since it is given by the nature of the interface once the enzyme is adsorbed. The substrate concentrations are static and discontinuous. A lipolytic enzyme in solution is faced with no substrate at all; adsorbed to the supersubstrate, it is met by the maximally possible concentration of substrate. It is to this concentration that V refers.

III. DETERMINATION OF THE K_m

The Michaelis constant K_m (Eq. III-4) can be read from the Lineweaver–Burk plot (Fig. III-1); the intercept of the $1/s$ axis is $-1/K_m$. From Eq. 5 it follows that K_m has the dimension of a concentration and equals numerically the substrate concentration at which $v = V/2$, or half the maximal velocity. The problem is how to measure the concentration of an insoluble substrate, an emulsion of oil in water. Obviously, at best only those molecules that are at the interface, on the surface of the oil droplets, are available to the enzyme. Furthermore, the substrate concentration is zero in the aqueous phase. It was first suggested by Schønheyder and Volqvartz (1944a,b, 1945) that the rate of lipolysis might depend on the available surface area. This area, rather than the weight or molarity

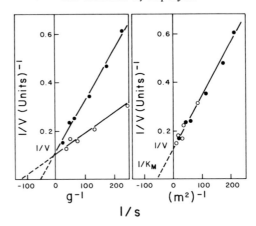

Fig. III-1 Lineweaver-Burk plot of the lipolysis of two oil emulsions, coarse (●) and fine (○). In the left diagram, the substrate concentration is plotted as gram⁻¹; in the right as (meter²)⁻¹. From Benzonana and Desnuelle (1965).

of the lipid, would then be a measure of the concentration s in the Michaelis–Menten equation. Schønheyder and Volqvartz determined the interfacial area of a trihexanoin emulsion by measuring and counting the oil droplets with a microscope. When they determined the rates with which pig pancreatic lipase hydrolyzed this emulsion they arrived at a K_m of 520 cm² per liter.

The concept of a surface-dependent interfacial reaction was further expounded by Sarda and Desnuelle (1958) who demonstrated the calculation of K_m with emulsions of tributyrin and triolein. These experiments led to the definition, by Desnuelle (1961), of lipases as esterases that act specifically on insoluble esters at the oil–water interface. The most painstaking measurements of an oil–water interface to date were made by Benzonana and Desnuelle (1965) with the help of an electronic particle counter. By measuring the rates of lipolysis of coarse and fine emulsions of the same substrate, the authors could demonstrate the role of the interfacial area. Figure III-1 shows, in the left diagram, the Lineweaver-Burk plots for a coarse emulsion (solid circles) and for a finer emulsion (open circles) with the substrate concentration plotted as the reciprocal of the weight in grams. It can be seen that both emulsions have the same maximal hydrolysis rates, expressed as $1/v$ at the intercepts with the ordinate. The Michaelis constants, however, measured as $-1/K_m$ at the intercepts with the abscissa, are different for both emulsions and not related to the weight of the substrate. However, if the concentration is expressed as the reciprocal of the surface area in square meters, as in the right diagram, a constant K_m as well as constant V is obtained.

These experiments show that lipolysis will conform to the Michaelis–Menten equation if the substrate concentration as well as the Michaelis constant are measured in units of interfacial area per volume.

IV. MEANING OF K_m

If formation and dissociation of the enzyme–substrate complex ES (Eq. III-1) proceed much faster than the catalytic reaction, i.e., k_1, $k_2 \gg k_3$, then

$$K_m \geq k_2/k_1 = K_{ES} \tag{III-7}$$

i.e., K_m approaches K_{ES}, the dissociation constant of the complex. For enzymatic reactions that take place in an aqueous phase, the Michaelis constant K_m can be, therefore, a measure of enzyme–substrate affinity, even though it is not usually identical with K_{ES}. A low K_m as compared with a high K_m normally indicates a greater ease of formation of the Michaelis complex, usually a better fit between enzyme and substrate. It can be expected, for instance, that the introduction of a bulky group into a water-soluble ester will augment this ester's K_m for an enzymatic hydrolysis; a reduced K_m would be taken as evidence that the enzyme possesses a special receptor for the bulky group of the substrate.

Lipolytic reactions, as has been shown, can formally be treated so that they follow Michaelis–Menten kinetics. It cannot be expected, however, that the calculated K_m will have similar importance and meaning as in reactions in an aqueous phase. In lipolysis, the formation of the Michaelis complex is preceded by adsorption of the lipase to the oil–water interface, which is the supersubstrate, S_s, in which the substrate, S, is imbedded.

$$\text{E} + \text{S}_s \underset{k_2}{\overset{k_1}{\rightleftharpoons}} [\text{E} + \text{S}]_{\text{adsorbed}} \underset{k_4}{\overset{k_3}{\rightleftharpoons}} \text{ES} \underset{k_6}{\overset{k_5}{\rightleftharpoons}} \text{E} + \text{P} \tag{III-8}$$

If we neglect the inequality, i.e., the difference between S_s and S, this equation describes formally a one-substrate enzymatic reaction with two intermediate complexes. It can be resolved for $v = dp/dt$ (Mahler and Cordes, 1971):

$$v = \frac{(k_1 k_3 k_5 s - k_2 k_4 k_6 p)e_0}{k_2 k_5 + k_2 k_4 + k_3 k_5 + k_1(k_3 + k_4 + k_5)s + k_6(k_2 + k_3 + k_4)p} \tag{III-9}$$

In the beginning of the steady state reaction $p = 0$, and

$$v = \frac{k_1 k_3 k_5 s e_0}{k_2 k_5 + k_2 k_4 + k_3 k_5 + k_1(k_3 + k_4 + k_5)s} \tag{III-10}$$

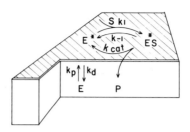

Fig. III-2 Proposed model for the action of a soluble enzyme at an interface. From Verger *et al.* (1973).

This equation can be converted into the form of the Michaelis–Menten equation by the appropriate substitution:

$$\frac{k_3 k_5 e_0}{k_3 + k_4 + k_5} = V \tag{III-11}$$

$$\frac{k_2 k_5 + k_2 k_4 + k_3 k_5}{k_1(k_3 + k_4 + k_5)} = K_m \tag{III-12}$$

$$v = Vs/(K_m + s) \tag{III-13}$$

The K_m can be determined as described, but it is seen to be composed of five individual rate constants (Eq. III-12) that cannot be calculated from only two experimental parameters, V and K_m. Two of the rate constants, k_1 and k_2, refer to the formation of the enzyme–supersubstrate (enzyme–interface) complex; k_3 and k_4 refer to the orthodox Michaelis complex; k_5 is k_{cat}. It is not possible with present methods to separate the formation of the supersubstrate–enzyme complex from the substrate–enzyme complex formation. It follows that a lipolytic K_m, despite the formal analogy, cannot be compared with the K_m of homogeneous reactions because its relationship to the active complex is much more remote, and it is not made any closer by expressing K_m in surface units.

Verger *et al.* (1973) have recently presented the problem in a different terminology, starting from the model of the interface shown in Fig. III-2. They distinguish between K_m, the conventionally measured Michaelis constant, and K_m^*, the interfacial or two-dimensional Michaelis constant, defined as

$$K_m^* = (k_{cat} + k_{-1})/k_1 \tag{III-14}$$

The equilibrium between dissolved and interfacial ("penetrated") enzyme, E and E*, is given by k_d/k_p, the desorption rate constant over the penetration rate constant. The Michaelis–Menten equation for lipolysis results as

$$v = \text{``}V_{\max}\text{''}s / (\text{``}K_m\text{''} + s) \tag{III-15}$$

with

$$\text{``}V_{\max}\text{''} = \frac{k_{\text{cat}}E_0 s}{K_m^* + s} \tag{III-16}$$

and

$$\text{``}K_m\text{''} = \frac{k_{\text{d}}}{k_{\text{p}}} \frac{K_m^* s}{K_m^* + s} \tag{III-17}$$

It can be seen that "K_m" and "$V_{\max}$" are functions of variables, k_1, k_{-1}, k_{d}, and k_{p}, which can neither be measured with present methods nor calculated from the experimental parameters presently available.

It has been argued (Brockerhoff, 1970) that when the interface has become large enough to adsorb half the available enzyme molecules, half the maximal velocity will be measured, and at that point the surface area concentration of the supersubstrate will equal "K_m". In other words, "K_m" could be considered the enzyme–interface dissociation constant. This interpretation was first advanced by Dixon and Webb (1964). Equation 12 seems to show that it cannot be correct, since only in the unlikely case that k_3 is very small compared to the other rate constants does "K_m" approach k_2/k_1. Nevertheless, the case can be argued for "K_m" indeed being the enzyme–interface dissociation constant. The two-complex Equation III-8 and its derivations (Eqs. III-9–III-13) as well as Eqs. III-15–III-17 have been developed with the assumption of equilibria existing between all successively written forms of the enzyme. It is possible, however, that this assumption is incorrect for lipolytic enzymes. The enzyme–supersubstrate equilibrium could possibly embrace all forms of the enzyme, whether unproductively bound or engaged in enzyme–substrate complexes. In other words, an increase in k_3 would increase ES and diminish (E + S) (Eq. III-8), but the supersubstrate, i.e., the interface, would still bind the same amount of protein and the concentration of E_{solution} would not be reduced. The following equation might express this concept:

$$E_{\text{solution}} + S_{\text{s}} \underset{k_2}{\overset{k_1}{\rightleftharpoons}} [E + S \underset{k_4}{\overset{k_3}{\rightleftharpoons}} ES \underset{k_6}{\overset{k_5}{\rightleftharpoons}} E + P]_{\text{interface}} \tag{III-18}$$

If it should be verified that most lipolytic enzymes have different sites for supersubstrates bonding and substrate bonding (Brockerhoff, 1973; Verger *et al.,* 1973; Pieterson *et al.,* 1974b), Eq. III-18 could indeed be expected to describe lipolytic reactions, and K_m would be k_2/k_1, or the surface concentration where half of the enzymatic protein is adsorbed. On the other hand, if Michaelis complex formation reinforces the binding of the protein to the interface, Eq. III-8 might be more appropriate. This situation might apply to those phospholipases that require a negative surface potential and also bind the negatively charged phosphate group in the active complex. The decision on the entire problem must await new experimental methods.

To summarize, the only reasonably meaningful constant available in the kinetics of enzymic lipolysis is the maximal reaction rate V; for example, all studies on the hydrolysis of different substrates have to rely on the measurement of this rate only. What is obviously needed is a method to assess the formation of the active complex in the interface; so far, no such method is in sight.

V. INHIBITION AT THE INTERFACE

On adding a solution of the poor substrate, triacetin, to an emulsion of the good substrate, tripropionin, it is found that the hydrolysis of the tripropionin by lipase is severely inhibited (Brockerhoff, 1969a). This seems to indicate that the poor substrate competes with the good substrate for a place at the active center of the enzyme. However, the enzyme is adsorbed to the interface and would not be available to the inhibitory water-soluble triacetin. A better explanation for this type of inhibition can be offered. Because of its partly hydrophobic character, the inhibitory substrate will distribute itself between the water and the tripropionin phase, and since it is more hydrophilic than tripropionin, it will accumulate at the interface and displace the better substrate. If this explanation in correct, the same effect of inhibition should be shown not only by esters, such as triacetin, but by any substance that can distribute itself between aqueous and nonaqueous phase. That this is indeed the case is shown in Figure III-3, where alcohols and ethers as well as esters are shown to be inhibitors. As might be expected, the more hydrophobic compounds are more effective.

Inhibition by displacement of the substrate at the interface has been quantitatively studied by Mattson *et al.* (1970) for pancreatic lipase and a series of aliphatic alcohols. Again, the longer, more hydrophobic compounds were the better inhibitors, and it could be shown that if the amount of long-chain alcohol was sufficient to occupy the total interfacial area, enzymatic hydrolysis was completely inhibited.

Interfacial substrate displacement also explains the inhibitory effects of bile salts and of proteins such as albumin. These agents also occupy the interface and prevent the mutual approach of enzyme and substrate (Desnuelle, 1961; Benzonana and Desnuelle, 1968; Schoor and Melius, 1969).

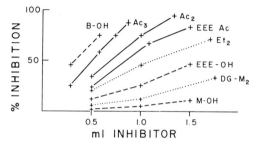

Inhibition of Lipolysis
of Tripropionin by Esters (———),
Alcohols (– – – –), and
Ethers (·········)

B-OH	Butanol
Ac₃	Triacetin
Ac₂	Glycol diacetate
EEE Ac	2-(2-Ethoxyethoxy)-ethyl acetate
Et₂	Diethyl ether
EEE-OH	2-(2-Ethoxyethoxy)-ethanol
DG-M₂	Diethylene glycol dimethyl ether
M-OH	Methanol

Fig. III-3 Inhibition of lipolysis of tripropionin, 0.3 ml in 15 ml 0.1 *M* NaCl, at pH 8. From Brockerhoff (1969a).

If they carry a charge, as the anions of the bile acids do, the enzyme may even be repelled from the interface.

Inhibition by substrate dilution at the interface is kinetically indistinguishable from competitive inhibition. On these grounds, the majority of "inhibition" experiments reported in the literature of lipolytic enzymes must be considered invalid.

VI. pH AND CATIONS AT THE INTERFACE

If the activity of an enzyme is plotted against the pH, a bell-shaped curve usually results, with either a sharp or broad pH optimum and slopes from which the apparent p*K* values for essential amino acid residues can be estimated (Dixon and Webb, 1964; Mahler and Cordes, 1971). The theory of activity–pH dependence has been worked out for enzymes and substrates reacting in aqueous solutions, and in principle it can be assumed to be valid for lipolytic enzymes also. However, a problem arises in the determination of the pH at the interface. The pH measured during an enzymatic lipolysis is the pH of the bulk phase; this pH need not be, and in most cases is not, identical with the pH at the interface. The problem can be exemplified by the influence of sodium chloride on the hydrolysis of long-chain triglycerides by pancreatic lipase.

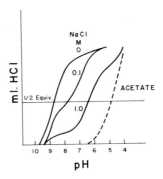

Fig. III-4 Titration curves of sodium oleate dissolved in 0, 0.1, and 1.0 M aqueous NaCl (solid lines), and of sodium acetate in water (broken line). From Mattson and Volpenhein (1966a).

The rate of pancreatic lipolysis can be measured by continuous titration of the liberated fatty acids. The reaction is usually carried out at pH 8 or 9, and since the pK_a of aliphatic acids is around 5 it would seem that the hydrolysis of fats could be quantitatively followed by the titration. It has been found, however, that the apparent pK_a of fatty acids that are water insoluble is around 9 (Schmidt-Nielsen, 1946). At pH 8–9, therefore, less than half of the insoluble fatty acids would be titratable. The apparent pK_a changes with the addition of sodium chloride. Mattson and Volpenhein (1966a) found that in a molar sodium chloride solution, 80–90% of the oleic acid present could be titrated at pH 8 or 9 (Fig. III-4). Benzonana and Desnuelle (1968) confirmed this shift of the titration curve and discussed the reasons for the effect; it may be explained by the existence of an interfacial pH gradient. In the equation

$$pH_s = pH_b + (\epsilon\psi/2.3kT) \qquad \text{(III-19)}$$

(Davies and Rideal, 1963) pH_s is the pH at the interface and pH_b the pH in the bulk phase, ϵ the charge of an electron, ψ the surface potential, and k the Boltzman constant. If the surface is negatively charged, as by fatty acid anions, ψ is negative and pH_s, at the interface, lower than pH_b; the interface attracts hydrogen ion. The pH in the bulk phase is consequently higher than the interfacial pH. In the presence of sodium ions, the following equation can be applied (Benzonana and Desnuelle, 1968):

$$pH_s = pH_b - \log\left([Na^+]_s/[Na^+]_b\right) \qquad \text{(III-20)}$$

where $[Na^+]_s$ is the concentration of Na^+ near the interface and $[Na^+]_b$ the concentration in the aqueous phase. Because of the electrostatic attraction, $[Na^+]_s$ is greater than $[Na^+]_b$ and pH_s lower than pH_b, but as the bulk concentration of sodium ion increases the difference between $[Na^+]_s$ and $[Na^+]_b$ decreases, their ratio approaches 1 and pH_s approaches pH_b. As Eq. III-20 and Fig. III-4 show, the interfacial pH gradient can be suppressed, but rather high ionic strength is required, and this may inhibit the enzymatic activity.

Charged interfaces are common to all supersubstrates of lipolytic enzymes because of the free fatty acids that are formed by their own action, and usually there are additional charge carriers present, such as the phosphate group of phospholipids or the bile acid anions in digestive lipolysis. A negative surface charge will attract cations. This attraction is maximal in the layer of water adjoining the interface and decreases in the direction of the bulk phase; the total electrostatic potential is called ζ potential. Magnitude and depth of the electrostatic field are influenced by the ionic strength of the solution. It has been found that in a buffer of ionic strength 0.02, the effective field will be about 20 Å deep (Heard and Seaman, 1960). Since the average diameter of a lipolytic enzyme may be 30–50 Å, it is clear that a portion of the enzyme, and certainly its reactive region, extends into a cationic gradient and consequently (Eq. III-20) also into a pH gradient. Since enzymes carry charged residues, this gradient can be expected to influence the orientation of the enzyme toward the lipid surface (Dawson, 1964). In any case, the hydrogen ion concentration around the center of enzymatic reaction cannot be identical with that of the bulk phase.

Interfacial phenomena are treated in books by Davies and Rideal (1963), Adamson (1967), and Gaines (1966) and in more recent collections (Brown, 1971; Hair, 1971). The action of lipolytic enzymes at interfaces, and in particular the influence of the ζ potential, has been studied extensively by R. M. C. Dawson, who has also written authoritative reviews on the subject (Dawson, 1964, 1968).

VII. MONOLAYERS

The monolayer technique introduced in enzymology by Hughes (1935) has been used extensively in studies of lipolytic enzymes. In this method, a monomolecular layer of a lipid is spread over water in a trough, and the enzyme is then added to the subphase. As the substrate molecules are hydrolyzed, the structure of the monolayer will change, especially if a substrate is chosen whose hydrolysis products diffuse into the water. Several experimental parameters can be measured in following the reaction (Bangham and Dawson, 1960; Dawson, 1969): the change in surface potential, in radioactivity, or in surface pressure. The surface potential is measured between the aqueous phase and an air-ionizing electrode mounted over the surface. The appearance of fatty acids in the layer will result in a change of its dielectric properties and therefore the potential; this change is recorded. Radioactivity is measured with a window counter. If the products of lipolysis, fatty acid or lysolipid, are water soluble and can diffuse into the subphase, a drop in radioactivity results. The method most often used is the measurement of the surface pressure with the help of a hydrophilic dipping plate suspended from a torsion balance. The initial drop in surface pressure can be measured (Garner and Smith, 1970b; Lagocki *et al.,* 1970, 1973), or, better, the surface pressure can be held constant while the surface area is constantly and automatically adjusted with a movable barrier, the decrease in the area being recorded (Dervichian, 1971; Olive and Dervichian, 1971). Thus, the substrate–surface concentration can be held constant, although the total available substrate decreases and the recorded decrease in area is therefore not linear. Verger and de Haas (1973) have described a trough of constant surface area in which the surface pressure is also held constant by replenishment of the hydrolyzed substrate, through a surface aperture, from a reservoir which also holds a substrate film but no enzyme. The change of the reservoir surface is then recorded. This device maintains substrate and enzyme concentration in the reaction trough constant and yields linear (zero-order) kinetics for extended periods of time. Brockman *et al.* (1973) also described a constant-area trough in which the surface pressure is kept constant by continuous (and recorded) administration of substrate.

Monolayer studies supplement the usual method of lipolysis in emulsion. They have some apparent advantages. First, all substrate molecules are exposed to the aqueous solution of the enzyme, not just a fraction as in emulsified oil droplets. Second, it is possible to dispense with some agents that are usually necessary in studies with emulsions but may interfere in unknown ways with the reaction; e.g., emulsifying agents such as gum arabic, or "activators" such as bile salts or calcium salts. Nevertheless, monolayer experiments are at present just as refractory to kinetic analysis as experiments with emulsions, or perhaps more so (Zografi *et al.,* 1971). In a monolayer system, it is not possible to vary the supersubstrate/volume concentration; even if the depth of the underlying aqueous phase should be changed, this would not change the effective supersubstrate concentration at the monolayer where the hydrolysis takes place. In emulsions, this problem is overcome by providing very large surface areas and vigorous agitation. In the monolayer technique, it is also difficult, though perhaps not impossible, to measure how much of the enzyme has been adsorbed to the monolayer from the aqueous phase. It is therefore not only impossible, as it is in emulsions, to determine a meaningful K_m, but the maximal rate V is also less meaningful. Only relative apparent hydrolysis rates of different substrates, or of one substrate at different degrees of surface pressure, can be obtained.

However, the lipolysis of monolayers does yield interesting information. It has been shown, for instance, that the reaction rates of phospholipase 2 against monolayered phospholipids are very much dependent on the surface pressure (Hughes, 1935; Shah and Schulman, 1967; Zografi *et al.*, 1971), and the same effect is found for substrates of lipases (Olive and Dervichian, 1971; Esposito *et al.*, 1973). This indicates that the packing or orientation of substrates in the monolayer is of great importance because of the changing degree with which the reactive groups of the substrate are exposed to solvent and enzyme. The great value of monolayer experiments lies, in fact, in the formation they can give on the structure of lipid layers and lipid–protein membranes (Colacicco, 1969). A recent study on the interaction of lipoprotein lipase with apolipoproteins has made successful use of a monolayer technique (Miller and Smith, 1973). The adsorption of a lipase to a lipid layer has been demonstrated by lateral transfer of the layer into a chamber that contained no enzyme (Dervichian *et al.*, 1973).

As for the study of lipolytic reactions, it seems that monolayers are presently less likely even than emulsions to yield significant information on the enzyme–substrate complex in lipolysis. On the other hand, future monolayer experiments can be expected to throw more light on the mechanism of the enzyme–supersubstrate association. A pre-steady state analysis of lipolysis may develop from such experiments.

VIII. WATER-SOLUBLE SUBSTRATES

The elucidation of the catalytic mechanism proper of lipolytic enzymes will best be achieved, in our opinion, with studies using water-soluble substrates. Such substrates can be hydrolyzed. Pancreatic lipase attacks dissolved triacetin and tripropionin, although there seems to be a correlation between reaction rate and micelle formation (Entressangles and Desnuelle, 1968). Completely water-miscible esters that are not likely to form micelles, such as 2-(2-ethoxyethoxy)ethyl acetate, are also hydrolyzed by the lipase (Brockerhoff, 1969a). Phospholipases 2 hydrolyze water-soluble dihexanoyllecithin (Roholt and Schlamowitz, 1958; de Haas *et al.*, 1971) and also monomolecular dibutyryllecithin (van Deenen and de Haas, 1963). The rates of hydrolysis are much lower for such substrates than for water-insoluble esters but sufficiently high for experimental determination; for dissolved triacetin and lipase, for example, k_{cat} is 2×10^4 min^{-1}.

Lipolysis in a homogeneous aqueous phase eliminates from the kinetic analysis all the complications that arise from the presence of an interface: the substrate concentration can be varied, the pH is defined, a meaningful K_m can be determined, and inhibition studies become possible. The potential of such studies has been demonstrated with the enzymatic hydrolysis of water-soluble dibutyryllecithin (Wells, 1972). An integrated description of interfacial lipolysis will probably emerge from a combination of water-phase and monolayer studies.

IV

Lipases

I. DETECTION AND ASSAY

For both detection and assay of lipases it is essential that substrates be chosen and reaction conditions arranged so that the definition of a lipase be met, i.e., hydrolysis of long-chain acylglycerols at an oil–water interface. With the proper procedures, nonlipolytic esterases will not be detected, and misleading reports on enzymes that are in the gray area between esterases and lipases can be avoided. Detection and assay of lipases have been briefly reviewed by Wills (1965), Jensen (1971), and Desnuelle (1972).

A. Substrates

Glycerol trioleate (triolein) is the most universal of substrates. It fulfills the definition of a lipase substrate by containing long-chain fatty acids only, and it is liquid at the usual assay temperatures; this is an important requirement for pancreatic lipase, which will hydrolyze solid triglycerides such as tristearin only very slowly; other lipases probably have a similar need for liquid substrates.

A good surrogate for triolein is olive oil, which contains over 70% oleic acid ester and has the advantage of being very cheap. Commercial samples usually contain varying amounts of free fatty acid and di- and monoglycerides. These contaminants must be removed. This can be achieved by filtration through active charcoal and chromatography on florisil (a magnesium silicate), but most easily by passing the oil through a column of neutral alumina in a solvent mixture of ether and petroleum ether 1:10 (Jensen *et al.*, 1966b). The solvent is then removed by evaporation, and the purity of the triglycerides can be assessed by thin-layer chromatography on silica gel with petroleum ether–ether–acetic acid 70:30:2. If the substrate contains substantial amounts of polyunsaturated acids, purification immediately before use is required. The products of oxidation, which inhibit both pancreatic and *Geotrichum candidum* lipases, are retained by the alumina (R. G. Jensen, unpublished data, 1973).

Tributyrin and all other esters can be purified in the same manner. Tributyrin is a convenient substrate, because it is easily dispersed in water by shaking or stirring without the addition of any emulsifiers. It is especially useful in lipase screening tests and in the quantitation of lipase activity by continuous automatic titration. These methods will be described below. It must be realized however, that tributyrin does not meet the requirements for a substrate of lipase as defined, and although most lipases may hydrolyze tributyrin and most enzymes hydrolyzing tributyrin may be lipases, exceptions must be expected. Therefore, if "lipase" activity is detected with tributyrin, it should be verified with triolein.

Chromogenic substances, such as the fatty acid esters of 4-methylumbelliferone, have been employed by many workers for ease and sensitivity of measurement. Esterases release an alcohol (eosin, umbelliferone, naphthol, etc.) from these substrates, which is then determined directly (Guilbault and Kramer, 1966; Jacks and Kircher, 1967) or after diazo coupling (Kramer *et al.,* 1963). Jacks and Kircher (1967) assayed pancreatic lipolytic activity with esters of 4-methylumbelliferone, but it has been found (Brockerhoff, 1969b; Melius and Doster, 1970) that umbelliferone caprylate is an extremely poor substrate, giving only about 0.2% of the activity observed with triolein. In fact, all phenolic esters appear to be unsuitable for assays of lipase activity (Barrowman and Borgström, 1968; Barrowman, 1969; Brockerhoff, 1969b). It is possible that the nonspecific esterase observed by Mattson and Volpenhein (1968) in rat pancreatic juice was responsible for the activity noted by Jacks and Kircher (1967) and by others with phenol esters.

Vinyl oleate has been tested as a substrate for lipolysis (Brockerhoff *et al.,* 1970) with about one-third of the activity toward triolein resulting. A very sensitive assay involves determination of the isomerization product of vinyl alcohol, acetaldehyde.

Many investigators (Chino and Gilbert, 1965; Greten *et al.,* 1968; Kelley, 1968; Belfrage and Vaughan, 1969; Kaplan, 1970; Schotz *et al.,* 1970; Tsai *et al.,* 1970; Marsh and Fitzgerald, 1972; Tornqvist *et al.,* 1972) have chosen radioactive substrates, mainly because of the extreme sensitivity that can be obtained. Use of these substrates may be mandatory in attempts to determine the low activity of nondigestive lipases, such as those of adipose tissue, or lipoprotein enzymes. Tornqvist *et al.* (1972) noted that radioactive triacylglycerols will spontaneously hydrolyze, even when stored at —20°C in pure solvents, forming partial acylglycerols of high activity. This effect, coupled with the presence, in all tissues, of monoacylglycerol hydrolase activity, necessitated frequent purification of the substrate in the assay of hormone sensitive lipase. To avoid these difficulties, the authors synthesized a glycerol diether ester with [³H]oleic acid in the 1(3) positions. Neither monoacyl glycerol hydrolases nor lipoprotein lipase would hydrolyze the diether substrate.

B. Conditions

The pH optima of most lipases lie between 8 to 9, although exceptions can be expected. Castor bean lipase is most active at pH 4.2, tissue lipases of liposomal origin have optima below 5, and the lipases in the microorganism *Mucor pusillus* has an optimum pH range of 5 to 6. Phosphate, NH_4OH–HCl, and tris have been used most often as buffers. A group of zwitterionic buffers, the Good series, is best for the pH range of 6 to 8 (Good *et al.,* 1966). These buffers are available from most biochemical supply firms. One of them, 2-(*N*-morpholine)ethanesulfonic acid (pH 6.5, 0.26 M) increased the activity of pregastric "esterase" (R. G. Jensen, unpublished data, 1973).

Lipolytic enzymes may be active over a very wide temperature range; e.g., some microbial lipases act at —20°C (Alford and Pierce, 1961) and the enzyme from *Vernonia anthelminthica* seed at 65°C. For substrates melting at higher temperatures, the temperature of incubation can be raised, for example, to 45°C (Luddy *et al.,* 1964), so as to ensure proper liquidity of the substrate. High-melting substrates can also be emulsified more readily if they are first dissolved in cosolvents such as hexane (Brockerhoff, 1965a), triolein, tripamitolein, methyl pentadecanoate or methyl oleate (Barford *et al.,* 1966).

Many lipolytic enzymes will require extended periods of incubation for the detection of activity. For comparison, 30 mg of the crude porcine pancreatic lipase available from laboratory supply firms will digest about 30–50% of the primary ester groups in 200 mg of purified olive oil in 5–15 minutes. This product, however, represents a concentrated source, and most tissue extracts will exhibit far less activity. Incubation periods longer than 15 minutes will result in acyl migration of the 1,2(2,3)-diacylglycerols and 2-monoacylglycerols, which may be products of lipolysis, to the 1,3-isomers and 1-isomers (Mattson and Volpenhein, 1962).

Vigorous agitation of the digestion mixture is required during incubation to constantly renew the surface of the oil droplet (Mattson and Volpenhein, 1966a). Very rapid agitation can be achieved with a modified dental amalgamator (Luddy *et al.*, 1969) or an orbital finishing sander with clamps for tubes mounted on the sanding surface (N. R. Bottino, personal communication, 1972).

Since the velocity of lipolysis is a function of the supersubstrate concentration, i.e., of the surface offered to the enzyme, it is important that the substrate be dispersed in as fine an emulsion as possible. For substrates such as olive oil, shaking or stirring are not sufficient, and even sonification, which is very effective in dispersing the oil, leads only to unstable emulsions. Stabilizing emulsifiers have to be added. The most common one is gum arabic (acacia) in concentrations between 2 and 10%; polyvinyl alcohol and methylated cellulose (Sémériva and Dufour, 1972) are also used. For routine assays, olive oil purified as described above and emulsified into a solution containing 10% powdered gum arabic with a Branson Sonicator or a Waring Blendor is satisfactory (Marchis-Mouren *et al.*, 1959). The solution should contain Na^+ ions (usually as 0.1–1.0 M NaCl or supplied by the buffer) which suppress enzyme inhibition by interfacial charge effects (Mattson and Volpenhein, 1966a,b; Benzonana and Desnuelle, 1968) and a fatty acid acceptor, usually Ca^{2+} ions (0.02–0.1 M $CaCl_2$) or albumin, since free fatty acids usually inhibit lipases. Many investigators also add bile salts, but these should probably be omitted because they may be inhibitory.

Certain lipolytic enzymes, such as lipoprotein lipase, require special conditions for activity. These will be discussed in the appropriate sections.

The digestion is usually terminated by the addition of acid to the incubation vessel. The acid inactivates the lipase, converts soaps to free acids, and prevents the formation of emulsions during subsequent extraction. However, exposure to the acid should be as brief as possible, or the acidification should be stopped at pH 4 with bromthymol blue as an indicator (Brockerhoff, 1965a), otherwise acyl migration of partial acylglycerols may occur. The mixture can then be extracted with organic solvents, such as a mixture of ether and petroleum ether. Short-chain acids or acylglycerols, which are difficult to extract, may be recovered with a technique involving silicic acid (Sampugna *et al.*, 1967). For up to 50 ml of digestion mixture, 250 ml of 9:1 $CHCl_3$–MeOH is used. The acidified digestion mixture is rinsed into a porcelain casserole with a small portion of the solvent, and enough silicic acid is added and mixed in with a pestle until a free-flowing powder results. The powder is slurried with another portion of the solvent, transferred into a chromatographic tube equipped with a fritted glass filter in the base, and the solvent is drawn through into a filter flask with the aid of a vacuum. The remainder of the solvent is then added to wash residual lipid through the silicic acid. Extraction is complete in a few minutes, many samples can be extracted in a short time, there are no problems with emulsions, the extract is dry, all products of lipolysis are extracted, and titration of the free acids can be done immediately in the filter flask. It should be noted, however, that monoglycerides will be isomerized by silicic acid.

C. Screening

Purified tributyrin mixed with agar and poured into a Petri dish is convenient for screening. Wells are cut in the agar and the extract being screened is added. If a lipolytic enzyme is present, the cloudiness will disappear because of the solubility in the water of the digestion products. Two pHs, such as 5.0 and 8.5, should be employed. Lawrence *et al.* (1967a,b) spread tributyrin–agar emulsions on a microscope slide, added the tissue extract to a hole in the agar, then incubated the slide in a Petri dish.

Fryer *et al.* (1967) developed two double-layer techniques for detection of lipase in microorganisms: one with tributyrin as a substrate and the other with milk fat plus Victoria blue. The organisms are grown in nutrient agar overlying the substrate agar. Advantages are that the colonies can be isolated after detection, lipolysis can be followed, and there is no toxicity due to the dye. The sensitivity of the tributyrin method can be increased by lowering the substrate concentration to 0.1% from 1.0%. Forty-five pure cultures of microorganisms were screened for lipolytic activity, with forty-two positive results. In addition, the method detected lipolytic microorganisms in raw milk. Berner and Hammond (1970a) screened preparations from nineteen animal species, representing nine phyla, with tributyrin agar clearing. Tributyrin agar was also used to detect activity in fractions obtained by gel electrophoresis.

Screening with the help of tributyrin may indicate lipolytic activity, but since tributyrin is not, by definition, a substrate of lipases only, all positive results should be confirmed by the hydrolysis of triolein.

Pancholy and Lynd (1972) isolated microorganisms from soil in agar plates containing the butyryl ester of 7-hydroxy-4-methylcoumarin. This nonfluorescent compound is converted to the highly fluorescent umbelliferone derivative. Hydrolysis of the ester occurred with nearly 1000-fold greater sensitivity than with the usual titration procedures. The method was also employed to visualize esterase loci in an *Aspergillus niger* mycelium. Although all of the esterase-producing microorganisms were subsequently shown to be lipase positive (they hydrolyzed olive oil emulsions), a caproylmethylcoumarin substrate has been found to be a very poor substrate for purified pancreatic lipase (Brockerhoff, 1969b; Melius and Doster, 1970).

D. Quantitation

Methods for the quantitation of lipolytic activity can be divided into three groups: (1) the measurement of physical changes in the reaction system, (2) assays of the liberated alcohols, and (3) assay of the freed fatty acids (Brockerhoff *et al.,* 1970).

As an example of group (1), we have already mentioned the use of tributyrin agar, with which the time required for clearance is inversely related to the lipolytic activity. A rapid and convenient physical method for the determination of pancreatic lipase activity in duodenal juice utilizes the clearing, measured photometrically, of an olive oil emulsion stabilized with monolein and sodium oleate (Borgström, 1957). The continuous decrease in optical density can be followed with a photometer. This procedure cannot be used below pH 8 because the emulsion collapses. Vogel and Zieve (1963) utilized turbidimetry as a rapid method to assay lipolytic activity in normal and pancreatitis serum.

Another physical method, stalagmometry, measures changes in surface tension brought about by the surface-active products of hydrolysis (Rona and Michaelis, 1911). Here, the number of drops in a given volume of fluid or the weight of each drop are determined. As an alternative, the changes in surface tension can be measured directly with a Du Nuoy tensiometer, and this has been done to detect milk lipase activity (Dunkley and Smith, 1951; Fredeen *et al.,* 1951). However, these methods do not give reliable quantitative estimates of lipase activity because changes in surface tension are dependent on the amounts and types of monoacylglycerols and free fatty acids present and on the ratio of soaps to free acids. The latter is in turn dependent upon pH and the presence or absence of Ca^{2+} ions and so on.

The assay procedures based on physical changes of the lipolytic system, convenient as they may be in special cases, have in common a relative lack of sensitivity.

The methods of the second group, involving assay of the liberated alcohol, are colorimetric or fluorometric and therefore sensitive and fast, but they suffer from the fact that most chromogenic alcohols are phenols, and their esters are therefore not proper substrates for lipases; vinyl ester may be an exception (Brockerhoff *et al.,* 1970).

The most generally useful techniques are found in the third group of assays, in which the fatty acids released by the enzyme are measured either directly by titration of the assay mixture after dilution with solvent (alcohol), or by nonaqueous titration of the solvent used to extract the mixture (Dole, 1956; Dole and Meinertz, 1960; Association of Official Agricultural Chemists, 1965). The recommended indicator, thymol blue, has a pK several units above that of fatty acids, and complete titration is thus assured. See Meinertz (1971) for an extensive review.

The sensitivity of fatty acid determination can be much improved with colorimetric procedures. The copper soap method of Duncombe (1963) is useful, and even more sensitive modifications have been reported (Mahadevan *et al.,* 1969; Mueller, 1972).

The release of fatty acid has also been monitored manometrically by the production of carbon dioxide from the reaction between a bicarbonate buffer and the free acids (Rona and Lasnitzki, 1924; Singer and Hofstee, 1948a,b) and by the color change of acid–base indicators (Massion and Seligson, 1967; Chakrubarty *et al.,* 1969; Nakai *et al.,* 1970; Kason *et al.,* 1972).

Radioactive fatty acids in a triacyl glycerol substrate afford the greatest sensitivity of any method. For example, nanogram quantities of [1-^{14}C]-oleic acid were easily detected after hydrolysis from [^{14}C]triolein, diluted with carrier olive oil, by purified pancreatic lipase (Marsh and Fitzgerald, 1972). The free fatty acids were recovered by chromatography with copper-impregnated ion exchange paper. The paper immobilizes the oleic acid, and the remaining tri-, di-, and monoacylglycerols are carried with the solvent front.

Several liquid–liquid extraction procedures have been applied to the recovery of labeled fatty acids present after lipolysis (Greten *et al.,* 1968; Belfrage and Vaughan, 1969; Kaplan, 1970). Since organic solvents that extract fatty acids will also extract the unreacted substrate triglyceride, a second extraction must be performed. For example, Kaplan (1970) extracted an acidified lipolysate with heptane and then recovered the free fatty acids from this extract with alkaline ethylene glycol. The fatty acids can also be separated from the glycerides by adsorption on an anionic resin (Kelley, 1968). Zieve (1973) has described a method using small disposable TEAE-cellulose columns.

In continuous titration, base is added as the pH decreases. This method is most conveniently carried out by a pH-Stat coupled with an automatic burette and a recorder. It has been used to measure the activity of pancreatic lipase (Marchis-Mouren *et al.,* 1959), a lipase from *Staphylococcus aureus* (San Clemente and Vadehra, 1967), a lipase from *Pseudomonas fragi* (Mencher and Alford, 1967), a rat adipose tissue lipase (Boyer, 1967), milk lipase (Parry *et al.,* 1966), and many others. The pH-Stat is valuable, indeed indispensible, in investigations of lipase kinetics, as initial velocities can be determined directly, but it is relatively insensitive and requires an alkaline pH of about 8 or 9. At pH levels below 8, less and less fatty acid can be titrated with base (Mattson and Volphenhein, 1966a; Benzonana and Desnuelle, 1968).

Tributyrin is very convenient for continuous titration (Erlanson and Borgström, 1970a) if the identity of the enzyme as a lipase has been ascertained. The substrate is kept in dispersion by vigorous stirring; 0.3 ml tributyrin in 10 or 15 ml of 0.1 M NaCl will give maximal reaction velocities for pancreatic lipase (H. Brockerhoff, unpublished). No further additives are required, and reactions can be followed at a lower pH than in the case of long-chain triglycerides; even at pH 6, 90% of the butyric acid released is titratable.

Fruchart *et al.* (1972) have developed a semiautomatic procedure for the determination of serum lipase which involves titration of a serum extract. Jaillard *et al.* (1972) applied the Technicon automated determination of glycerol to the estimation of lipoprotein lipase activity in serum. The adaption offered advantages in speed and precision.

The measurement of "serum lipase" activity has been described in detail by Tietz and Fiereck (1972). Their procedure, intended for use in clinical laboratories, involves a potentiometric titration of the fatty acids hydrolyzed by serum lipase from olive oil that has been purified by slurrying with alumina followed by filtration.

E. Specificity

The substrate specificity of a lipase is defined by its positional specificity, i.e., the ability to hydrolyze only the primary or both primary and secondary ester bonds of a triglyceride; by its stereospecificity, i.e., the (hypothetical) ability to hydrolyze only ester 1 or only ester 3 or the triglyceride; by its preference for longer or shorter, saturated or unsaturated acids; and, in general, by the dependence of the reaction rates on the structure of the substrate.

Positional specificity can be readily determined. Thin-layer chromatography on silica gel impregnated with boric acid will separate 1,2- from 1,3-diacylglycerols and 1- and 2-monoacyl glycerols (Thomas *et al.,* 1965). These products can be quantitated directly by gas–liquid chromatography after conversion to trimethylsilyl ether derivatives(Tallent *et al.,* 1966; Tallent and Kleiman, 1968). Alternatively, a mixed triglyceride with known positional distribution of fatty acids can be used as a substrate. After lipolysis, the reaction products are separated and analyzed. For example, from a column of silicic acid, the solvent mixture hexane–ether 70:30 will elute triglycerides, fatty acids, and diglycerides (Barron and Hanahan, 1958); the monoglyceride can then be eluted with ether, and its fatty acid composition can be determined by gas–liquid chromatography. By thin-layer chromatography on silica gel with petrol ether–ether–acetic acid 70:30:2, all four products can be separated. Appropriate substrates can be prepared by synthesis (Mattson and Volpenhein, 1962; Quinn *et al.*, 1967; Jensen, 1972). In many cases, natural fats can be employed instead; lard, for example, is known to contain almost all its palmitic acid in position 2.

The stereospecificity of a lipase could be similarly detected with a substrate of known steric fatty acid distribution, which can be prepared by total synthesis or by acylation, in position 3, of a 1,2-diglyceride prepared from a phospholipid with the help of phospholipase 3 (Tattrie, 1959; Hanahan *et al.*, 1960; Brockerhoff, 1965b). Morley and Kuksis (1972) tested lipoprotein lipase for stereospecificity by isolating the diglycerides from the lipolysis mixture and converting them to phosphatidylphenols, which were then subjected to phospholipase 2. Only the phospholipid derived from the 1,2-diglyceride is hydrolyzed by this enzyme (Brockerhoff, 1965a,b); thus, the extent of hydrolysis, if it should deviate from 50%, would point to a stereospecific action of the lipase.

In determining the preference of a lipase for acids of certain chain lengths it must be ascertained that changes in reaction rates that are the result of changes of the interface are not mistaken for chain-length specificity. There are numerous reports on such specificity, especially concerning pancreatic lipase, in which the possibility of interfacial effects has not been taken into account. First of all, the relative velocities of hydrolysis of monoacid triglycerides cannot be used to determine chain-length specificity (which refers to the fatty acid part of an ester), because the alcoholic moiety is changed together with the fatty acid. Mixed esters of glycerol (Clément *et al.*, 1962a) or ethylene glycol (Mani and Lakshminarayana, 1970) in which the two primary groups are esterified with different fatty acids have been used in attempts to overcome this difficulty; the relative percentage of hydrolysis of the acids has been interpreted as relative chain length specificity. However, these assays fail to consider the possibility of substrate orientation; it is likely that the shorter end of the molecule is more often exposed to the interface. In fact, such experiments tend to show higher specificity of pancreatic lipase for short-chain acids.

When synthetic mixed triglycerides are mixed with carrier triglycerides such as triolein, it is found that a low molecular weight of the total molecule facilitates hydrolysis by the lipase, but that both long and short acids of such triglycerides are hydrolyzed at approximately equal rates (Jensen *et al.*, 1964b); this may be explained by the higher hydrophilicity or diffusional mobility of the lower triglycerides.

A more direct method of determining chain-length specificity makes use of monoesters (Mattson and Volpenhein, 1969; Brockerhoff, 1970). All objections that apply to triglycerides or glycol esters that can thus be eliminated. Such experiments do not, of course, settle the question of whether a chain-length specificity that may have been found is the result of interfacial mobility, orientation, hydration, and similar factors, or the result of enzyme–substrate binding and geometrical fit.

Similar arguments apply to the specificity of lipases against acids of different unsaturation. For example, a resistence of tripalmitin and tristearin as compared to triolein is more likely to be the result of the crystalline structure of the supersubstrate than the result of the structure of the fatty acid. Only in one lipase, that of the mold *Geotrichum,* has a definite specificity (for *cis* Δ9 unsaturated acids) been established (Alford and Pierce, 1961).

Substrate specificity in the most general sense, i.e., dependence of reaction rates on the chemical, steric, inductive, or mesomeric properties of substrates must be determined with synthetic substrates other than triglycerides (Brockerhoff, 1968).

II. PANCREATIC LIPASE

A. Introduction

Pancreatic lipase is the best known and most often investigated of all lipolytic enzymes, and it is for this reason that it occupies a disproportionately large section in this presentation. This intensive coverage can be further justified on the ground that pancreatic lipase can serve as a model and prototype for other digestive lipases, and for the lipases of plants and microorganisms as well, and probably also, to a considerable extent, for the mobilizing lipases of tissues. In addition, the enzyme has found extensive use as a reasearch tool in lipid chemistry and biochemistry owing to the fact that it specifically hydrolyzes the esters of primary alcohols. This property has been much exploited in the analysis and synthesis of fats and other glycerides; several reviews are available on the subject (Coleman, 1963; Litchfield, 1972). Authoritative reviews on pancreatic lipase from the standpoint of an enzymologist have been presented by Desnuelle (1961, 1971, 1972). Bibliographic information on the history of lipase research can be found in the book on lipids by Deuel (1955) and in the review by Wills (1965).

Mammals cannot assimilate intact dietary glycerides; as far as known, neither can other vertebrates nor the arthropods. The triglycerides must be partially hydrolyzed before they can be absorbed in the intestinal tract, and this hydrolysis is performed by digestive lipases which are produced in the pancreas of mammals or in the corresponding organs of lower animals, e.g., the "diffuse pancreas" of bony fishes and the hepatopancreas of arthropods. In the mammalian organ, the enzyme is formed in the acinar cells and accumulates in the zymogen granules that contain the complement of digestive enzymes, or their proenzymes (Hokin, 1967). From these granules, the lipase is released into the ducts that carry the pancreatic juice to its destination in the small intestine. No evidence has been found for a proenzyme or zymogen of lipase, and there is no reason to expect its existence. In contrast to proteinases or phospholipases, whose active forms might attack constituents of the pancreatic cells, lipase has a pronounced specificity for its proper substrate, triglyceride, emulsified as it occurs in the intestine but not in cells. There is, therefore, no need to protect the pancreas from the enzyme.

Pancreatic lipase was one of the earlier enzymes to be recognized; Claude Bernard described its action in 1856. Subsequently the enzyme received much attention from physiologists. The first intensive effort by biochemists to purify and characterize it was made by Willstätter and his group in the nineteen-twenties (1923; Willstätter and Memmen, 1928). Chromatographic methods were then, and have since been, employed, but the purification of the enzymes proved exceedingly difficult and has fully succeeded only very recently with the isolation of the lipase of the pig (Sarda *et al.,* 1964; Verger *et al.,* 1969) and of the rat (Gidez, 1968; Vandermeers and Christophe, 1968.)

B. Isolation of Pancreatic Lipase

The attempts of Willstätter's group to isolate pancreatic lipase (Willstätter *et al.,* 1923; Willstätter and Memmen, 1928) led to some enrichment of the enzyme but were on the whole unsuccessful, and so were attempts by others during the following thirty years. Willstätter sought to exploit the tendency of lipase to adsorb to its substrate, triglyceride, or to other lipid materials. This tendency is still used in some modern procedures to prepare highly enriched lipase ("fast" lipase), although it has proven to be a stumbling block in the preparation of absolutely pure lipase.

Until recently, the most effective method of purification (Marchis-Mouren *et al.,* 1959) employed adsorption to calcium phosphate and aluminum hydroxide followed by zone electrophoresis to yield a nearly pure enzyme. This method, however, yielded no more than a milligram of enzyme per run, and it has now been superseded by simpler and more effective procedures.

1. Fast and Slow Lipases

The first step in the purification of lipase has usually been the preparation of the dehydrated and defatted acetone powder of pig pancreas. When

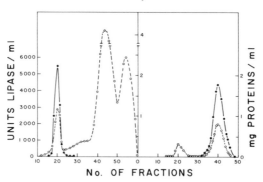

Fig. IV-1 Behavior of porcine pancreatic lipase on Sephadex G-200. Lipase activity (solid line); protein (broken line). (*Left*) Pancreas extract, fast lipase in first peak. (*Right*) Rechromatography of the fast lipase after treatment with ether–alcohol–deoxycholate solution. From Sarda *et al.* (1964).

aqueous extracts of this powder are chromatographed on the molecular-sieving gel Sephadex, the lipase appears in the first protein fraction that emerges from the column (Gelotte, 1964; Downey and Andrews, 1965b); from the elution volume an apparent molecular weight of 300,000 or more can be calculated. This preparation has been called the fast lipase (Sarda *et al.*, 1964). Under certain conditions, a second, more slowly migrating peak of lipase appears (Gelotte, 1964; Berndt *et al.*, 1968). A fast lipase can also be obtained from rat pancreas (Ramachandran *et al.*, 1970), whereas rat pancreas juice yields a preparation of lower apparent molecular weight (Morgan *et al.*, 1968). Pancreatic juice from humans has yielded a fast lipase as well as a slow lipase (Fraser and Nicol, 1966; Erlanson and Borgström, 1970b). Figure IV-1 (left) shows the appearance of the fast lipase peak in the chromatography of pancreatic proteins on Sephadex-200. The starting material was an aqueous extract of fresh pig pancreas, stabilized with diisopropyl fluorophosphate (DFP) against digestion by proteinases and concentrated by precipitation of the protein with ammonium sulfate. From the proportion of the peak to the remaining protein peaks, it is apparent that a considerable concentration of the enzyme has taken place. A comparable preparation from rat pancreas has been described as nearly pure (Ramachandran *et al.*, 1970).

Since the fast lipase does not occur in pig pancreatic juice, and the yield of slow lipase from the pancreas can be increased at the expense of the fast enzyme (Gelotte, 1964), it has been proposed that the fast enzyme is a multimolecular aggregate of the slow monomolecular form (Gelotte, 1964; Sarda *et al.*, 1964; Downey and Andrews, 1965b). Figure IV-1 (right) shows that the fast lipase can indeed be converted into the slow lipase by treatment with a mixture of aqueous deoxycholate, ether, and alcohol.

A further concentration of the enzyme is thereby achieved, and subsequent ion-exchange chromatography of this material on DEAE-cellulose has led to a preparation of high activity (Sarda *et al.*, 1964), which was, however, still contaminated with other proteins and contained about 6% of extractable lipid (Verger *et al.*, 1969). Furthermore, the treatment of the fast lipase with solvents usually entails a large loss in activity.

The nature of the fast multimolecular enzyme complex is suggested by its breakdown on treatment with organic solvents. It is an aggregate of lipase molecules and lipids, especially phospholipids, which are known to be insoluble in acetone and therefore not removed during the preparation of pancreas powder. Fast lipase is an artifact formed during the extraction of the powder (and also of fresh pancreas) with water, and may contain fats, fatty acids, phospholipids, and lysophospholipids (Schoor and Melius, 1969). An acidic phospholipid forms an especially tenacious association with the enzyme (Verger *et al.*, 1969).

2. *Purification of Porcine Pancreatic Lipase*

The complete separation of the tightly bound lipids from the fast lipase requires drastic conditions which lead to an extensive loss of enzymatic activity; however, a thorough delipidation of the pancreas before the extraction of the lipases does not affect the enzyme. From such pancreas powder, Verger *et al.* (1969) could isolate pure lipase, and their procedure will be described here.

The fresh pancreas is homogenized and extracted several times with chloroform–butanol mixtures, then with acetone and ether. These extractions remove 98% of the lipid. The dried pancreas powder is extracted with buffer at pH 9, which contains DFP to inactivate the proteinases, and from the extract those proteins are collected that precipitate with ammonium sulfate between 0.32 and 0.52 saturation (Table IV-1). At this

TABLE IV-1

Purification of Porcine Pancreatic Lipase[a]

Method	Number of lipase units ($\times 10^3$)	Specific activity	Yield (%)
1. Extract of pH 9.0 of 40 gm of delipidated powder	1000	150	100
2. Ammonium sulfate fractionation	700	220	70
3. *n*-Butanol treatment and dialysis	540	120	54
4. DEAE-cellulose chromatography	350	2000	35
5. Sephadex G-100 filtration	300	4500	30

[a] From Verger *et al.* (1969).

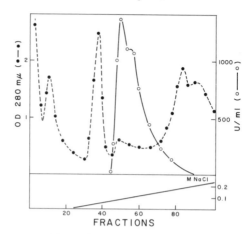

Fig. IV-2 Chromatographic purification of porcine pancreatic lipase on DEAE-cellulose. Elution with a linear concentration gradient of sodium chloride. Lipase (solid line); protein (broken line). From Verger *et al.* (1969).

stage, the remaining acidic phospholipid is removed by agitating an ammonium sulfate-containing buffer solution of the proteins with butanol. This treatment, described by Melius and Simmons (1965), breaks the protein–lipid association and yields a lipid-free cream of protein which accumulates at the butanol–water interface. This material can now be subjected to the chromatographic separations on basic ion-exchange (DEAE) cellulose and molecular sieving gel (Sephadex) that had earlier been described by Benzonana *et al.* (1964). Figure IV-2 shows that a highly concentrated lipase is eluted from DEAE-cellulose by gradient solution of sodium chloride. The shape of the lipase peak indicates the presence of two enzymes; this will be discussed later. At this stage, the enzyme is still accompanied by deoxyribonuclease and by part of the amylases of the pancreas. These are removed in the final purification on a column of Sephadex G-100 (Fig. IV-3). Table IV-1 summarizes the complete procedure. The course of purification is followed throughout by determinations of specific activity and by analytical disk gel electrophoresis. In the disk gel, the final preparation shows only two bands running closely together, corresponding to the lipase species L_A and L_B. The sedimentation pattern obtained on ultracentrifugation also attests the purity of the enzyme. The specific activity of the lipase against olive oil is 4500–5000.

A similar procedure reported by Garner and Smith (1970a) also starts with a thorough solvent extraction of porcine pancreas. Phenylmethyl sulfonylfluoride, urea, and 2-mercaptoethanol are employed to inhibit the proteinases; the treatment with butanol and the chromatography on DEAE-

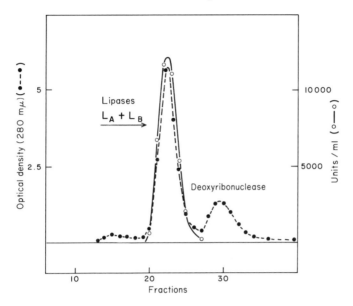

Fig. IV-3 Filtration of a porcine lipase fraction obtained by chromatography on DEAE-cellulose (Fig. IV-2) through Sephadex G-100. From Verger *et al.* (1969).

cellulose are omitted. The final preparation contains about 50% lipase, probably still associated with some lipid material. In a later study, Garner and Smith (1972) used the method of Verger *et al.* (1969).

Bovine pancreatic lipase has been purified in a yield of 35% by a similar procedure (Julien *et al.*, 1972). The preparation showed one band on disk gel electrophoresis and had a specific activity of 3000 (pH 9, 25°C, triolein). The molecular weight was 48,500.

3. Purification of Rat Pancreatic Lipase

Gel filtration of rat pancreatic juice yields a lipase with a molecular weight of about 40,000; however, the purity of this preparation has not been assessed (Morgan *et al.*, 1968).

Extraction of rat pancreas with a phosphate buffer at pH 7.2 containing 0.1% deoxycholate has yielded a fast lipase (MW > 3,000,000) that can be purified by repeated gel filtrations (Ramachandran *et al.*, 1970). The final product probably contains considerable amounts of lipid.

Slow lipase is extracted from the gland with phosphate buffer at pH 8.2 or 8.5 without the addition of a surfactant. Gel filtration and ion-exchange chromatography of this extract result in the isolation of the enzyme. Two essentially similar procedures have been described, one of them (Gidez, 1968) apparently leading to a product of somewhat higher specific

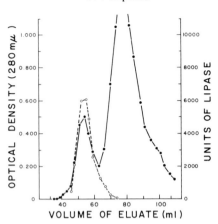

Fig. IV-4 Chromatography of a rat pancreas protein fraction (Table IV-2) on Sephadex G-100. Lipase (broken line); protein (solid line). From Gidez (1968).

activity and lower molecular weight than the other (Vandermeers and Christophe, 1968). Again, it is lipid material that clings tenaciously to the enzyme; the lipase of Vandermeers and Christophe (MW 43,000) contained 14% of lipid, of which 37% were free fatty acids, 35% phospholipids, 20% cholesterol esters, and 8% glycerides.

In the procedure of Gidez (1968), glands are homogenized with phosphate buffer of pH 8.2 (with added DFP). The supernate from the following centrifugation is dialyzed and passed through a column of DEAE-cellulose at pH 8.2. Since rat pancreatic lipase is a "cationic" protein it appears in the breakthrough peak. The pancreatic amylase is then precipitated by complex-formation with glycogen and addition of ethanol to a concentration of 12%. Subsequent gel filtration (Fig. IV-4) results in a threefold increase of specific activity, and when the most active fractions of Fig. IV-4 are subjected to chromatography on the anionic gel carboxy-methyl-Sephadex, a lipase with the specific activity of 5300 is obtained in the central fractions. A summary of the purification procedure is given in Table IV-2. The enzyme is pure as judged by disk gel electrophoresis. It has an approximate molecular weight of 32,000. It is possible that the ethanol precipitation of the glycogen–amylase complex removed the lipid that is usually associated with the lipase; however, the lipid content of the final preparation has not been determined.

Bradshaw and Rutter (1972) have found two lipases, A and B, in rat pancreas and pancreatic juice. The enzymes differed in electrophoretic mobility. Their relation to the lipase purified by Gidez (1968) and Vandermeers and Christophe (1968) is unknown. The level of pancreatic lipase in the rat can be increased by feeding a diet rich in fat (Gidez, 1973).

TABLE IV-2

Purification of Rat Pancreatic Lipase[a]

Fraction	Activity recovered (%)	Mean specific activity
Homogenate	100	126 (80–182)
40,000 rpm supernate	59 (48–73)	166 (86–370)
DEAE eluate	93	326
Amylase supernate	—	1010
G-100 eluate	52	3280
Carboxymethyl-Sephadex eluate	60	5330

[a] From Gidez (1968).

C. Chemistry of Pancreatic Lipase

The primary structure of pancreatic lipase, i.e., its amino acid sequence, is as yet unknown. We are restricted to discussing the amino acid composition, the reaction of lipase with various inhibitors, and the possible occurrence and importance of some amino acids in the active center of the enzyme and elsewhere. A large part of our present knowledge of the biochemistry of pancreatic lipase originated in the laboratory of P. Desnuelle and his colleagues in Marseille. While this is true for all aspects of research on the enzyme, the present chapter is nearly exclusively based on the recent findings of the research group in Marseille.

1. Porcine Pancreatic Lipase L_A and L_B

The uneven peak of porcine lipase that emerges from a DEAE-cellulose ion-exchange column (Fig. IV-2) indicates that there is more than one species of the enzyme present. Disk gel electrophoresis shows that two components are also present in the purest lipase obtained after Sephadex filtration. These two species can be cleanly separated by chromatography on the anionic carboxymethylcellulose (Verger *et al.,* 1969). They emerge consecutively with a pH gradient increasing from pH 5.0 to pH 5.2 (Fig. IV-5). The first molecular species has been named L_A, because it is more acidic; the second, more basic species, is L_B. Both species have the same specific activity, around 4500 against olive oil. When lyophilized, they are quite stable for long periods in the refrigerator. The molecular weight is obviously nearly the same in both species, since both have been obtained from a single Sephadex fraction; it is around 45,000 according to retention on Sephadex G-200, and 50,000 according to ultracentrifugation.

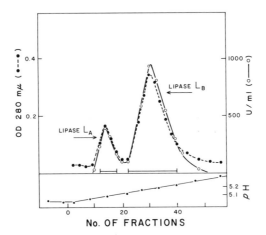

Fig. IV-5 Separation of porcine pancreatic lipases L$_A$ and L$_B$ by chromatography on carboxymethyl-cellulose with a linear acetate pH gradient. Lipase (solid line); protein (broken line). From Verger *et al.* (1969).

2. Amino Acid Composition of Pig and Rat Lipase

The amino acid compositions of L$_A$ and L$_B$ are indistinguishable except perhaps for a slight difference in the isoleucine content (Table IV-3) (Verger *et al.,* 1969; Verger, 1970). Since the isoleucine residue is uncharged, this difference cannot explain the different behavior of L$_A$ and L$_B$ on ion-exchange chromotography.

The amino acid composition of rat pancreatic lipase has been calculated from the analysis given by Vandermeers and Christophe (1968) with the assumption of a molecular weight of 34,500, intermediate between that determined by these authors and that of Gidez (1968). The composition shown in Table IV-4 is similar to that of the porcine lipases of Table IV-3.

A surprising feature of these analyses is the presence of only "normal" amounts of apolar amino acids. Since lipases attack hydrophobic substrates and form strong associations with lipids, it might have been expected that they contain a surplus of hydrophobic amino acids to facilitate these associations. However, if, according to Hatch (1965), the residues Asx, Glx, Lys, Arg, Ser, and Thr are considered as polar, Val, Leu, Ile, Met, Pro, and Phe as apolar, the ratios of apolar to polar amino acids are found not to be especially high. They are, for lipase L$_A$, 0.74; for L$_B$, 0.72; for rat lipase, 0.72 (Verger, 1970). For comparison, the ratio for porcine chymotrypsinogen A is 0.80; for porcine ribonuclease, 0.47; carboxypeptidase A$_I$, 0.61; amylase I, 0.67 (Hatch, 1965).

Although the lipase molecule as a whole does not have an excessively hydrophobic composition, it is, of course, possible that its secondary and

TABLE IV-3

Composition in Amino Acids of the Porcine Lipases L_A and L_B[a]

Amino acids	Number of residues per mole			
	Lipase L_A experimental value	Lipase L_B experimental value	Lipase L_A nearest whole number	Lipase L_B nearest whole number
Ala	20.4	20.3	20	20
Arg	19.6	20.0	20	20
Asx	56.4	56.4	56	56
½Cys	11.4	11.0	12	12
Cys	1.95	1.95	2	2
Glx	37.9	38.1	38	38
Gly	39	36.6	39	37
His	10	10.1	10	10
Ile	27.6	23.6	28	24
Leu	28.8	28.6	29	29
Lys	21.4	21.8	21	22
Met	4.1	4.0	4	4
Phe	22.0	22.7	22	23
Pro	24.9	23.9	25	24
Ser	28.5	28	28	28
Thr	26.3	24.9	26	25
Try	7.3	6.9	7	7
Tyr	15.4	15.4	15	15
Val	34.2	33.0	34	33

[a] From Verger (1970).

TABLE IV-4

Amino Acid Composition of Rat Pancreatic Lipase

Amino acid	Moles/ 10^5 gm[a]	Number of residues/mole[b]	Amino acid	Moles/ 10^5 gm[a]	Number of residues/mole[b]
Ala	49.0	17	Lys	52.4	18
Arg	36.8	13	Met	14.9	5
Asx	118.4	41	Phe	47.9	17
Cys	23.6	8	Pro	37.7	13
Glx	69.5	24	Ser	61.7	23
Gly	83.4	29	Thr	53.2	18
His	22.3	8	Try	16.2	6
Ile	48.3	17	Tyr	29.6	10
Leu	58.4	20	Val	70.9	24

[a] Calculated from Vandermeers and Christophe (1968).
[b] Assumed molecular weight 34,500.

tertiary structure are such that a considerable number of apolar amino acids are exposed at the surface of the molecule rather than buried in the inside, and that they form a hydrophobic head of affinity to interfaces (Brockerhoff, 1973).

3. The Carbohydrate Moiety

Both porcine lipase L_A and L_B contain carbohydrate. Verger (1970) found 3.1 and 2.7 residues of glucosamine, Plummer (1970) found in addition 2.2 residues of mannose; these data have not been published except in a thesis by Verger (1970). Garner and Smith (1972) found 3.8 moles of mannose and 2.9 moles of N-acetylglucosamine in both lipases.

It has been suggested that the carbohydrate moieties are remote from the active site, like those of ribonucleases (Jackson and Hirs, 1970), and that they form a hydrophilic tail opposite the hydrophobic head of the enzyme and thus assist in the proper interfacial orientation of lipase (Brockerhoff, 1973).

4. Sulfhydryl Groups of Porcine Pancreatic Lipase

Those sulfhydryl groups of cysteine residues that are not involved in disulfide linkages can be detected by a number of reagents. Among these, iodoacetate and iodoacetamide have been found not to inhibit pancreatic lipase (Barth, 1934), whereas p-chloromercuribenzoate causes only partial inhibition (Barrón and Singer, 1943). It has been concluded that the lipase is not a sulfhydryl enzyme but that the blocking of sulfhydryl groups may interfere with the enzymatic activity by a steric effect (Wills, 1960). With the isolation of a pure lipase from porcine pancreas (Verger *et al.,* 1969) and the introduction of new sulfhydryl reagents such as 5,5'-dithiobis(2-nitrobenzoic acid) (DTNB) (Ellman, 1959), quantitative studies of the sulfhydryl groups have become possible. Such studies have been carried out by Verger *et al.* (1971) on the porcine lipase L_B.

The lipase contains two sulfhydryl groups. One of them (SH_I) reacts readily with DTNB, p-chloromercuribenzoate, N-ethylmaleimide, and phenylmercuric ion. The second group (SH_{II}) is buried and reacts only with the last reagent in the native enzyme. After denaturation of the enzyme with 0.3% sodium dodecyl sulfate or 8 M urea, all reagents, and also iodoacetate and iodoacetamide, react with both groups. The reaction of DTNB with the native or denatured enzyme can be conveniently followed by photometric titration (Fig. IV-6). The mono-5-thio-2-nitrobenzoic acid (TNB) lipase can be prepared, as can the di(phenyl-mercuic) (DPM) lipase, and also a mixed derivative carrying TNB on the SH_I group and phenylmercury on SH_{II}. The native lipase can be recovered with thiol

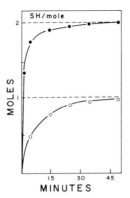

Fig. IV-6 Titration of the thiol groups of porcine pancreatic lipase with DTNB. Ordinate: moles thionitrobenzoate released as determined photometrically. Native lipase (open circles); lipase in the presence of 0.3% sodium dodecylsulfate, (solid circles). From Verger *et al.* (1971).

reagents like cysteine. From the mono-TNB lipase, the TNB can be removed by denaturation to yield a (nonactive) lipase with an additional disulfide bridge and no more free sulfhydryl groups.

Lipase $\begin{array}{c} SH_I-S \\ \\ SH_{II} \end{array}$ —⟨NO_2 ring⟩—COO$^-$ $\xrightarrow[\text{dodecyl sulfate, } O_2]{\text{urea,}}$ Lipase $\begin{array}{c} S_I \\ \\ S_{II} \end{array}$

Mono-TNB-lipase   Oxidized lipase

Lipase $\begin{array}{c} S_I-Hg-⟨\text{ring}⟩ \\ \\ S_{II}-Hg-⟨\text{ring}⟩ \end{array}$

DPM-Lipase

The substituted compounds prepared from native lipase are still enzymatically active. Mono-TNB lipase displays the original maximal reaction rate, *V*, with emulsified triglycerides, but its "K_m" is increased ten times. In DPM lipase the "K_m" remains unchanged, but *V* is reduced by 40%. On the basis of these results, Verger *et al.* (1971) suggest that SH_I is near a site that is responsible for the hydrophobic attachment of lipase to interfaces, and that the substitution with the negatively charged TNB group interferes with this attachment. The SH_{II} group would be near the catalytic site; its substitution with phenylmercury introduces steric hindrance at the active center of the enzyme. At any rate, the results show clearly that pancreatic lipase is not a "sulfhydryl enzyme."

5. The Reactive Serine

Many hydrolytic enzymes are "serine enzymes": they contain a reactive serine residue that attacks the substrate with its alkoxy oxygen as a nucleophile (Bender and Kézdy, 1965). These enzymes are usually inhibited by a number of organophosphoric compounds that phosphorylate the reactive serine. The most common of these compounds is DFP, diisopropyl fluorophosphate. Lipase, however, is not usually inhibited by DFP (Maylié et al., 1972) nor by aliphatic sulfonyl halides; in fact, DFP may advantageously be added to lipase because it protects the enzyme from the attack of proteinases (Verger et al., 1969).

Recently, an inhibition of pancreatic lipase by incubation with highly concentrated DFP solutions was reported by Maylié et al. (1969). After digestion of the DFP-labeled enzyme with chymotrypsin, a peptide could be isolated that had the structure Thr–Asn–Glu–(Asx_{1-2}, Glx_{0-1}, Tyr)–Leu and in which, according to several tests, the phosphate was esterified to the hydroxyl of the tyrosine. However, this phosphorylation was later found to have no direct relation to the inactivation of the enzyme (Sémériva et al., 1971; Maylié et al., 1972). The labeled tyrosine is, therefore, not a part of the active center.

The organophosphate diethyl-p-nitrophenyl phosphate had been found in 1960 by Sarda et al. to inhibit lipase. It appeared that only emulsions but not solutions of this compound reacted with the enzyme, and this was interpreted as a further example of lipolytic action at an interface (Desnuelle, 1961) and seemed to offer an explanation why the water-soluble DFP did not inhibit. However, eventually it could be shown that diethyl-p-nitrophenyl phosphate could also be made to react with pancreatic lipase in an aqueous solution in the presence of bile salt (Maylié et al., 1969). Figure IV-7 shows that the amount of p-nitrophenol released during the reaction corresponds closely to the deactivation of the enzyme.

Proteolytic digestion of the lipase labeled with diethyl phosphate has yielded a peptide of the formula Leu–Ser(P)–Gly–His, in which the phosphate is indeed bound to a serine residue (Desnuelle, 1971; Maylié et al., 1972). On the basis of this result, lipase can be assumed with considerable probability to be a serine enzyme. The fact that DFP does not inhibit lipase is probably due not so much to its not being presented at an interface as to an effect of steric hindrance; this will be discussed later.

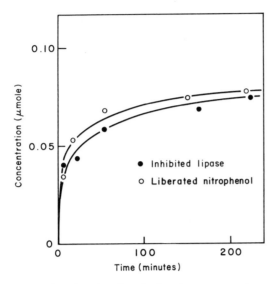

Fig. IV-7 Action of diethyl-*p*-nitrophenyl phosphate on porcine pancreatic lipase. Lipase, 0.097 μmole, in a solution of DNP (4.6 m*M*), bile salt (0.1%), NaCl (0.1 *M*), CaCl₂ (0.05 *M*), at pH 6 and 0°C. From Maylié *et al.* (1969).

The amino acid sequence around the reactive serine is remarkably different from the sequences known for other hydrolases (Table IV-5). The second amino acid from the carboxyl end of the serine is histidine rather than the usual glycine; the significance of this difference remains unknown. The replacement of the acidic aspartic or glutamic acid, or the polar threonine, with the apolar leucine elicits a number of speculations. First, the

TABLE IV-5

Amino Acid Sequence around the Reactive Serine of Several Hydrolases

Hydrolase	Amino acid sequence
Ox liver esterase[a]	–Gly–Glu–Ser(P)–Ala–Gly
Pig kidney esterase[b]	–Gly–Glu–Ser(P)–Ala–Gly
Subtilisin[c]	–Gly–Thr–Ser(P)–Met–Ala
Bovine chymotrypsin[d]	–Gly–Asp–Ser(P)–Gly–Gly
Porcine pancreatic lipase[e]	–Leu–Ser(P)–Gly–His

[a] Augusteyn *et al.* (1969). [d] Blow (1971).
[b] Heymann *et al.* (1970). [e] Desnuelle (1971).
[c] Oosterbaan and Cohen (1964).

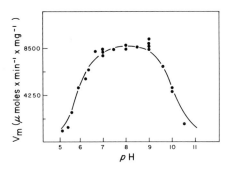

Fig. IV-8 pH dependence of V of porcine pancreatic lipase. Experimental values are represented by filled circles. The solid line is a theoretical curve calculated using values of $pK_1 = 5.8$ and $pK_2 = 10.1$. Lipase activity was measured on tributyrin emulsions at 25°C. From Sémériva et al. (1971).

aspartic acid of pancreatic chymotrypsin is essential in establishing the confirmation of the active enzyme after the hydrolysis of the proenzyme (Blow, 1971); (subtilisin has no proenzyme). The absence of the aspartic residue in lipase is in accord with the nonexistence of a prolipase. In chymotrypsin, the aspartic acid forms a salt linkage with a remote isoleucine residue, and its CH_2COO^- moiety consequently points away from the serine and does not encumber it. For the $(CH_3)_2CH_2CH_2$ group of the leucine, however, there is no *a priori* reason why it should not exert an effect of steric hindrance on the neighboring serine; such an effffect would go far to explain some differences in substrate specificity. Finally, the leucine residue must increase the hydrophobicity of the active center considerably; this is in harmony with the hydrophobic nature of lipolysis.

6. Reactive Histidine; Carboxyl Functions

In the serine proteinases, a histidine residue is known to be involved in the catalytic mechanism (Blow, 1971). Evidence for the involvement of a histidine in lipolysis has come from the photooxidation studies of Sémériva et al. (1971). First, it was found that the dependence of the maximal rate of hydrolysis, V, of tributyrin on pH fitted a bell-shaped curve with inflection points at pH 5.8 and pH 10.1 (Fig. IV-8); and imidazole is the only residue in proteins whose ionic state could vary around pH 5.8. Second, the lipase was subjected to photooxidation. Tryptophan, cysteine, methionine, and histidine residues were found to be oxidized. Several tests indicated that none of the first three groups of amino acids was essential for the activity of the enzyme. The photooxidation of the histidines followed a pattern that could be separated into three first-order reactions (Table IV-6). One histidine residue was oxidized with a rate constant of

TABLE IV-6

Calculated First-Order Rate Constants of the Photooxidation of Porcine Pancreatic Lipase[a]

Reactions	Rate constants (min^{-1})
Lipase inactivation	0.160
Tryptophan oxidation	
Reactive	0.490
Unreactive	0.015
Methionine oxidation	
Reactive	0.290
Unreactive	0.042
Cysteine oxidation	
SH_I	0.008
SH_{II}	0.120
Histidine oxidation	
Reactive I	0.690
Reactive II	0.150
Unreactive	0.021

[a] From Sémériva *et al.* (1971).

$k_1 = 0.69$ min^{-1}, the second with $k_2 = 0.15$ min^{-1}, and the remaining eight histidines with $k_3 = 0.021$ min^{-1}. Since the first-order deactivation of the lipase proceeded with $K = 0.16$ min^{-1}, and K_2 was the only rate near to this value, it was tentatively concluded that the histidine involved was an essential part of the active center of the enzyme (Sémériva *et al.*, 1971).

Modification of the free carboxyl groups of pig pancreatic lipase by amide formation with glycine or norleucine esters and carbodiimides as condensing agents leads to inactivation (Sémériva *et al.*, 1972). Only one of the carboxyl groups is essential for the activity of the enzyme; however, in contrast to the buried aspartic carboxyl that is involved in the catalytic mechanism of chymotrypsin (Blow, 1971), the carboxyl group modified in the lipase appears to be exposed.

D. Substrate Specificity

The specificity of an enzyme is normally defined by the chemical structure of its substrates. In the case of lipase, however, it is also the physical form of the substrate that has been of much interest to investigators; namely, the question of whether lipase needs an insoluble substrate, or whether it also attacks esters in aqueous solution. A discussion on the substrate specificity of pancreatic lipase can, therefore, be discussed with advantage under two aspects, one on the physical state of the substrates and one on their chemical structure.

1. Action on Insoluble Substrates

Willstätter and his co-workers (1923) discovered that pancreatic lipase could be adsorbed by insoluble lipids, and Schønheyder and Volqvartz (1945) described the action of lipase on the interface of an insoluble substrate, tricaproin, and water. At the time, however, soluble esters were also believed to be substrates for lipase. The clarification of the problem had to wait for the preparation of a reasonably pure enzyme. The hypothesis that action at an interface was not only typical but essential for lipase was expressed by Sarda et al. (1957), Sarda and Desnuelle (1958), and Desnuelle (1961).

Sarda and Desnuelle (1958) gave a convincing demonstration of the action of pancreatic lipase on insoluble substrates at the oil–water interface. In Fig. IV-9A, to the left of the vertical broken line, the substrate triacetin is dissolved in water (the saturation point is 1.0); to the right of the line, the substrate is offered as an emulsion with an increasing interfacial area. The difference is especially dramatic if the hydrolysis rates of the highly purified enzyme ($\triangle$) are considered. Figure IV-9B, for comparison, presents the hydrolysis of the same substrate by horse liver esterase. Here, the maximal rate is reached long before the solution becomes oversaturated, and the appearance of an emulsion does not change it.

These experiments showed the dependence of lipase on an interface, and they led to a widely accepted definition of lipases as carboxylesterases that act not on dissolved but only on emulsified substrates. Although recent work has shown that lipase can also act on dissolved substrates, in the

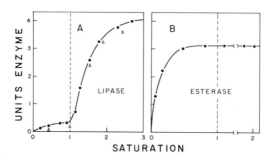

Fig. IV-9 Hydrolysis of triacetin by pancreatic lipase and horse liver esterase. Reaction rates as function of substrate concentration, which is expressed in multiples of saturation. The broken vertical lines indicate the saturation point. From Sarda and Desnuelle (1958).

absence of an interface, the specific ability of lipase to act at interfaces remains undisputedly one of its outstanding properties. It might be mentioned, here, that lipase acts on carboxyl esters but not on thioesters (Brockman *et al.*, 1973).

2. Action on Soluble Substrates

Entressangles and Desnuelle (1968) found that triacetin in aqueous solution may be attacked by pancreatic lipase, and at a much higher rate than that found previously (Sarda and Desnuelle, 1958) if the solution contained sufficient sodium chloride. The rates of lipolysis in 0.1 M aqueous NaCl are shown in Fig. IV-10 for triacetin and tripropionin. There is increasing hydrolysis with increasing substrate concentration even before saturation is reached, and the appearance of an emulsion causes no sharp break in the rate profiles. An explanation for this unexpected result has been sought in the known ability of sodium chloride and other salts to promote the molecular aggregation of poorly soluble compounds in water; instead of the multimolecular substrate aggregates at interfaces, smaller micellar aggregates might fulfill the minimum requirements for proper substrates. When the onset of micelle formation (critical micellar concentration) is measured for the two substrates, triacetin and tripropionin (Fig. IV-11), it appears, indeed, that the enzymatic hydrolysis depends, more or less, on the presence and concentration of micelles. The size of the micellar aggregates has been determined by light-scattering experiments; tripropionin micelles contain about fifteen molecules, triacetin micelles are much smaller, containing perhaps three molecules.

Interestingly, the maximal rate V of pancreatic lipolysis remains unchanged whether the substrate is emulsified or dissolved and micellar. The enzyme does not appear to require aggregates of very many substrate mole-

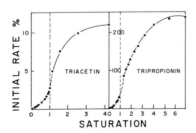

Fig. IV-10 Hydrolysis of triacetin and tripropionin by porcine pancreatic lipase in 0.1 M NaCl as a function of substrate concentration. The total concentration of the substrate (solution + emulsion) is expressed in multiples of the saturation. The vertical dashed line indicates the saturation point. From Entressangles and Desnuelle (1968).

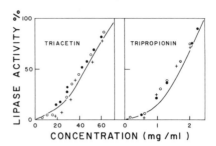

Fig. IV-11 Correlation between lipase activity and the formation of aggregates in triacetin and tripionin solutions. The continuous curve indicates lipase activity versus substrate concentration. The other signs indicate the responses of three different techniques used for the detection of aggregates. From Entressangles and Desnuelle (1968).

cules. The question must be raised, then, whether monomolecularly dispersed esters may also be hydrolyzed. A monoester that is completely miscible with water and very unlikely to form any micelles, 2-(2-ethoxyethoxy)ethyl acetate, is in fact attacked by lipase (Brockerhoff, 1969a). The rate is very low in comparison to that of good substrates, but the turnover number is still around 1000 min⁻¹.

3. Hydrophilic and Hydrophobic Substrates

When esters of different solubility in water are offered to pancreatic lipase *as emulsions* it is found that the more soluble esters are the poorer substrates (Table IV-7). This is especially apparent if corrections are applied to the rates considering the different electrophilic structures of the esters; the rationale for these corrections will be discussed later. Figure IV-12 presents graphically the maximal (corrected) rates V for a number of substrates against their solubility. It must again be stressed that the substrates are in emulsions. Therefore, the rates do not depend on the presence or absence of an interface but on the solubility, or hydrophilicity, of the substrates. It appears that the more hydrophobic esters are the better substrates regardless of their physical state. This result led to the hypothesis that the hydrophobicity of an ester group rather than its presence at an interface controls the rate of lipolysis (Brockerhoff, 1969a).

How can this hypothesis be reconciled with the indisputable fact that only insoluble esters at interface give the extremely high reaction rates typical for pancreatic lipase? First, hydrophobicity, micelle formation, and formation of an interface are parallel; very hydrophobic esters are, for that reason, insoluble and form interfaces. Triolein, for instance, is extremely hydrophobic; therefore it forms an interface with water, and is also a good

TABLE IV-7

Maximal Velocities of Hydrolysis of Emulsified Esters by Pancreatic Lipase

Ester	Velocity (tripropionin = 1.0)	Velocity corrected	Solubility (%)	Conditions
Triolein	0.5	0.5	0.00	pH 9; 37°C
Tributyrin	1.04	1.04	0.01	pH 9; 37°C
Tripropionin	1.0	1.0	0.25	
2-Methoxyethyl hexanoate	0.032	0.20	0.4	pH 8; 37°C
1,3-Dibutyrin	0.065	0.32	1.0	pH 7; 25°C
2-Methoxyethyl pentanoate	0.018	0.11	1.3	pH 8; 37°C
Glycol dipropionate	0.022	0.08	1.7	pH 8; 37°C
Cyanomethyl propionate	0.039	0.05	3	pH 8; 37°C
Triacetin	0.05	0.05	7	pH 7; 25°C
Cyanomethyl acetate	0.033	0.04	8	pH 7; 25°C
Glycol diacetate	0.004	0.02	12.5	pH 7; 25°C
2-(2-Ethoxyethoxy)ethyl propionate	0.009	0.03	15	pH 7; 25°C
2-(2-Ethoxyethoxy)ethyl acetate	0.0004	0.001	∞	6 % solution at pH 7; 25°C

[a] From Brockerhoff (1969a).

substrate; but the one is not the result of the other. Second, lipase is adsorbed to interfaces and finds there its substrates at extremely high concentration, higher than could possibly be achieved in an aqueous solution; in addition, the enzyme is oriented so as to bring its active site into the closest vicinity of the substrate molecules in the supersubstrate (Brockerhoff, 1973). This leads to high measured reaction rates which cannot be

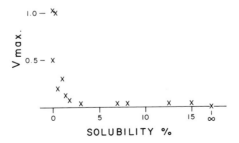

Fig. IV-12 Solubility of esters in water and relative maximal rates of lipolysis. V of tripropionin = 1.0. From Brockerhoff (1969a).

achieved in a solution when the substrate concentration is extremely low and the orientations of enzyme and substrate haphazard.

4. Sequence of Triglyceride Hydrolysis

Triglycerides are the natural substrates of pancreatic lipases, and their digestion is a subject of great interest for physiology. Triglycerides have therefore been used as substrates for lipase in the overwhelming majority of investigations. Although it has recently become clear that this has obscured rather than cleared the view for the enzymologist, triglycerides still deserve a separate treatment in our discussion.

Since triglycerides carry three esterified fatty acids, it is clear that their hydrolysis, enzymatic or chemical, must proceed in steps via diglyceride and monoglyceride to glycerol and free fatty acids or soaps. Hydrolysis in the intestine was originally assumed to be complete; Frazer and Sammons, however, in 1945, found that the principal products of *in vitro* digestion were lower glycerides rather than glycerol. Desnuelle *et al.* (1947) found that diglycerides were formed rapidly, monoglycerides more slowly, and glycerol very slowly. A quantitative study of the course of hydrolysis of olive oil was undertaken by Constantin *et al.* (1960). Figure IV-13 shows how the first product, diglyceride, gives rise to the second one, monoglyceride. Glycerol starts to appear only when almost all the triglyceride, and more than 50% of the total ester bonds, have been hydrolyzed. Monoglycerides, in contrast to triglycerides and diglycerides, are very poor substrates for pancreatic lipase. Whether they are attacked at all, or whether their hydrolysis is due to an enzyme other than pancreatic lipase will be answered later.

Mention must be made of the physiological significance of the resistance of monoglycerides. Monoglycerides, together with fatty acids and bile salts, form the micelles that are vehicles of fat absorption through the intestinal wall (Hofmann and Borgström, 1964). The monoglycerides (precisely,

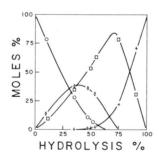

Fig. IV-13 Hydrolysis of olive oil by pancreatic lipase, in the presence of taurocholate and calcium chloride. Triglyceride (○); diglyceride (▲); monoglyceride (□); glycerol (+). From Constantin *et al.* (1960).

2-monoglycerides) then serve as the backbone for the resynthesis of the triglycerides that are then released into the lymph as chylomicrons. Thus, while triglycerides cannot be assimilated intact, the organism is spared the caloric expense of a total resynthesis.

5. *Positional Specificity*

Glycerol has two primary and one secondary hydroxyl group (disregarding, for the present, the prochiral properties of glycerol). The question which ester groups, primary or secondary or both, are hydrolyzed by pancreatic lipase was in effect answered by Balls and Matlack in 1938 in favor of the hydrolysis of primary esters. Nevertheless, the demonstration that the outer, primary ester groups of triglycerides are the only ones attacked came much later, and in the wake of some confusing reports. Schønheyder and Volqvartz in 1952 concluded from kinetic studies that tripropionin was degraded to 1,2-dipropionin; in 1954, they suggested that trilaurin was degraded to 1,2-dilaurin and 2-monolaurin. Borgström, in 1953, oxidized the diglycerides with CrO_3 and found that they were the 1,2-isomers. Final proof for the hydrolysis at the α-positions 1 and 3 was presented by Mattson and Beck (1955, 1956) and Savary and Desnuelle (1956). These authors used triglycerides with different fatty acids in the α- and β-positions and isolated and analyzed the monoglycerides obtained with pancreatic lipase, according to Scheme IV-1 (Mattson and Beck, 1955); in which

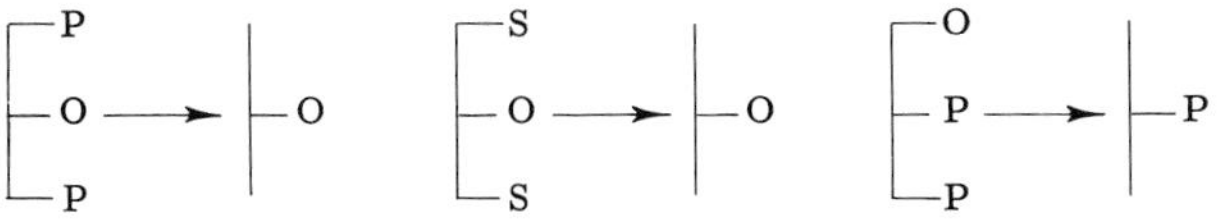

SCHEME IV-1

P stands for the ester of palmitic acid; O is oleic acid; S, stearic acid. The result proved the positional specificity of pancreatic lipase, and that this specificity is independent of the nature of the fatty acid. However, some α-monoglyceride is always found among the reaction products; this was explained as due to acyl migration of the acid from position 2 to position 1 or 3 of the monoglycerides (Mattson and Beck, 1956). Traces of 1,3-diglycerides are found for the same reason. Recent reinvestigations of the question have shown that the α-specificity of the lipase is, in fact, almost absolute (Entressangles *et al.,* 1966; Brockerhoff, 1968; Mattson and Volpenhein, 1968).

The breakdown of triglycerides by pancreatic lipase proceeds, then, in the manner depicted in Fig. IV-14. In this figure, we have anticipated the nonstereospecificity of the lipase. The major pathways and products appear

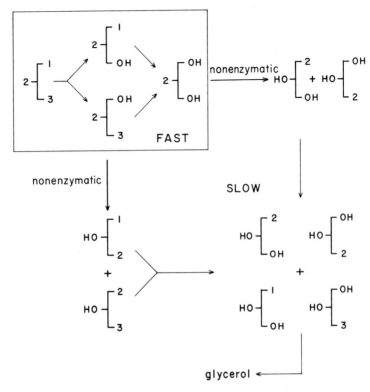

Fig. IV-14 Stepwise hydrolysis of triglycerides by pancreatic lipase. The quantitatively predominant reactions are in the rectangle.

in the rectangle; 1, 2, and 3 stand for the fatty acids in these positions; the liberated fatty acids are not displayed.

6. Absence of Stereospecificity

The two primary carbinol groups of glycerol are not sterically equivalent, they can be distinguished as positions 1 and 3 of *sn*-glycerol (Chapter II, Section I). Since stereospecificity is a typical feature of enzymes, it might be expected that pancreatic lipase makes a distinction between these positions; but the fact that the enzyme hydrolyzes both of the primary esters argues against this assumption. It would still be possible, however, that one ester group is hydrolyzed before the other, or faster than the other. The diglycerides formed should then be completely or partly optically active; however, this is not the case. Strict proof that pancreatic lipase is not stereospecific was supplied by Tattrie *et al.* (1958) and by Karnovsky and Wolff (1960). Tattrie *et al.* synthesized *sn*-glycerol 1,2-dipalmitate

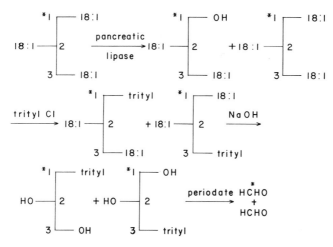

Fig. IV-15 Method used by Karnovsky and Wolff (1960) to show lack of stereospecificity of pancreatic lipase. The triolein glycerol was labelled in the *sn*-1 position with ^{14}C. If the lipase had been stereospecific one or the other of the diglyceride isomers would have been produced and the HCHO would have contained all or none of the label. Instead it contained half as shown.

3-oleate and *sn*-glycerol 1-oleate 2,3-dipalmitate and subjected them to pancreatic lipase. The diglycerides formed were in both cases equal mixtures of glycerol palmitate oleate; this proved random hydrolysis at positions 1 and 3. Similar experiments by Jensen *et al.* (1970) confirmed these results. The procedure used by Karnovsky and Wolff is explained in Fig. IV-15. Triolein prepared from *sn*-[1-^{14}C]glycerol was digested with the lipase, the diglycerides were hydrolyzed, and their free hydroxyl groups were tritylated. Saponification yielded trityl glycerols; these were oxidized with periodate. The ^{14}C in the formaldehyde formed by this oxidation was then counted. Had the lipase attacked only the 1 or the 3 position, the formaldehyde should have contained all or none of the activity; in fact, half the activity was observed.

Why is it that pancreatic lipase lacks steric specificity, whereas phospholipases are strictly stereospecific? Is it to be expected that digestive triglyceride hydrolases will be found that do possess stereospecificity? Probably not. Explanations for the lack of specificity can be given from an evolutionary or from molecular viewpoint.

Organisms use lipases to obtain free fatty acids for the synthesis of other lipids or as a source of energy, and the steric position in which the fatty acid had originally been bound does not matter. In contrast to phospholipids, triglycerides have no structural function; they occur in isolated droplets rather than as essential parts of biological membranes, and their configuration, or that of their breakdown products, is probably of little concern to the organism. This does not imply the nonexistence of species-specific or organ-specific triglycerides of nonrandom steric structure—indeed we know that they exist (Brockerhoff, 1971b)—but when they are broken down by lipases, their structures become irrelevant. Therefore, there has been no need for organisms to develop stereospecific digestive or mobilizing lipases.

Explanations in molecular terms are as follows: While the specificity of phospholipases results from the binding of the enzyme to the phosphate group in the natural *sn*-glycerol 3-phosphate structure, no such binding is necessary or possible for pancreatic lipase. Pancreatic lipase, as will be discussed later, requires only a single ester group that should be hydrophobic, unhindered, and activated, and other lipases are likely to have similar requirements. There is little probability that a stereospecific digestive lipase will be discovered; it is somewhat less unlikely that intracellular lipases may be found that are stereospecific. Such lipases might, for instance, use a pool of triglyceride as a source of the 1,2-diglyceride needed in the biosynthesis of phospholipids.

7. Synthetic Substrates: Influence of the Alcohol

Esters are composed of two parts, the alcohol and the acid, and in order to access the contribution of each part to the substrate specificity it is necessary to vary one and keep the other constant. With triglycerides as substrates, this is impossible. For instance, if the specificity of lipase against acetic and oleic acid should be compared, it must be remembered that in the respective triglycerides not only the acids but also the alcohols differ; diacetin and diolein have vastly different physicochemical properties. Unambiguous studies on substrate specificity can be carried out only with monoesters.

The first comparison of different alcohols was made by Balls and Matlack (1938). They found that all esters of primary alcohols were digested by pancreas extract, whereas secondary and tertiary esters were hardly attacked. After their report, no systematic or quantitative studies were made for three decades, but Sarda and Desnuelle reported in 1958 that methyl oleate was attacked twenty-five times more slowly than triolein. In 1968, oleates of different alcohols were compared as substrates for pancreatic lipase (Brockerhoff, 1968). The alcohols had been systematically selected for their chemical structures and properties.

The action of pancreatic lipase on its substrate can be assumed to proceed by nucleophilic attack on the carbonyl carbon of the ester. Nucleophilic esterolysis by chemical means has been extensively investigated. Among the factors that regulate hydrolysis rates, an inductive effect has

TABLE IV-8

Hammet Constants σ for Meta-bound Substituents[a]

Substituent	σ	Substituent	σ
O—	−0.71	Cl	+0.37
CH$_3$	−0.07	Br	+0.39
OH	−0.002	CO$_2$C$_2$H$_5$	+0.40
OCH$_3$	+0.12	CF$_3$	+0.43
C$_6$H$_5$	+0.22	CN	+0.68
F	+0.34	NO$_2$	+0.71

[a] Compiled from Hine (1956).

been recognized as important; if the alcohol moiety of an ester contains an electrophilic substituent, electrons are withdrawn from the carbonyl carbon and nucleophilic attack is facilitated. The inductive effect can be quantitatively described with the Hammett equation (Hammett, 1935):

$$\log (k/k_0) = \sigma\rho$$

in which k_0 and k are rate constants for the reaction of the unsubstituted (hydrogen) and the substituted compound; σ is the Hammett constant that characterizes the substituent; and ρ is a constant for the given reaction under given conditions. The substituent constant σ can be obtained by measuring the dissociation constants of substituted benzoic acids. Ortho and para substitution contribute resonance as well as inductive factors, but meta substitution yields Hammett constants σ that measure only the inductive effect. Table IV-8, compiled from a textbook of organic physical chemistry (Hine, 1956), shows these constants for some common substituents. The σ for hydrogen is zero. A negative σ signifies donation, a positive σ abstraction of electrons.

Table IV-9 shows the maximal rates V with which pancreatic lipase hydrolyzes the oleates of different alcohols, compared to the rate with which triolein is hydrolyzed. On comparing the values of Tables IV-8 and IV-9, the inductive effect of electrophilic substituents on lipolysis becomes apparent. The deactivating hydroxyl and alkyl groups result in low activities. The rates increase with an alkoxy and, significantly, with an ester group. Halogens promote lipolysis, and so does the phenyl radical of the benzyl group, and if this phenyl group carries further electrophilic substituents, very high reaction rates are achieved. The reaction constant ρ is obviously positive as for all nucleophilic reaction. A precise value for ρ cannot be calculated from the data because the measurements are not very accurate, the influence

TABLE IV-9

Speed of Hydrolysis of Esters of Oleic Acid[a] by Pancreatic Lipase

Ester	Relative speed of hydrolysis (triolein = 1.00)	Ester	Relative speed of hydrolysis (triolein = 1.00)
Triolein	1.00	FCH_2CH_2O-	1.2
$HOCH_2CH_2O-$	0.05	$ClCH_2CH_2O-$	0.25
Oleyl—	0.08	$BrCH_2CH_2O-$	0.16
CH_3O-	0.07	$NCCH_2O-$	0.80
$C_6H_{13}OCH_2CH_2O-$	0.16	$C_6H_5CH_2O-$	0.27
CH_2O-		$p-ClC_6H_4CH_2O-$	1.3
$\vert$	0.27	$p-NO_2C_6H_4CH_2O-$	1.4
CH_2O-			

[a] From Brockerhoff (1968).

of steric factors has not been assessed, and the special conditions of hydrolysis at the interface cannot be evaluated; steric factors, for instance, are likely to play a role in the decrease of activation from F to Cl to Br. Nevertheless, there can be no doubt as to the existence and importance of the inductive effect in pancreatic lipolysis.

The inductive effect has been invoked to interpret the hydrolysis of phospholipids by pancreatic lipase. It had been found (van den Bosch *et al.,* 1965) that pancreas extract attacks phospholipids not only at position 2 (the activity of phospholipase 2) but also, to some extent, at position 1. De Haas *et al.* (1965) showed that pancreatic lipase is responsible for this action, but that the hydrolysis of a phospholipid such as lecithin by the lipase is much slower than that of a triglyceride. Even slower is the attack of β-lecithin. These results agree well with the concept of electrophilic induction (Brockerhoff, 1968; Slotboom *et al.*, 1970a). Figure IV-16 shows that in lecithin one activating ester group of a triglyceride has been replaced by a phosphate group, which has a high electron density and a negative charge. Such a group is electron-donating and therefore deactivating. In the β-lecithin, the phosphate has moved into the immediate neighborhood of the ester group to be attacked. Consequently, the hydrolysis rate is even lower. If, on the other hand, the negative charge of the phosphate is cancelled, as in the diphenyl phosphatidate, the rate is increased. Slotboom *et al.* (1970a) subjected a large number of synthetic phospholipids to pancreatic lipase and found the rate of hydrolysis of the α-ester bond dependent on the type of substituent at the β-position, and the effect was in agreement with the concept of electrophilic induction.

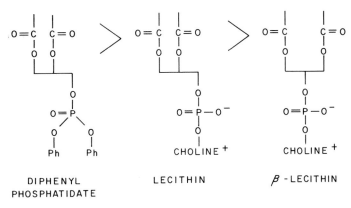

Fig. IV-16   Hydrolysis of phospholipids by porcine pancreatic lipase.

The stepwise hydrolysis of triglycerides, with its decreasing reaction rates for each step (Constantin *et al.,* 1960), can be interpreted in the light of the inductive effect (see Scheme IV-2). Compared to the triglyceride, the

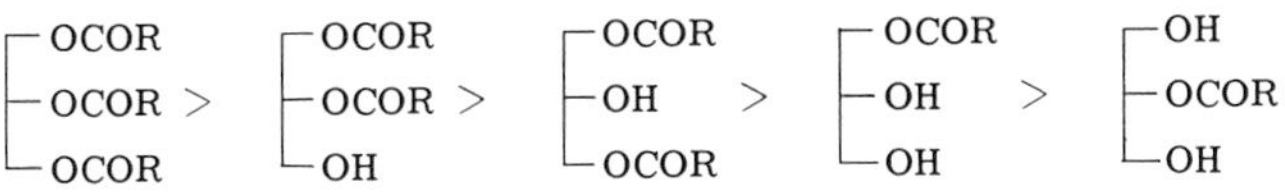

$$
\begin{array}{ccccc}
\left[\begin{array}{l}\text{OCOR}\\\text{OCOR}\\\text{OCOR}\end{array}\right. > &
\left[\begin{array}{l}\text{OCOR}\\\text{OCOR}\\\text{OH}\end{array}\right. > &
\left[\begin{array}{l}\text{OCOR}\\\text{OH}\\\text{OCOR}\end{array}\right. > &
\left[\begin{array}{l}\text{OCOR}\\\text{OH}\\\text{OH}\end{array}\right. > &
\left[\begin{array}{l}\text{OH}\\\text{OCOR}\\\text{OH}\end{array}\right.
\end{array}
$$

SCHEME IV-2. Reaction rates of lipolytic attack.

1,2-diglyceride is somewhat less activated with only one neighboring electrophilic ester group. In the 1,3-diglyceride, the activating group is farther removed; the 1-monoglyceride is not activated at all; and the 2-monoglyceride ester is flanked by two nonactivating groups. (However, steric hindrance is more important in this compound). It should be mentioned, however, that the above sequence of reaction rates, while probably generally correct, has not been strictly established, particularly not for the relative position of the two diglycerides in the sequence. The orientation of the substrate in the interface is also likely to influence the relative reaction rates (Mattson and Volpenhein, 1969; Garner and Smith, 1970b). In the case of phospholipids, the charge of the substrate molecule may also have to be considered (Slotboom *et al.,* 1970a).

Savary (1971) found that ethyl esters were five times less rapidly hydrolyzed than methyl esters; hexyl esters were the most active substrates. Mattson and Volpenhein (1969) determined lipolysis rates for a whole spectrum of esters of normal alcohols (C_1–C_8, C_{12}, C_{16}, and $C_{18:1}$) with C_3–C_{18} fatty acids. They found that methyl esters (C_1) were usually attacked about twice as fast as the corresponding ethyl esters (C_2). The rate

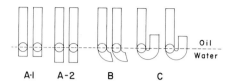

Fig. IV-17 Proposed orientation of esters at an oil–water interphase, A-1, methyl ester; A-2, ethyl ester; B, esters C_3-C_5, C, esters C_6 and longer. From Mattson and Volpenhein (1969).

increased again for the *n*-propyl esters and reached a maximum for the *n*-heptyl esters. The authors suggest that these activity profiles might result from orientation effects at the interface (Fig. IV-17). A methyl ester group would stick out into the aqueous phase, an ethyl group even more so, but, starting with the propyl group, the hydrophobic alcohol chains would be bent back into the lipid phase and expose the carboxyl moiety to the on-slaught of the enzyme. Alternative explanations are possible (Brockerhoff, 1968). The drop of activity from the methyl to the ethyl ester may correspond to an increase in steric hindrance; the increase for C_3 and higher esters may be due to an increased hydrophobicity around the ester group (Brockerhoff, 1969a); the inductive effect, however, cannot be expected to play a large role within a homologous series. Probably none of these explanations is exclusively correct, or, to put it more positively, all of these effects are probably involved.

Garner and Smith (1970b) tested a number of substrates of pancreatic lipase with the help of the monolayer technique. Their results are compatible with the concept of induction; for instance, a 1,2-propanediol diester is a better substrate than a 1,3-propanediol diester. However, orientation effects also seem to play a role, as might be expected in monolayer experiments.

Next to induction, the second important effect in the nucleophilic hydrolysis of esters is steric hindrance. Table IV-10, which is taken from a textbook (Hine, 1956), shows the increase of steric hindrance in going from methyl to ethyl ester, but especially from primary to secondary to tertiary ester. Table IV-11 gives the relative rates of pancreatic lipolysis for some secondary and phenolic esters, in comparison to the rate of tri-olein. More accurate rates for phenolic esters have been supplied later (Brockerhoff, 1969b). The hindrance effects are much greater than those in Table IV-10, probably because the complicated catalytic center of the enzyme is much more sensitive to steric effects than the small hydroxyl ion of the chemical hydrolysis. That steric hindrance is the cause of the isopropyl ester being a resistant substrate, rather than the deactivating effect of the two methyl groups, is shown by the fact that introducing the

TABLE IV-10

**Relative Rates of Alkaline Ester Hydrolysis
in Water at 25°C[a]**

Ester	Rate of hydrolysis
$CH_3CO_2CH_3$	1.00
$CH_3CO_2C_2H_5$	0.60
$CH_3CO_2CH(CH_3)_2$	0.15
$CH_3CO_2C(CH_3)_3$	0.008

[a] Compiled from Hine (1956).

TABLE IV-11

Lipolysis of Secondary and Phenolic Oleic Esters[a]

Ester	Rate of lipolysis	Ester	Rate of lipolysis
Triolein $\equiv$	1.00	CH_3O-	0.07
$(CH_3)_2CHO-$	0.00	C_6H_5O-	0.00
	(<0.003)	β-naphthyl—	0.02
$(ClCH_2)_2CHO-$	0.01	p-ClC_6H_4O-	0.01
$(FCH_2)_2CHO-$	0.05	p-$NO_2C_6H_4O-$	0.07
$(F_3C)_2CHO-$	0.01	$CH_2{=}CHO-$	0.45

[a] From Brockerhoff (1968).

two highly electron-negative fluorines does not lead to a very active substrate. The resistance of the 2-monoglycerides to digestion, and consequently the "positional specificity" of pancreatic lipase, can now be explained by their steric hindrance. Since the two hydroxyl groups of 2-monoglyceride, contrary to the fluorines, have no power of electrophilic induction (Table IV-8), it becomes clear from Table IV-11 that the hydrolysis rate of the 2-monoglyceride can at best be only one-five hundredth of that of a triglyceride (Brockerhoff, 1968, 1969b).

Steric hindrance effects by branching on the side of the alcohol are most severe at the first carbon; triglyceride, which is branched at the second carbon, is a good substrate. However, further branching at carbon 2, as in trioctanoyl-2-methylglycerol, leads again to almost complete resistance (Garner and Smith 1970b; Derbesy and Naudet, 1972). It is not the absence of hydrogen in the β-position of this substrate that is responsible for this resistance, since benzyl oleate, which also possesses no hydrogen in position 2 of the alcohol, is a good substrate. However, the benzyl group has a flat structure, and it must be an umbrella-like spreading of the substituents at 2 that prevents the formation of an enzyme–substrate complex. More surprising is the resistance of esters of alcohols with four or more hydroxyl groups. Erythritol tetraoleate, for instance, is hardly attacked by pancreatic lipase (Mattson and Volpenhein, 1972a,b,c). Since the orientation of the tetraoleate ester groups at the interface probably does not differ much from that of a triglyceride, we must conclude that the enzyme in its active complex cannot accommodate more than two aliphatic chains on the side of the alcohol.

8. Influence of the Fatty Acid

Considering the fatty acid moiety of a substrate, it can be expected that the same factors that explain the role of the alcohol will also explain much of the role of the acid. However, one of these factors, electrophilic induction, can hardly be assessed in the case of the acid because the second factor, steric hindrance, is very much more in evidence. Experiments to measure the inductive effect would require, for substrates, α- or β-substituted acids, with the substituents, for example, being halogens; but such substrates would be severely sterically hindered.

The influence of the fatty acid structure on lipolysis can be discussed in three parts: first, the effect of branching or bulky substituents; second, the effect of unsaturation; third, the effect of the chain length. This last effect, which might also be counted among the steric effects, has received much attention, mainly because of its practical importance in the analysis of triglycerides.

The inhibitory effect of chain branching was first noted in esters of isobutyric acid (Morel and Terroine, 1909), then in esters of α,α-dimethylstearic acids (Bergström et al., 1954; Blomstrand et al., 1956); these are not digested by pancreatic lipase. A systematic study on the influence of branching on substrate specificity dates from 1970 (Brockerhoff, 1970). Table IV-12 gives maximal lipolysis rates of different monoesters of methylbranched or phenyl- or cyclohexyl-substituted acids by porcine pancreatic lipase. The inhibition, which is almost absolute for α-substitution, disappears gradually as the chain is lengthened to carbon 6. The ω-phenyl octanoates have an abnormally high rate of lipolysis, for reasons as yet unexplained. Multiple branching reinforces the steric hindrance. Complete resistance of α-branched fatty acid esters has also been found by Garner and Smith (1970b) and Derbesy and Naudet (1972). The resistance of phytanic esters $(3,7,11,14\text{-Me}_4\text{-C}_{20}$; Table IV-12) was confirmed by Ellingboe and Steinberg (1972).

TABLE IV-12

**Maximal Rates of Hydrolysis of Branched Fatty Acid Esters by
Porcine Pancreatic Lipase[a]**

Acid		V_{rel}[b]	Conditions
Methyl-branched acids			
2-Me-C_3[c]	ClBzl[d]	0.001	29°C, pH 8
3-Me-C_4	ClBzl	0.04	37°C, pH 9
3,7,11,14-Me_4-C_{20}	ClBzl	0.006	37°C, pH 9
4-Me-C_6	ClBzl	0.04	29°C, pH 8
5-Me-C_7	ClBzl	0.37	29°C, pH 8
ω-Phenyl and ω-cyclohexyl acids			
Cyclohexyl-C_2	FEt	0.003	25°C, pH 8
Cyclohexyl-C_3	FEt	0.01	25°C, pH 8
Phenyl-C_3	FEt	0.04	25°C, pH 8
Phenyl-C_4	ClBzl	0.15	37°C, pH 9
Phenyl-C_5	ClBzl	0.20	37°C, pH 9
Phenyl-C_6	FEt	0.59	37°C, pH 9
Phenyl-C_8	FEt; ClBzl	3.4; 2.2	37°C, pH 9
Phenyl-C_{10}	ClBzl	0.98	37°C, pH 9

[a] From Brockerhoff (1970).
[b] Relative to oleic esters, $V = 1.00$.
[c] Chain length.
[d] ClBzl, *p*-chlorobenzyl ester; FEt, 2-fluoroethyl ester.

The effect of fatty acid unsaturation on substrate activity is not quite as clear cut. It was noted by Brockerhoff (1965a) and Bottino *et al.* (1967) that the polyunsaturated fatty acid esters of marine oils were hydrolyzed at much slower rates by pancreatic lipase than the monounsaturated and saturated esters. The most important of these polyenoic acids are eicosapentaenoic acid, all-*cis*-5,8,11,14,17–20:5, with the first double bond between carbon 5 and 6; and docosahexaenoic acid, all-*cis*-4,7,10,13,16,19–22:6, with the first double bond between 4 and 5. Retarded hydrolysis by pancreatic lipase has also been found for seed oils with *trans*-3-unsaturation (Kleiman *et al.,* 1970). Table IV-13 presents relative lipolysis rates for the monoesters of some polyunsaturated acids (Brockerhoff, 1970). The last two esters, a linoleate and linolenate, with the first double bond in position 9, are almost as readily hydrolyzed as oleic esters, but the other acids are hindered. Hindrance effects of similar magnitude are found in Table IV-14 (Brockerhoff, 1970) for monoenoic acids with the double bond in the same position as the first double bonds of the polyenoic acid. Accumulation of double bonds beyond the first one

TABLE IV-13

Maximal Rates of Pancreatic Lipolysis of Polyenoic Fatty Acid Esters[a]

Polyunsaturated acid		V_{rel}[b]	Conditions
t2,t4–6:2[c]	ClBzl[d]	0.008	37°C, pH 9
c4,c7,c10,c13,c16,c19–22:6	FEt; ClBzl	0.20; 0.34	37°C, pH 9
c5,c8,c11,c14–20:4	ClBzl	0.09	37°C, pH 9
c5,c8,c11,c14,c17–20:5	FEt; ClBzl	0.13; 0.09	37°C, pH 9
c9,c12–18:2	ClBzl	0.85	37°C, pH 9
c9,c12,c15–18:3	ClBzl	0.89	37°C, pH 9

[a] From Brockerhoff (1970).
[b] Relative to oleic ester, $V = 1.0$.
[c] Nomenclature after Holman (1966): configuration + position of double bond − chain length: number of double bonds.
[d] ClBzl; *p*-chlorobenzyl ester; FEt, 2-fluoroethyl ester.

does not introduce further hindrance except in the case of the conjugated *t2,t4*–6:2 (sorbic acid) (Table IV-13). Between unsaturation at $\Delta^{2,3}$ and $\Delta^{5,6}$, the resistance does not depend in a coherent manner on the actual position of the double bond; in fact, $\Delta^{2,3}$ acids and $\Delta^{4,5}$ acids seem to be slightly better substrates than $\Delta^{5,6}$ acids. Cis and trans unsaturation and the one triple bond tested inhibit lipolysis by comparable degrees. Heimermann *et al.* (1973) studied the pancreatic lipolysis of isomeric octadecanoates and found that the inhibition was greatest for the Δ^5-isomer (Fig. IV-18).

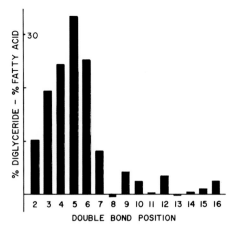

Fig. IV-18 Degree of discrimination by porcine pancreatic lipase against isomeric *cis*-19:1 acids in random mixed triglycerides. The ordinate shows percent 18:1 in DG-FFA. Absence of discrimination results in zero. From Heimermann *et al.* (1973).

II. *Pancreatic Lipase* **67**

TABLE IV-14

Maximal Rates of Pancreatic Lipolysis of Monoenic Fatty Acid Esters[a]

Monounsaturated acids		V_{rel}[b]	Conditions
t2-6:1[c]	ClBzl[d]	0.04	25°C, pH 8
t2-8:1	ClBzl	0.10	25°C, pH 8
2y-8:1[c]	ClBzl	0.21; 0.22	25°C, pH 8; 37°C, pH 9
c2-8:1	ClBzl	0.19	37°C, pH 9
t2-12:1	FEt	0.08	37°C, pH 9
t3-6:1	ClBzl	0.04	25°C, pH 8
c4-12:1	FEt	0.18	37°C, pH 9
c5-14:1	ClBzl	0.13	37°C, pH 9
c5-20:1	FEt	0.08	37°C, pH 9
c6-18:1	FEt	0.34	37°C, pH 9
c9-18:1	Ester of reference	1.00	—, —
10-11:1	ClBzl	0.81	37°C, pH 9
c11-20:1	FEt; ClBzl	0.72; 0.82	37°C, pH 9
c11-22:1	FEt	0.61	37°C, pH 9

[a] From Brockerhoff (1970).
[b] Relative to oleic ester, $V = 1.0$.
[c] Nomenclature after Holman (1966); y, triple bond.
[d] ClBzl, *p*-chlorobenzyl ester; FEt, 2-fluoroethyl ester.

The effect of the double bond on substrate specificity is not likely to be one of induction or resonance, since both would fall off rapidly or abruptly as the bond recedes from the carboxyl group. The possibility of a folding effect that would bring the methyl end of the polyenoic acids into the vicinity of the head group for steric interference has been discussed by Bottino *et al.* (1967); however, multiple double bonds are not more inhibitory than a single one, and an acid with maximal and rigid refolding, *cis*-2-octenoic acid, is not more resistant than two similar acids in which the chain points straightly away from the carboxyl group, *trans*-2-octenoic and 2-octynoic acid (Table IV-14). The effect of the double bonds is probably a consequence of rotational restrictions. With the introduction of a double bond, the fatty acid chain can no longer freely bend away from the enzyme, as may be required for the optimal formation of the enzyme–substrate complex (Brockerhoff, 1970).

The influence of fatty acid chain length on substrate specificity has been the subject of more studies than have been devoted to any other aspect of the specificity of pancreatic lipase. The reason is the occurrence of short-chain acids (especially butyric acid) in milk triglycerides, and the many efforts by lipid biochemists to determine the positional distribution of these acids. Many studies have been carried out with triglycerides as

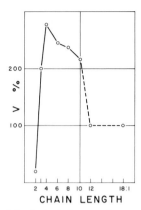

Fig. IV-19   Relative rates of hydrolysis by porcine pancreatic lipase of triglycerides of fatty acids of different chain lengths. From Entressangles *et al.* (1961b).

substrates; the more direct approach of comparing monesters has only recently been taken.

Among triglycerides, triacetin is a poor substrate, whereas tripropionin (Wills, 1961b) and tributyrin (Entressangles *et al.*, 1961b) are hydrolyzed faster than other triglycerides. Figure IV-19 shows a comparison of *V* for different triglycerides. From such results it has been concluded that butyrates are preferred substrates for pancreatic lipase. This conclusion seemed to be born out by experiments with mixed glycerides. When *rac*-glycerol 1-palmitate 3-butyrate was subjected to pancreatic lipase the mixture of the free fatty acids found consisted of 72% butyric and 28% palmitic acid (Entressangles *et al.*, 1961a,b). In such an experiment, however, effects of substrate orientation at the interface may well play a significant role; for instance, the butyric ester part of the molecular might be oriented toward the aqueous phase, and the palmitoyl part away from it. It has also been found that even if the butyric acid is attached in position 2, it will still appear among the free fatty acids, and the hydrolysis of 2-butyryldipalmitin has yielded 40% monopalmitin in its monoglyceride fraction (Clément *et al.*, 1962a,b). The migration of the short-chain acids on the glycerol seems to proceed much faster than that of long-chain acids, and butyric acid is probably split from the intermediate diglyceride after migration to the 1 or 3 position (Entressangles *et al.*, 1966).

Evidence has recently been presented that indicates that the total size of a triglyceride rather than the length of a particular fatty acid regulates the rate of lipolysis (Sampugna *et al.*, 1967). In a mixture of glycerol 2-butyrate 1,3-dipalmitate and triolein, the first triglyceride is digested faster, although the butyric ester is not, in the beginning, attacked by the lipase (Jensen, 1971). Glycerol 1-butyrate 2,3-dipalmitate, with a fatty

TABLE IV-15

**Quantities of Free Fatty Acids Derived from Pancreatic
Lipolyses of Equimolar Mixtures of Butyrate
Glycerides and Triolein**[a]

Triglyceride mixture	Free fatty acids (mole %)		
	Palmitate	Butyrate	Oleate
PBB + 000	28.2	28.8	43.0
Theory[b]	26.4	26.4	47.2
PBP + 000	55.7	—	44.3
Theory	51.8	—	48.2
PPB + 000	23.6	28.2	48.2
Theory	24.7	24.7	50.6

[a] From Sampugna *et al.* (1967).
[b] Theoretical values calculated assuming specificity for primary positions, but none for butyrate.

acid carbon number of 36, is digested more rapidly than triolein, with a carbon number of 54. The presumed preference of pancreatic lipase for short-chain acids was also tested with mixtures of well defined triglycerides; glycerol 1-palmitate 2,3-butyrate (PBB), glycerol 1-butyrate 2,3-palmitate (BPP), glycerol 2-butyrate 1,3-dipalmitate (PBP), and triolein (000) (Table IV-15). The mixtures of fatty acids obtained by lipolysis had almost theoretical compositions, i.e., the compositions calculated for positional specificity but absence of fatty acid specificity of lipase. The data showed no evidence for a preferential hydrolysis of butyrate. When the diglyceride glycerol 1-palmitate 3-butyrate was treated with lipase (Table IV-16), the results showed that palmitic acid, rather than butyric acid,

TABLE IV-16

Products from Pancreatic Lipolysis of Glyceryl 1-Palmitate 3-Butyrate[a]

Time (minutes)	Monoglycerides (mole %)		Free fatty acids (mole %)	
	Butyrate	Palmitate	Butyrate	Palmitate
2.5	58.0	42.0	39.3	60.7
15.0	66.1	33.9	39.1	60.9
Theory	50.0	50.0	50.0	50.0

[a] From Sampugna *et al.* (1967). The 240 mg sample was incubated at 37°C.

had been preferentially released; this was probably in part due to the resistance of the water-soluble monobutyrin against lipase, but it tends to confirm the nonpreference of the enzyme for the short-chain acid. Perhaps the earlier experiments with the same substrate had given the opposite result (Entressangles *et al.,* 1961a,b) because of the longer incubation periods which may have favored acyl migration. The more rapid hydrolysis of triglycerides of lower molecular weight may be caused by the accumulation of these molecules in the surface of the oil droplets, or by their greater diffusional mobility.

The chain-length studies with glycerides as substrates are of great interest for lipid research, but for the enzymologist they leave much to be desired. They do not study the effect of the chain length in isolation, because the alcoholic group is changed together with the acid and it is impracticable to disentangle the contributions of the two. This difficulty is compounded with our ignorance of orientation effects or of interfacial substrate displacements. Mani and Lakshminarayana (1970) endeavored to reduce these problems by using mixed diesters of ethylene glycol as substrates for pancreatic lipase. They found that oleic, stearic, lauric, elaidic, stearolic, and elaidohydnocarpic acid were all released at nearly the same rate, arachidic (C_{20}) acid a little more slowly, but caproic (C_6) acid twice as fast.

A more direct approach to the problem was taken by Mattson and Volpenhein (1969) who determined the rates with which several series of *n*-alkyl esters of different chain lengths were hydrolyzed by rat pancreatic lipase. The first graph of Fig. IV-20 represents their results: butyric esters are indeed hydrolyzed faster than the next homologous esters, but not much faster than oleic and somewhat slower than dodecanoic esters. Interesting is the sudden drop in substrate activity between the C_4 and the C_5 acids. Mattson and Volpenhein have interpreted their findings as examples of orientation effects at the interface, as depicted in Fig. IV-17. The second, third, and fourth graphs of Fig. IV-20 show relative lipolysis rates (porcine enzyme) of *p*-chlorobenzyl, 2-hexoxyethyl, and vinyl esters of a sequence of fatty acids (Brockerhoff, 1970). The exceptional position of butyric acid is again apparent, although the hexoxyethyl esters series shows a shoulder rather than a peak at this chain length. In all series, the substrate activity shows a minimum for the C_5 fatty acid. The interpretation of the results can be given in terms that disregard orientation effects. The lower activities of C_2 and C_3, for example, might result from hydration of the ester group, the minimum at C_5 from steric hindrance of the end methyl group in the formation of the substrate–enzyme complex. Longer chains might bend away from the enzyme and reduce this interference (Brockerhoff, 1970). Immersion of the ends of longer chains into the lipid phase might assist in the bending away of the chains. (At this point, the hypotheses arguing with orientation effects and those arguing with active complex effects overlap.) Butyric acid may be located at a point where the hydrophobic effect is still overpowering the steric hindrance effect.

From Fig. IV-20, it becomes clear that although butyric esters are indeed favored substrates of pancreatic lipase, they are not necessarily hydrolyzed faster than long-chain esters. The influence of the alcohol is also important; while tributyrin and vinyl butyrate are better substrates than the corresponding oleates, such is not true for the alkyl or chlorobenzyl esters, and certainly not for hexoxyethyl esters. The longstanding arguments that were based on studies with glycerides should be reconsidered with these considerations in mind; many seemingly contradictory results could probably be reconciled.

The series of p-chlorobenzyl esters has been extended to C_1, the formate (Fig. IV-20); this ester is found to have about 60% of the substrate activity of the oleate (Brockerhoff, 1970). Another ester, cinnamoyl formate (represented by the asterisk in the p-chlorobenzyl graph of Fig. IV-20), was 2.6 times as active a substrate as cinnamoyl oleate (Brockerhoff, 1970). The high reactivity of formate esters has been confirmed by Savary

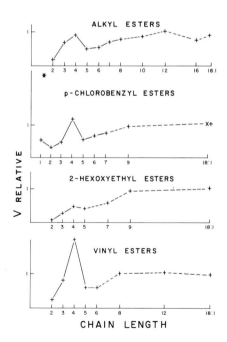

Fig. IV-20 Rates of hydrolysis of esters of fatty acids of different chain lengths by pancreatic lipase. Oleic ester, $V = 1.0$. Top curve, rat lipase, from Mattson and Volpenhein (1969); other curves, porcine lipase, from Brockerhoff (1970).

SUBSTRATES   INHIBITORS

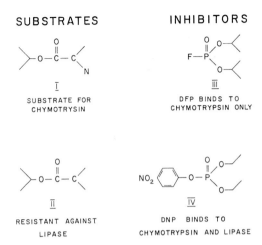

I

SUBSTRATE FOR
CHYMOTRYSIN

III

DFP BINDS TO
CHYMOTRYPSIN ONLY

II

RESISTANT AGAINST
LIPASE

IV

DNP BINDS TO
CHYMOTRYPSIN AND LIPASE

Fig. IV-21 Substrates and irreversible inhibitors for chymotrypsin and pancreatic lipase. From Brockerhoff (1973).

(1972). It is clear, therefore, that pancreatic lipase has no special need for substrates with a fatty acid chain.

9. The Enzyme–Substrate Complex

The results of the studies with synthetic substrates permit some conclusions concerning the structure of the enzyme–substrate complex, and they suggest an explanation for the behavior of pancreatic lipase toward organophosphates. Diisopropyl fluorophosphate (DFP) does not inhibit the lipase, but diethyl-*p*-nitrophenyl phosphate (DNP) does, though slowly. Both phosphates inhibit other serine enzymes, for instance, chymotrypsin. Figure IV-21 shows, in I, part of the skelton of a typical chymotrypsin substrate; compound II, however, with an analogous backbone, would be completely resistant towards lipase. Lipase has a much more stringent requirement for an unencumbered substrate structure than the proteinase, especially on the acid side of the substrate. A comparison of the inhibitors DFP (III) and DNP (IV) suggests a reason why the one might be resistant to lipase and the other not—DNP is much less bulky on the "acid" side than DFP.

The experiments with branched substrates have made it clear that the steric effects are different at the side of the alcohol and the side of the acid. Branching in the fatty acid is much more inhibitory than equidistant branching in the alcohol. The formulas of Fig. IV-22 illustrate this point. The ester on the left is a very good substrate, but the ester on the right has only a few percent of its activity, although the inductive powers of

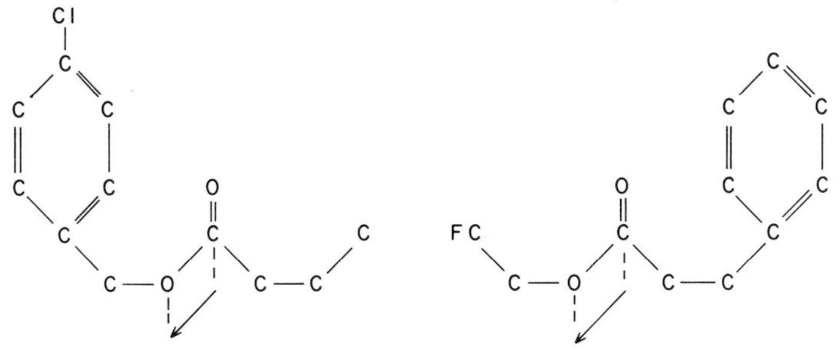

p−Chlorobenzyl butyrate

Fluoroethyl−
3−phenyl propionate

Fig. IV-22 Directional fixation of substrates to pancreatic lipase. From Brockerhoff (1970).

the alcohols and the shape of the molecules (except for the far-away chlorine) are very similar. This means that the substrate cannot turn around on the enzyme; it must be fixed at least in two points (Brockerhoff, 1970). Since it is not necessary for a good substrate to possess a fatty acid chain, and since the alcohol can be varied in many ways (or be replaced by hydrogen) and still leave an active substrate, the fixation must be on the —O—C=O structure. The place of the carbonyl carbon is, of course, fixed, because it is there that the enzyme attacks. The second fixation cannot be on the carbonyl oxygen, =O, because then the substrate molecule could still rotate around the double bond; it is, of course, possible, in fact, probable, that the carbonyl oxygen is also hydrogen-bonded to the enzyme, like the analogous oxygen of substrates of chymotrypsin (Fersht *et al.*, 1973). The directional fixation, however, must be along the C—O bond as shown in Fig. IV-22.

It has been suggested (Brockerhoff, 1970) that the fixation of the ether oxygen of the substrate might result from hydrogen bonding to the reactive histidine. However, if the active center of lipase is similar to that of other serine enzymes, e.g., chymotrypsin, the histidine N_{ε_2} will be hydrogen-bonded to the reactive serine (Blow *et al.*, 1969), and the proton involved cannot also bind to the substrate. This proton is assumed, in chymotrypsin-catalyzed hydrolysis of esters, to accept the leaving R—O— group, which as a powerful nucleophile has to be neutralized immediately in order not to reverse the reaction. Figure IV-23 shows that the two-dimensional fixation of an ester in the active substrate–lipase complex may result from the necessity to place the ether oxygen within reach of the proton (Brockerhoff, 1973). Thus, the fixation of the substrate is not an actual pretransitional bonding but an obligatory orientation.

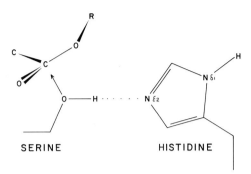

SERINE   HISTIDINE

Fig. IV-23 Hypothetical alignment of the "reactive" serine and histidine and the substrate in the lipase–substrate complex. From Brockerhoff (1973).

E. Factors Affecting Pancreatic Lipolysis

The number of compounds that can influence the rate of hydrolysis of a given substrate by pancreatic lipase is legion. Assay systems for lipase often contain sodium chloride, calcium chloride, bile salts, proteins, and emulsifiers of all descriptions, and all of these compounds, and many more, can influence the course of lipolysis. One of the major problems of lipase research has been the difficulty of sorting out the effects of any one of these agents on the degree of emulsion, the interface, the substrate, the products, or perhaps the activation of the enzyme itself. This field of research is covered with the corpses of precocious theories. Only recently have some longstanding problems been successfully attacked, but as yet not two researchers agree on the function of all factors, and there are still differences of opinion on each single factor. The following expositions are, of course, strongly biased toward our own opinions.

The most important activators of lipolysis are sodium chloride, or salts in general, calcium ions, bile salts, and proteid cofactors. We shall discuss them in this order.

1. Inorganic Salts

Sodium chloride activates the lipolysis of insoluble long-chain and short-chain triglycerides as well as the lipolysis of soluble short-chain triglycerides. The apparent rates of enzymatic hydrolysis of triolein, tributyrin, and triacetin all increase with the addition of the salt. However, the mechanisms of this acceleration are different; at least two factors have been recognized.

In the case of long-chain triglycerides such as triolein, the apparent stimulation of pancreatic lipolysis is largely the result of the peculiar conditions in lipolytic assay systems, as they are described in Chapter III, Section

VI. In short, cations suppress the pH gradient at the interface and lower the apparent pK_a of the fatty acids, which can then be more quantitatively titrated.

However, the shift of the titration curve alone is not an exhaustive explanation of the role of sodium chloride in the lipolysis of long-chain triglycerides. In the complete absence of salts, lipolysis will not proceed at all (Benzonana and Desnuelle, 1968), even at pH 9 when at least a portion of oleic acid is titratable. It may be assumed that, in the absence of any counter ion, electrostatic repulsion prevents either the enzyme–supersubstrate binding or the enzyme–substrate binding entirely.

The promoting effect of sodium chloride on the action of pancreatic lipase on dissolved short-chain glycerides (Entressangles and Desnuelle, 1968) has been ascribed to the promotion of micelle formation by sodium chloride and the subsequent action of lipase on micellar aggregates; but the issue remains open to questioning, especially since sodium chloride also accelerates the enzymatic hydrolysis of short-chain but insoluble glycerides, in particular, tributyrin (Verger *et al.,* 1970; Erlanson and Borgström, 1970a). Tributyrin is only sparsely soluble in water (0.02% at 30°C) (Schønheyder and Volqvartz, 1944b) and the enzymatic hydrolysis of its emulsion undoubtedly takes place at the interface and not in a micellar solution. On the other hand, butyric acid is water soluble and does not give rise to interfacial charge effects and shifts of titration curves; its pK_a is essentially the same, around 5, in aqueous solution or in a tributyrin emulsion, with or without sodium chloride present (H. Brockerhoff, unpublished). Nevertheless, the hydrolysis of tributyrin by pancreatic lipase is at least four times faster in 0.1 *M* NaCl than in distilled water. It must be suspected that salts exert an effect other than promoting micelle formation or shifting titration curves, a more immediate effect on enzyme or substrate or both.

2. Calcium Ions

The activation of lipolysis by calcium chloride was already employed by Willstätter *et al.* (1923) and has since then been confirmed on numerous occasions. In the first report, it was noted that the calcium ion was not absolutely essential but could be replaced in part by albumin or bile salt. Triacetin hydrolysis was independent of calcium (Willstätter and Memmen, 1928). Schønheyder and Volqvartz suggested in 1945 that the function of calcium was the removal of the inhibitory long-chain acids as calcium salts; and this hypothesis has been, by and large, confirmed.

Borgström (1954) found that the lipase-catalyzed reacylation of partial glycerides is inhibited by calcium; evidently, the free fatty acids needed for the resynthesis are removed from the system as calcium salts. Constan-

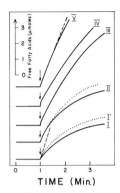

TIME (Min.)

Fig. IV-24 Modification of the kinetic curves of olive oil lipolysis by Ca^{2+} and deoxycholate. Curve I, no addition. Curves II–IV, $CaCl_2$ at 0.011 mM, 0.083 mM, 0.33 mM. Curve V, 0.17 mM $CaCl_2$ and 2.5 mM deoxycholate. Dotted corrected curves for incomplete ionization of oleic acid. Conditions: pH 9, 37°C, 0.1 M NaCl, trace sodium oleate. From Benzonana and Desnuelle (1968).

tin *et al.* (1960) found that Ca^{2+} did not increase the initial reaction rate of lipolysis, and Benzonana and Desnuelle (1968) demonstrated this fact thoroughly. Figure IV-24 shows that the lipolysis of olive oil starts with the same velocity whether Ca^{2+} is present or not, but in the absence of Ca^{2+}, the reaction is very soon inhibited. Increasing concentrations of calcium lead to increasingly linear reaction rates. In principle, the action of calcium is not much different from that of sodium or other cations; it acts as a counter ion to the fatty acid anion accumulating at the interface. But whereas with sodium an equilibrium is soon reached with the surface still negatively charged, with calcium the surface charge is actually cancelled because the calcium soaps are insoluble. In different words, the supersubstrate binds calcium much better than sodium (Benzonana, 1968). This is also true for strontium and barium, but not for magnesium (H. Brockerhoff, unpublished). The effect of these ions can be expected to vary with the nature of the substrate; the results may be different for triglycerides forming soaps other than oleates and with different solubilities.

   While it is generally agreed that calcium acts through the formation of calcium soaps, it is not at all clear how the lipolysis at the interface is affected by these soaps and whether soap formation is the only mechanism by which calcium affects the reaction. In *in vitro* experiments, the calcium soaps eventually form crystalline aggregates and precipitate from the emulsion, but in the early stages of lipolysis they will stay attached to the interface, probably with their head group toward the aqueous phase and their tail buried in the oil droplets. The lipase first approaching a supersubstrate of esters is soon confronting a supersubstrate of esters mixed with calcium soaps. A drop in the apparent hydrolysis rate owing to substrate displacements might be expected; but the initial rate of lipolysis, before any calcium soaps are formed, is in fact only slightly higher than the later reaction rate (H. Brockerhoff, unpublished). It might be speculated that the presumptive displacement effect is compensated by an actual activation of the ester groups of the remaining substrate by the calcium soaps. Because of the lipophilic association of the long chains of soaps and esters, calcium might come into very close contact with ester groups and activate them by polarization of the $C\!\!=\!\!O$ bond. Again, although calcium is not an obligatory cofactor of lipase, the high concentration of calcium at the interface might result in inductive effects modifying the enzyme. Furthermore, it must be remembered that Ca^{2+} is a bivalent ion while fatty acid anions are monovalent. The possible existence of the species fatty acid anion–Ca^+ must be considered. To couch it in other terms, the anion may attract more positive charges to the interface than it can balance. In solution, the cationic species would not long survive because of the speed of ionic reactions; but the slower diffusion in the lipid phase might extend its life span. We cannot predict what influence this ion, and its counter ions, would have on the interfacial potential and other interfacial properties. This question, however, might well be amenable to theoretical or experimental treatment.

3. Bile Salts

It has been known for a long time that bile salts can accelerate the digestion of fats (Rachford, 1891). Listing and discussing the experiments on this phenomenon and the hypotheses proposed would alone fill a book. Some references to the older literature can be found in the review by Wills (1965); here we must restrict ourselves to discussing those studies which appear best suited to illustrate our present state of knowledge of the subject.

It is generally conceded that the principal physiological function of bile acids is their role in the absorption of fats. It was shown by Verzar and his school (Verzar and Laszt, 1934) that fat, even after digestion with lipase, will not pass the intestinal wall unless bile salts are present, and Hofmann and Borgström (1964) have shown that the mixed micelles of bile salts, fatty acids, and monoglycerides are the vehicles of fat absorption Hofmann and Small (1967) have published an exhaustive review on this subject.

As for the role of bile salts in the lipolysis of triglycerides, it can be stated categorically that it is not an obligatory one; the enzymatic hydrolysis of triacetin (Entressangles and Desnuelle, 1968) or tributyrin (Erlanson and Borgström, 1970a) is not stimulated by bile salts, and in the hydrolysis of long-chain triglycerides, bile salts can be replaced by such divers agents as calcium ions and albumin (Willstätter and Memmen, 1928) or hexadecylpyridinium bromide (Wills, 1955). Bile salts are also often described as inhibitors rather than activators of lipase; this "inhibition" can almost certainly be attributed to substrate displacement effects and charge effects at the interface. Recently it has been shown by Morgan *et al.* (1969) that bile acids as such are not bound by pancreatic lipase and do not measurably change the conformation of the enzyme.

While bile salts cannot be regarded as true cofactors of pancreatic lipase, they often do promote lipolysis *in vitro,* and this activity calls for an explanation. One theory is based on the fact that bile salts form mixed micelles with the reaction products of lipolysis—fatty acids and monoglycerides (Borgström, 1964b). It is significant that bile salts have been found most active at slightly acidic conditions that would also favor the formation of undissociated fatty acids. In the absence of bile salts, the lipolysis-resynthesis equilibrium of triolein, as catalyzed by rat pancreatic lipase, yielded 80% triolein at pH 6; in the presence of bile salt under otherwise equal conditions, only 25% triolein was left (Borgström, 1964b). It can be concluded that the species needed for resynthesis is the undissociated fatty acid rather than the anion. The incorporation of fatty acids into micelles with bile salt will lower their apparent pK_a (Hofmann and Small, 1967), shift the dissociation equilibrium toward the anion and therefore inhibit the resynthesis of glycerides. Recently, Borgström and Erlanson (1971) have suggested that the shift of the pH optimum in the presence of bile salt is connected with the formation of a bile salt–lipase–co-lipase complex.

The bile salt used in Borgström's (1964b) study was the conjugated taurodeoxycholate; this salt does not precipitate as an insoluble acid at acidic pH. Benzonana (1969) has studied the micelle formation of the unconjugated deoxycholate at pH 9 and finds that a micelle of about thirteen molecules can "bury" one molecule of oleic acid anion. In this manner, the bile salts would remove the reaction product, fatty acid, from the interface. Benzonana and Desnuelle (1968) also showed that (in the presence of some sodium oleate and calcium chloride) deoxycholate does not augment the initial lipolysis rate but improves the linearity of hydrolysis with time; this is shown in the uppermost curve of Fig. IV-24. In Fig. IV-25, which records initial reaction rates of triolein lipolysis, it is seen that the optimal activating concentration, or rather the starting concentration of inhibition, is very different for different bile salts; relatively high for the unconjugated acids of the left diagram but low for the conjugated acids at the right, although the conjugated taurocholate, at the left, is somewhat less inhibitory than its relatives (Benzonana and Desnuelle, 1968). Since the conjugated compounds are much stronger acids, their stronger inhibi-

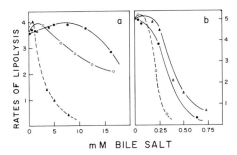

Fig. IV-25 Effect of bile salts on the pancreatic lipolysis of olive oil *in vitro,* at pH 9, 37°C, 0.1 *M* NaCl, and 0.5mM CaCl₂. (*Left*) cholate (○); deoxycholate (●); taurocholate (▲). (*Right*) taurodeoxycholate (○); glycodeoxycholate (●); glycocholate (△). From Benzonana and Desnuelle (1968).

tory faculty points to an electrostatic action of their anions at the interface, although it is not clear if this is a case of a repulsion or of misorientation of enzyme at the interface. Measuring the "K_m" in the presence of the different bile salts would give some information on this subject.

It can be considered as probable that bile salts, under certain *in vitro* conditions, will accelerate, i.e., linearize lipolysis by complexing fatty acids in micelles and removing them for the interface. This mechanism is certainly essential *in vivo* as one step in the digestive assimilation of fats, though not essential for the mere degradation of triglycerides by lipase.

An additional hypothesis of bile salt action (Brockerhoff, 1971a) is founded on the long-known fact that enzymes, and proteins in general, are adsorbed to oil–water interfaces and denatured by hydrophobic bonding and unfolding (Bull, 1947; James and Augenstein, 1966; McLaren and Packer, 1970); the effect is a function of the interfacial tension. In this respect lipase does not differ from other proteins. Figure IV-26 shows the denaturation of lipase by the shaking of an aqueous solution (A) and the much faster denaturation at a hexadexane–water interface (B). It can be seen that, after substantial denaturation within the first 2 minutes, further denaturation of a similar magnitude takes about a day. This interval is the turnover time of the release of the adsorbed and denatured enzyme into the aqueous phase, a release that must take place for another batch of enzyme to be adsorbed and denatured. This turnover is obviously very slow; the interface is effectively blocked by the denatured protein. Taurocholate (Fig. IV-26C) prevents denaturation and blocking. (The increase of activity after 2 minutes is only apparent; it is due to the introduction of some bile salt to the final assay system). Figure IV-27 shows that the protective action of taurocholate on lipase starts at concentrations much lower than the physiological, which is, for man, slightly lower than 10^{-2} *M*.

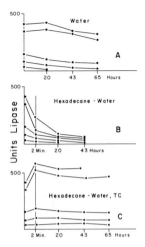

Fig. IV-26 Denaturation of lipase at the air–water (A) and hexadecane–water (B) interphase, and its protection by taurocholate (TC) (C). From Brockerhoff (1971a).

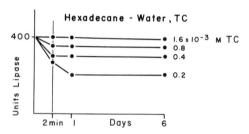

Fig. IV-27 Protecting effect of taurocholate (TC) at different concentrations on lipase. From Brockerhoff (1971a).

The denaturation of the lipase takes place not only with hexadecane but also at the interface of a natural substrate, olive oil, and water. Figure IV-28 shows the titration of fatty acids released from olive oil emulsified with gum arabic. Calcium chloride and sodium chloride are present, taurocholate is introduced where indicated. Curve C shows the lipolysis in the complete system; curve A shows that the action of the lipase stops soon in the absence of taurocholate. This "inhibition" is the result of enzyme denaturation. When taurocholate is introduced after $\frac{1}{2}$ minute or 2 minutes (curves C and B) the enzyme is not reactivated to its original activity in D; but the further inactivation is halted and the residual activity conserved.

Long-chain surfactants like sodium dodecyl sulfonate (Brockerhoff, 1971a) or dodecyl sulfate (Wills, 1954) will not protect the enzyme but denature it, obviously because they form hydrophobic bonds with the protein and cause themselves the unfolding that should be prevented. Bile salts,

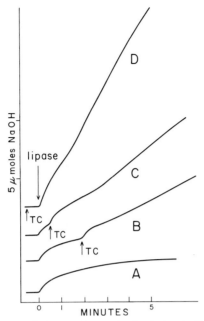

Fig. IV-28 Effect of taurocholate (TC) on the lipolysis of olive oil. One milliliter of 0.02 *M* taurocholate, pH 8, added at arrow. From Brockerhoff (1971a).

on the other hand, have a bulky, kinked structure and possess hydrophilic hydroxyl groups. They are lipophilic enough to adsorb to the interface and prevent hydrophobic enzyme–substrate interaction, but hydrophilic enough not to establish themselves such bonding with the enzyme.

An earlier hypothesis (Benzonana and Desnuelle, 1968; Schoor and Melius, 1970), which stated that bile salt facilitates the adsorption and desorption of the lipase, can now be confirmed and reinterpreted; bile salts prevent the unfolding of proteins, which would block the interface.

The denaturation experiments were carried out with pure porcine lipase, and it can be asked if they have any bearing on *in vivo* lipolysis, or even on *in vitro* assays of lipase under usual practical conditions. In the intestine, the lipase is accompanied by a large excess of other pancreatic proteins, not to mention the dietary proteins, which will also protect the native structure of the lipase. There would seem to be no need for a protection by bile salt. However, all the proteins will crowd the oil–water interfaces. Proteolytic enzymes will eventually break down these barriers, but bile salts may be helpful in preventing the initial blockage of the interfaces. That bile salts can even reverse such blockage was indicated by experiments of Fraser and Nicol (1966) with human pancreatic lipase. The enzyme was severely inhibited by bovine albumin. Prior addition of deoxycholate prevented this inhibition, addition of the bile salt after the albumin did not. But as the lipolysis proceeded, a titration curve with a concave slope was obtained, indicating a slow reversal of the inhibition caused by the albumin.

Apart from explaining an aspect of bile salt function, the experiments on the denaturation of lipase are relevant to the problems of the specificity and the enzyme–substrate complex formation of lipase, because they show that rather than bonding to the lipid chains of its substrates the enzyme has to be protected from them.

4. Albumin

The activating effect of albumin in incubations of long-chain triglycerides with pancreatic lipase (Willstätter *et al.*, 1923) is probably a result of the fact that albumin can bind fatty acids and thus cancel their inhibitory effect. A different effect of albumin at low concentrations has recently been described (Brockerhoff, 1971a). The addition of 0.1 mg of bovine albumin to an emulsion of 0.5 gm tributyrin in 15 ml 0.1 *M* aqueous NaCl (Curve C in Fig. IV-29) resulted in a steady rate of lipolysis. In the absence of albumin (Curve A), denaturation occurred that could be halted but not

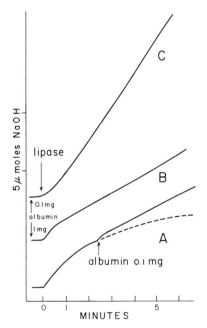

Fig. IV-29   Influence of bovine albumin on the lipolysis of tributyrin. Albumin was added as a 1% solution. See text for details. From Brockerhoff (1971a).

reversed by the addition of albumin. Obviously, the albumin acts in the same manner as bile salts, preventing the unfolding of the lipase at the interface. Other proteins act in the same way. If the amount of albumin is too high, as in Curve B of Fig. IV-28, the interface will be blocked and the rate of lipolysis reduced, and the albumin acts as an inhibitor. A protein from rat liver, with the molecular weight of 100,000, has been described as an inhibitor of pancreatic lipase (Machovich *et al.*, 1970).

5. Co-lipase

The first evidence for a co-lipase was presented by Baskys *et al.* (1963). It was found that porcine pancreatic lipase had lost its activity against an olive oil emulsion containing deoxycholate after passage through a DEAE-cellulose column. Addition of boiled pancreatic extract restored the activity. Morgan *et al.* (1969) found that a heat-stable cofactor with a molecular weight of around 12,000 could be separated from rat pancreatic lipase by chromatography on Sephadex G-100. In the presence of bile salt, the cofactor did not separate from the lipase. Co-lipase has also been found in human duodenal contents (Figarella *et al.*, 1972) and in bovine pancreas (Julien *et al.*, 1972).

The co-lipase of porcine pancreas has been isolated and characterized in two laboratories. Erlanson and Borgström (1972), starting from an acidic extract of pancreas, used heat-treatment, ammonium sulfate fractionation, treatment with ethanol, and chromatography on SP-Sephadex, DEAE-cellulose, and Sephadex G-75 to prepare the cofactor with a yield of 12%. Maylié *et al.* (1971, 1973) extracted the proteins from defatted pancreas, delipidated them further by butanol treatment and ammonium sulfate precipitation, and then chromatographed on QAE-Sephadex and SP-Sephadex. Two forms of the factor, co-lipase I and II, were obtained, with an overall yield of 30%. Co-lipase II differs from I by being about ten amino acid residues smaller. The amino acid composition of both forms is given in Table IV-17. An analysis published by Erlanson and Borgström (1972) for their preparation is almost identical with that of co-lipase I. This protein has a molecular weight of 10,200–10,500, with 94–95 amino acid residues and five disulfide bridges. It contains no methionine or tryptophan, but has a rather high content in acidic and hydroxyl-containing amino acids, which may serve to knit the molecule together by hydrogen bonding. The terminal amino acid is reported as isoleucine (Erlanson and Borgström, 1972) or glycine (Maylié *et al.*, 1973). The amino acid sequence of the amino end chain (Maylié *et al.*, 1973):

$$\overset{1}{\text{Gly}}\text{–Ile–Ile–Ile–Asn–Leu–Asp–Glu–Gly–Glu–}\overset{10}{\text{Leu}}\text{–Cys–Leu–Asn–Cys–Ala–}$$
$$\overset{20}{\text{Gln}}\text{–Cys–Lys–Ser–Asn}$$

TABLE IV-17

Amino Acid Composition of Porcine Co-lipase[a]

Residue	Nearest integral number		Residue	Nearest integral number	
	Co-lipase I	Co-lipase II		Co-lipase I	Co-lipase II
Ala	5	4	Lys	4–5	3–4
Arg	4	3	Met	0	0
Asx	11	10	Phe	2	1
Cys	10	10	Pro	2	1
Glx	9	8	Ser	7	7
Gly	9	8	Thr	5	3–4
His	2	2	Trp	0	0
Ile	7	6	Tyr	3	2
Leu	11	9	Val	3	1

Residues: (I) 94–95, (II) 84–85; calculated weight: (I) 10,157–10,285, (II) 9,103–9,190.

[a] From Maylié *et al.* (1973).

is remarkable for the sequence of three isoleucine residues at the end and a clustering of three cysteine bridges between residues 12 and 19.

Co-lipase may form a complex with pancreatic lipase, and many earlier preparations of the enzyme may have been contaminated with the cofactor. After separation from the co-lipase, lipase still showed about 30–40% of its full activity against a triglyceride emulsion stabilized with hydroxymethylcellulose and containing deoxycholate (Maylié *et al.*, 1971). However, Sephadex filtration of a mono-TNB lipase (Chapter IV, Section II, A, 4), in which the "accessible" sulfhydryl group (SH_I) of lipase is blocked with a thionitrobenzoic acid residue, led to an almost inactive mono-TNB lipase preparation which was not reactivated by conversion to the native enzyme with dithiothreitol. Addition of co-lipase to the deactivated mono-TNB lipase, at a molar ratio of co-lipase to lipase between 1 and 2, restored its activity. Maylié *et al.* (1971) have concluded that the SH_I group of the lipase molecule, or the region around it, controls the binding of the enzyme to the coenzyme.

Borgström and Erlanson (1971) and Morgan and Hofmann (1971), using fractions enriched in co-lipase by Sephadex filtration of rat pancreatic juice, found that co-lipase had no effect on lipase unless bile salts were present. In that case, co-lipase restored the activity lost by bile salt inhibition and caused some additional stimulation of lipolysis at acidic pH. Figure IV-30 shows the inhibition of corn oil hydrolysis by taurodeoxycholate and taurocholate and the effect of co-lipase at pH 6. The lipase

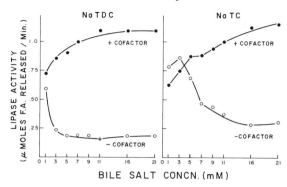

Fig. IV-30 Effect of bile salts on human lipase activity in the presence and absence of cofactor. Lipase source: 200 μl pooled lipase from Sephadex G-100 column (protein, 0.35 mg/ml). Cofactor: 400 μl pooled cofactor from same column (protein, 0.16 mg/ml). Cofactor added (●); no cofactor added (○). NaTDC, Na-taurodeoxycholate; NaTC, Na-taurocholate. From Morgan and Hoffman (1971).

is that of human pancreas. It can be seen that inhibition and reactivation start at different concentrations for each salt. These concentrations are near the micellar points, and it seems that it is the micelles of the bile salts that inhibit lipolysis, and that this inhibition is cancelled by the co-lipase. Maylié *et al.* (1973), however, found that the inhibition of lipase started below the critical micellar point of taurocholate.

When the hydrolysis of tributyrin by rat lipase is inhibited by bile salt and restored with co-lipase, it is found (Borgström and Erlanson, 1971) that the pH optimum of the reaction has shifted from 9 to 6 (Fig. IV-31). This points to an interaction of lipase, co-lipase, and bile salt at this pH. In fact, when co-lipase is chromatographed on Sephadex in the presence

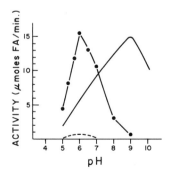

Fig. IV-31 Influence of bile salt and co-lipase on the pH optimum of rat pancreatic lipase. Substrate, tributyrin; lipase + co-lipase (solid line); lipase + taurodeoxycholate (broken line); lipase + co-lipase + sodium-taurodeoxycholate, 1 m*M*. (solid circle). From Borgström and Erlanson (1971).

of 4 m*M* taurodeoxycholate, it appears as a dimer of molecular weight 26,000, at pH 9 as well as at pH 5.8. Lipase and co-lipase in taurocholate separate at pH 9 into lipase (MW 36,000) and co-lipase dimer (MW 26,000), but at pH 5.8 a species of molecular weight 62,000 emerges from the column, corresponding to 1 mole of lipase and 2 of co-lipase.

It appears, then, that it is the function of co-lipase to overcome the inhibitory effect that bile salts will have when they occur in the micellar state, as in the intestinal juices. Borgström and Erlanson (1971) suggest that the co-lipase emerged in the evolution of higher animals as an adjustment of lipase to bile salts in weakly acidic solutions (invertebrates have no bile salts).

The dramatic downward shift of the pH of lipolysis in the presence of bile salt and co-lipase is extremely puzzling. On the one hand, it is hard to believe that the same catalytic mechanism should be at work at two peaks of efficiency three pH units apart; on the other hand, it is unlikely that bile salt and co-lipase should induce a completely different kind of catalytic mechanism. It may be suspected that the shift of the optimal pH is apparent rather than real. Bile salt micelles are known to incorporate fatty acids and shift their apparent pK_a from 9 to 6.5 (Hofmann and Small, 1967). Bile acid anions will also adsorb to the interface and their own pK_a will change. It must be realized that the optimal pH that has been measured in the studies discussed is the pH of the aqueous bulk phase; the interfacial optimal pH may, quite possibly, remain constant.

The suggestion has been made that the co-lipase may act on the oil–water interface rather than directly on the enzyme, (Brockerhoff, 1971a, 1973), and possess a hydrophobic head and a hydrophilic tail, as they have been postulated for lipase, which would bind and orient the protein at the interface. As a small, heat-stable protein with five disulfide bridges, the co-lipase would probably be immune against interfacial unfolding. It might in some manner neutralize the bile anions, or assist the lipase in reaching the interface, or support the proper orientation of the enzyme.

A determination of the apparent Michaelis constant "K_m" of an olive oil emulsion showed that, without co-lipase, "K_m" increased from 0.6 to 14 (relative values) as the taurocholate concentration increased from 0 to 0.8 m*M*; with co-lipase present, no such increase in "K_m" was apparent (Maylié *et al.*, 1973). This result can be interpreted as inhibition of enzyme–supersubstrate binding by the bile salt and counterinhibition by the co-lipase. However, whether the co-lipase forms a complex with lipase and bile salts or influences the catalytic mechanism or acts at the interface, an entirely intelligible model for its action cannot yet be proposed. A forthcoming structural analysis of the protein will perhaps shed more light on the problem.

F. Conclusion

Pancreatic lipase has always been a difficult enzyme, and despite a long history of research it is still much less well understood than other hydrolases, for instance, phosphatases or proteinases. The reasons have been the problems of isolation, caused mainly by the association of lipids with the enzyme, the relatively high molecular weight, and especially the peculiarity of the substrates of lipase to be insoluble. However, in the last few years great advances have been made concerning all aspects of the enzyme.

1. Isolation and Structure

The pancreatic lipases of beef (Julien *et al.,* 1972), of the rat (Gidez, 1968; Vandermeers and Christophe, 1968) and of the pig (Verger *et al.,* 1969) have been isolated. The porcine lipase with a molecular weight of 48,000 occurs in two equally active forms L_A and L_B, both containing a few carbohydrate residues. The amino acid compositions are known for the lipases of both rat and pig, but the amino acid sequences are unknown. The lipases have not yet been crystallized.

The lipase of the pig consists of one chain with six disulfide linkages and two free sulfhydryl groups, SH_I and SH_{II} (Verger *et al.,* 1970). SH_I may be in a molecular region that regulates the affinity of the enzyme to its supersubstrate; SH_{II} seems to be near the active center of the enzyme; neither group is involved in the catalytic mechanism. Lipase, like many other hydrolases, appears to be a serine enzyme. After inactivation of the enzyme with diethyl-*p*-nitrophenyl phosphate and subsequent tryptic digestion, a peptide can be isolated in which the phosphate is attached to a serine flanked by a glycine and a leucine residue (Desnuelle, 1971). It is reasonable to expect that the hydrophobic leucine is involved in the steric hindrance effects and the hydrophobic nature of lipolysis. There are also indications that a histidine residue is involved in the catalytic reaction (Sémériva *et al.,* 1971). It seems that a histidine has to be deprotonated for catalytic activity, and the destruction of pancreatic lipase by photooxidation is kinetically parallel to the loss of one of the enzyme's histidines. The qualifications of serine and histidine as parts of the reactive center are not yet definitively established, but it would be surprising if they should not soon be confirmed. It remains to be seen if the buried aspartic acid which is part of the "charge relay system" of proteinases plays a role in lipases also.

2. Cofactors

The outcome of almost a century of research on activators of lipase is that no cofactor is necessary for the catalytic action of the enzyme. There is, however, a multitude of agents that can influence the speed and course of lipolysis. Sodium chloride provides counter ions at the interface and thereby influences the apparent dissociation constants of long-chain fatty acids; this results in an increase in the speed of lipolysis. A more direct effect of cations on the enzyme–substrate complex must also be considered as a possibility. Calcium ions are essential in the hydrolysis of long-chain triglycerides, because the free fatty acids (or anions) must be removed as their insoluble calcium salts.

Bile acids will in general inhibit lipolysis, most likely by molecular and electrostatic blocking of the supersubstrate–water interface. Only at low concentrations, usually below their critical micellar point, and *in vitro* may they function as activators, by preventing the unfolding and denaturation of lipase, and by unblocking the oil–water interface. The physiological function of bile salts, the micellar removal of fatty acids and monoglycerides, is of decisive importance *in vivo* for the intestinal absorption of fats but not for the action of lipase.

Co-lipase is a small protein with a molecular weight of 10,000. It is not an obligatory factor *in vitro,* but it may be essential for lipolysis in the intestine of vertebrates. Its function seems to be to counteract the inhibitory effect of bile salts on lipolysis. The co-lipase could possibly serve to counteract the blocking of oil–water interfaces by bile salts. Further information on structure and function of the co-lipase must be awaited.

3. Substrate Specificity and Active Complex

Studies with synthetic substrates indicate that pancreatic lipase prefers esters that are unhindered, hydrophobic, and activated by an electrophilic alcohol moiety. Steric hindrance explains the "positional specificity" of the enzyme for the primary α-esters of glycerides; electrophilic induction by neighboring ester groups explains the high rate of triglyceride hydrolysis. Lipase is most active on insoluble esters at oil–water interfaces, but it can also hydrolyze water-soluble esters, and the speed of hydrolysis then depends on the hydrophobic nature of the substrate.

Enzyme–substrate complex formation does not appear to involve lipophilic binding, i.e., hydrophobic bonding of aliphatic chains. Fixation of the substrate occurs along the C—O axis of the ester group (although hydrogen bonding to the carbonyl oxygen is also possible). Borrowing from the research results on chymotrypsin we may postulate that the reactive serine forms a hydrogen bond, C—O—H $\cdots$ N$_{\varepsilon_2}$ to the reactive histidine, and that the two-dimensional fixation of the substrate is the orientation necessary for the leaving alkoxy group to capture the serine proton (Fig. IV-32). The N$_{\delta_1}$ of the histidine may perhaps, as in the proteinases, be hydrogen-bonded to an aspartic acid carboxyl group. The leucine resi-

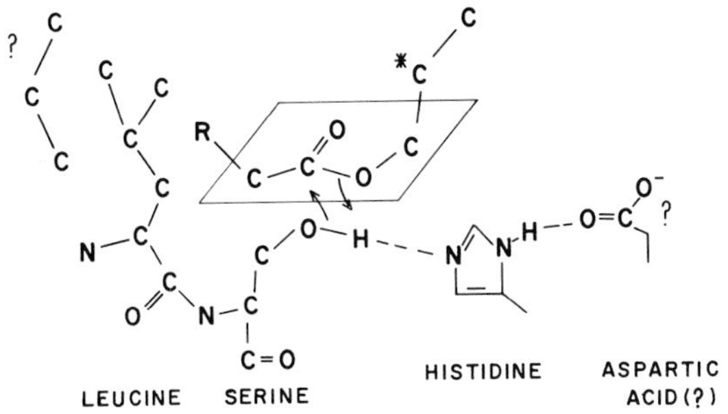

Fig. IV-32 Hypothetical alignment of amino acids in the reactive center of lipase. From Brockerhoff (1973).

due next to the serine may confer hydrophobicity to the active center, and also steric hindrance. This might, for instance, explain the difference in the hydrolysis rates of butyroyl and valeroyl esters. In Fig. IV-32 it is further speculated that the leucine chain is held in an unburied position by hydrophobic bonding or steric restriction from another part of the enzyme. The figure shows clearly that C*, the anomeric carbon of the glyceride, is not involved in bonding or catalytic reaction; therefore the "nonstereospecificity" of the lipase. It must be stressed that the model of Figure IV-32, plausible as it is, is still almost entirely hypothetical. Future research is likely to modify the picture considerably.

4. Lipase at the Interface

The outstanding characteristic of pancreatic lipase is its affinity to oil–water interfaces and the high rates of hydrolysis it catalyzes when adsorbed. A great affinity to lipids is needed if the enzyme is to compete for interfaces with the multitude of other proteins found in the intestine during digestion. Paradoxically, however, there appears to be no lipophilic binding, i.e., hydrophobic (van der Waals) binding of aliphatic chains, in the enzyme–substrate complex. A tentative resolution of the paradox is offered in Fig. IV-33. The model proposes that the attachment to the supersubstrate (the surface of the oil droplet) is achieved by a supersubstrate binding site, the hydrophobic head. This site not only binds the enzyme to the supersubstrate but also ensures its correct orientation (Brockerhoff, 1973).

The reactive site is near to the hydrophobic head but separate from it. By this arrangement, substrate and active center are brought together, but lipophilic enzyme-substrate binding is avoided, and a path of access if left open for the water that is needed to hydrolyze the acyl enzyme. The proper orientation of the lipase molecule may be further stabilized by polar or

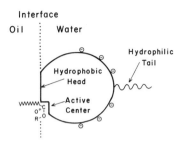

Fig. IV-33 Hypothetical model of the orientation of pancreatic lipase at the super-substrate–water interface.

charged amino acid or carbohydrate residues which constitute a hydrophilic tail. The figure explains why the lipase is so much larger, two to four times, than the digestive proteinases and nucleases, although its reactive center is probably more simply constructed; a larger infrastructure is needed to stabilize the separate existence of reactive site, hydrophobic head, and hydrophilic tail.

The model of Fig. IV-33 suggests a new definition of lipases, and lipolytic enzymes in general, which is an elaboration of that of Desnuelle (1961). These enzymes are hydrolases that have developed supersubstrate binding and orientation sites that enable them to bring their reactive sites toward their insoluble substrates. The inability of the substrate molecules to move in the third dimension is countered by the enzymes by fixating their active center, too, in a two-dimensional plane.

III. GASTRIC, PREGASTRIC, AND LINGUAL LIPASES

The existence of nonruminant gastric lipases has yet to be definitely established, although lipolytic activity has been observed in gastric juice, dietary fat has been found to be hydrolyzed in the stomach, lipase or esterase activity has been detected histochemically in gastric mucosa, and various gastric mucosal preparations have been shown to contain lipolytic activity (see Cohen *et al.*, 1971, for a comprehensive list of references).

Cohen *et al.* (1971) collected data that might prove the occurrence of a lipase in gastric juice not identical with regurgitated pancreatic lipase. Lipolytic activity was found in human gastric aspirates which were essentially free of duodenal reflux. On Sephadex G-200 the gastric lipase appeared in an elution volume corresponding to a molecular weight of 40,000 to 50,000. The lipolytic activity of duodenal juice was excluded from the column, with an apparent molecular weight of >500,000. The authors,

TABLE IV-18

Comparison of Human Gastric and Duodenal Lipases[a]

Characteristic	Gastric	Duodenal
Apparent molecular weight	40,000–50,000	> 500,000
Inactivation at pH 3	Little	Marked
Resistance to trypsin	Little	Moderate
Inhibition by 8:0	Moderate	Little
Chain-length specificity	4 > 6 > 8 ≫ 18	4 > 6 > 8 > 18
Positional specificity	1 > 2	1 ≫ 2
pH optimum in presence of bile acids	4–8	6–9
Esterification of fatty acid acceptor	Moderate	Marked
Hydrolysis of milk triglyceride	Moderate ($\cong$ that of triolein)	Virtually absent unless bile acids present

[a] Cohen *et al.* (1971).

however, mentioned that the polymeric form of pancreatic lipase could be dissociated into a monomolecular form of 35,000 to 42,000 molecular weight. However, other observed differences, listed in Table IV-18, suggested the separate existence of a gastric lipase; the greater stability of the gastric enzyme at pH 3 is especially persuasive. Healthy adults had relatively little lipolytic activity in the gastric juice, and Cohen *et al.* (1971) postulated that gastric lipase could aid in the digestion of milk in infants, but might contribute little to normal lipolysis in adults. Barrowman and Darnton (1970) found that gastric mucosa of rats hydrolyzed medium-chain triglycerides (trioctanoin) but not long-chain triglycerides. Engström *et al.* (1968) and Blum and Linscheer (1970) found trioctanoin hydrolase activity in canine gastric juice.

Information on gastric lipases of ruminants has become available from studies on enzyme preparations used in making Italian cheese varieties. These lipases provide the free fatty acids that contribute to the distinctive flavor of the cheeses. In the past, rennet paste was prepared from tissue as well as contents of the fourth stomach of kid goat, lamb, or calf. More recently, partially purified pregastric esterase preparations from stomach contents only have been utilized. Richardson *et al.* (1971) found that lamb gastric lipase had a pH optimum of 7.0 to 7.6 and hydrolyzed tributyrin more rapidly than β-naphthyl laurate and milk fat. Both di- and monobutyrins were digested.

The saliva of calves and other young ruminants contains a pregastric esterase which hydrolyzes milk triglycerides in the stomach (Otterby *et al.,* 1964). The enzyme preferentially released 4:0 and 6:0 from milk fat (Siewert and Otterby, 1968) and hydrolyzed tributyrin very much faster than longer-chain triglycerides (Grosskopf, 1965). The optimum pH range of the enzyme was 4.5 to 6.0; it was inactivated below pH 2.4 and above 7.8; and its activity was reduced by sodium taurocholate. Pitas and Jensen (1970) observed relative rates of hydrolysis of synthetic racemic triglycerides as follows: 16:0–4:0–4:0, 100; 16:0–8:0–4:0, 36; 16:0–4:0–16:0, 32; 16:0–16:0–4:0, 28; 16:0–18:1–4:0, 25. The triglycerides of the lower carbon number were hydrolyzed more rapidly.

Hamosh and Scow (1973) found a lipase in the serous glands of rat tongue which hydrolyzed triglycerides at pH 4.5 to 5.4. Homogenates of the linqual serous glands of one animal digested 40–70 mmole of triglyceride per hour. The saliva also contained lipolytic activity. The hydrolysis produced mainly diglycerides and free fatty acids from milk and corn oil triglycerides. The authors proposed that partial hydrolysis by lingual lipase of triglycerides in the stomach is the first step in the digestion of dietary fats in the rat.

It is tempting to speculate that in some instances lipolytic activity designated as gastric may have been due to mouth glandular secretions into the saliva which traveled to the stomach. Purification and characterization of these activities will be required before definite differentiation can be made.

IV. DIGESTIVE LIPASES OF NONMAMMALIAN ANIMALS

There are few references in the literature to digestive lipases of vertebrates other than mammals and fish. Reports on lipases of invertebrates are more numerous. However, no studies have been published that are comparable in depth to the studies on mammalian pancreatic lipases. In most cases, only the presence of lipolytic activity in a fish or in an invertebrate has been reported; quantitative studies are rare. It is usually impossible to decide from the published data whether the lipolytic activity was of the digestive or the intracellular types, "digestive lipases" being defined as extracellular enzymes that act in the digestive tract of the animal and assist in the incorporation of the fat. This question will be discussed in the conclusion of this chapter. Furthermore, many investigators have employed unsuitable substrates, especially tributyrin, and this makes the relevance of many reports suspect.

In keeping with the purpose of this book, we shall not list all the references to lipases, as they can be extracted from *Biological Abstracts,* but discuss only a few studies on fish and invertebrates in which the lipase was undoubtedly of the digestive type and where some quantitative data are given.

A. Fish

Cartilaginous fish possess a discreet pancreas, and this organ contains a lipase. The enzyme of a skate, *Raia radiata,* was found to hydrolyze the primary ester bonds of a triglyceride of known fatty acid distribution without regard to stereospecificity (Brockerhoff and Hoyle, 1965). The specific activity of an acetone powder of the gland approached that of commercial pig pancreatic powder. Dogfish, *Squalus acanthias,* pancreas has lower lipolytic activity (H. Brockerhoff, unpublished). Berner and Hammond (1970a) could not find any lipase in this animal.

In bony fishes, the pancreas is diffuse and difficult to collect. In a feeding experiment with cod, *Gadus morrhua,* it was shown that triglycerides are hydrolyzed nonstereospecifically in the primary positions (Brockerhoff, 1966a). The lipase of a trout, *Salmo gairdneri,* has recently been partially purified (Leger *et al.,* 1970; Leger, 1972). The diffuse pancreas had a specific activity of 0.4–0.6 at 25°C and pH 8.0. Ammonium sulfate precipitation from an extract of the defatted lyophilized tissue, followed by gel filtration on Sephadex G-100, yielded a preparation with the specific activity of 15, in a yield of 8.5%. A "fast" and a "slow" enzyme emerged from the Sephadex column. Increasing the sodium chloride concentration of the eluting buffer converted the fast lipase into the slow one. The enzyme was specific for primary esters, but also appeared to show a preference for oleic acid esters regardless of position (Leger and Bauchart, 1972).

B. Invertebrates

Berner and Hammond (1970a) tested a number of invertebrates for lipolytic activity. In a crayfish (*Cambarus virilis*) hepatopancreas, they found a lipase partially, but not entirely, specific for the α-position of triglycerides. The specific activity of the pancreas powder was in the same range as that of terrestrial vertebrates (man, rhesus monkey, pig, opossum, chicken, turtle). In other invertebrates, the lipolytic activity was either very low (*Dendrostrum pyroides,* peanut worm; *Lumbricus terrestris,* earthworm; *Chaetopterus variopedatus,* pardiment tube worm; *Aplysia californica,* sea hare; *Strongylocentrotus purpurtus,* sea urchin; *Dysidea amblia,* sponge; *Tetrahymena pyriformis,* protozoan) or undetectable (*Pisaster giganteus,* sea star; *Parastichopus parvimensis,* sea cucumber; *Metridium senile,* anemone), and there was no strong positional specificity.

The best-characterized invertebrate lipase is that of the American lobster, *Homarus americanus* (Brockerhoff *et al.,* 1967, 1970). The enzyme is found in the gastric juice of the animal. It hydrolyzes the primary ester bonds of triglycerides. Its optimal pH is around 7; its molecular weight, determined by gel filtration, 43,000. The specific activity of the gastric juice proteins at pH 7.5 and 37°C was 1.2. Similar to porcine pancreatic lipase, the enzyme hydrolyzes tributyrin twice as fast as triolein; according to their behavior on gel filtration and ion-exchange chromatography, tributyrinase and lipase are identical. The enzyme is also insensitive to DFP and very sensitive to surface denaturation by the emulsified substrate; it can be protected by bile salts and albumin (Brockerhoff, 1971a).

A comparison of the total lipolytic activity in the gastric juice with the food intake of the lobster showed that there is sufficient enzyme to hydrolyze all ingested fat (Brockerhoff *et al.,* 1970); the digestive nature of the enzyme is also attested by its occurrence in the gastric juice. The lipase is probably produced in the hepatopancreas of the animal, but its activity in this organ is more than ten times lower than in the juice (H. Brockerhoff, unpublished). The locus of triglyceride hydrolysis is somewhat of a mystery, since the pH of the gastric juice is around 5, a region where *in vitro* the lipase is hardly active. The lobster's intestine is too short to be considered as the place of breakdown and adsorption of fats. It is perhaps possible that the ducts between stomach and hepatopancrease fulfill this function; the hepatopancreas is also the organ of fat storage.

In the insects *Periplaneta americana* and *Blabenus cranifer* a lipase that hydrolyzes long-chain triglycerides is secreted by the midgut. The enzyme specifically hydrolyzes the primary ester bonds (Bollade *et al.,* 1970).

A lipase has been found in the crystalline style of the surf clam, *Spisula solidissima* (Patton and Quinn, 1972). The specific activity of the lyophilized organ was less than 1% of that of hog pancreatic powder. The pH optimum was 8.0. The enzyme was specific for primary esters. Esters of polyenoic acids resisted hydrolysis by the enzyme.

C. Conclusion

The lipolytic activities in the digestive organs or juices of bony fish and invertebrates are usually very low, perhaps one thousand times lower than those in mammalian pancreas or pancreatic juice. This is perhaps due to lower quantities of enzyme rather than to lower specific activities of the enzymic proteins. For example, a lipase from lobster, obtained by preparative electrophoresis of gastric juice proteins, had the respectable specific activity of 300 (toward tributyrin) although the preparation was still quite impure (H. Brockerhoff and R. J. Hoyle, unpublished). A pure preparation might not be much less active than porcine pancreatic lipase (specific activity 10,000). In any case, it will be very difficult to purify the nonpancreatic lipases and isolate them in amounts sufficient for intensive examinations. Until this becomes possible, the most significant information on these enzymes is obtained by comparison of their activity and specificity with that of the mammalian pancreatic lipases, and by extrapolating from our knowledge of these better known enzymes.

The results on skate, lobster, *Periplaneta,* and *Spisula* suggest that all animal digestive lipases are specific for the 1- and 3-position of triglycerides. Judging from the lobster lipase, their molecular size may be similar to that of the mammalian enzymes, and they may be equally sensitive to surface denaturation. Possibly they are also serine enzymes though not inhibited by DFP.

The enzymes found in fish pancreas and in the digestive juices of invertebrates are undoubtedly digestive lipases rather than tissue lipases. However, fat might also be ingested by phagocytosis and stored without initial hydrolysis until it is hydrolyzed by an intracellular lipase for the purpose of lipid synthesis or energy generation. The question may be asked at what level of evolution the intracellular digestion of fats gives way to extracellular digestion and how the two types of lipase differ. The enzymes of phagocytes might be expected to act at acidic pH, with a mechanism quite different from that of the neutral or basic serine lipases. On the other hand, both types of lipases might be similar in structure and mechanism, the extracellular evolving from the intracellular enzymes as the entire process of digestion evolves from intra- to extracellular. In those animals, such as many bivalves, which have a mixed-type digestion, the extra- and intracellular enzymes might be very similar or even identical. In this view, there is no fundamental difference between intracellular and extracellular lipases. The lipases of the exocrine pancreas are perhaps descendants of intracellular enzymes, specialized for the competitive environment of the intestinal lumen by a more potent affinity for oil–water interfaces and by a more developed polar structure that ensures the proper orientation at the interface. The greater activity, or higher concentration, of the extracellular enzymes of higher animals may be necessitated by the greater quantitites of food stuff that have to be assimilated in a shorter time of digestion.

V. LIPOPROTEIN LIPASE

Triglycerides circulate in the blood as components of aggregates with high molecular weights in combination with phospholipids, cholesterol and cholesterol esters, and proteins. From these lipoproteins, the triglycerides are hydrolyzed by lipoprotein lipases (LPL). Such hydrolysis is necessary because intact triglyceride molecules cannot cross biological membranes.

The lipoproteins of the blood can be divided into four (overlapping) groups according to increasing density, which is inversely related to their triglyceride content: (1) chylomicrons, (2) very low density lipoproteins (VLDL), (3) low density lipoproteins (LDL), and (4) high density lipoproteins (HDL). Several recent reviews describe structure, metabolism, and interrelations of these compounds (Schumaker and Adams, 1969; Fredrickson and Levy, 1972; Fredrickson *et al.,* 1972a,b; Robinson, 1970; Scanu and Wisdom, 1972; Skipski, 1972). We must restrict ourselves to a short characterization of those lipoproteins that primarily serve as substrates for the enzymes.

Chylomicrons are assembled in the intestinal mucosa and appear in the blood after ingestion of fat. They contain around 90% triglyceride and 1–2% protein, and they are large enough (generally 1000–5000 Å) to give the blood a milky appearance (lipemia). Very low density lipoproteins originate mostly in the liver; they contain, on the average, 55% triglyceride and 10% protein, 20% phospholipid, and 15% free and esterified cholesterol. Several of the proteins of this fraction have been isolated and characterized. These "apolipoproteins" are classified according to their N-terminal and C-terminal amino acids, e.g., as Thr-apoVLDL-Glu.

The triglycerides contained in low density and high density lipoproteins (10% or less of triglyceride in the aggregate) are probably not hydrolyzed by LPL unless they are incorporated into fractions of lower density.

In 1943, Hahn found that an injection of heparin caused accelerated clearance of lipemic blood in dogs. Korn (1955a) established that the "clearing factor" was identical with an enzyme that he could extract from rat heart and that he named lipoprotein lipase. He found that the enzyme was inhibited by protamine, pyrophosphate, and 1 *M* NaCl, and that an emulsion of coconut oil was hydrolyzed only after incubation with serum, and in particular with an HDL fraction of serum (Korn, 1954, 1955a,b). These three criteria: (a) activation by heparin, (b) activation by a protein cofactor in serum, and (c) inhibition by sodium chloride and protamine, have been applied by most investigators to identify lipoprotein lipase activity.

Recently it has become apparent that there are at least two triglyceride lipases that act on the lipoprotein triglycerides of blood, one of hepatic and the other of extrahepatic origin (La Rosa *et al.,* 1970a; Greten *et al.,* 1972; Assmann *et al.,* 1973a; Krauss *et al.,* 1973). Contrary to the extrahepatic enzyme, the hepatic lipase does not require activation by serum lipoproteins and is quite resistant against sodium chloride and protamine and thus lacks some of the critical properties listed by Korn. Nevertheless, since enzymes are usually classified according to their natural substrates, the hepatic lipase deserves to be included among "lipoprotein lipases"; this name should be used for all enzymes that degrade the triglycerides of the chylomicrons and other lipoproteins of the blood.

Reviews on LPL research have been supplied by Korn (1959) and Robinson and French (1960). A shorter, but more recent, account is given in a review by Fredrickson and Levy (1972). These authors treat the enzyme in the context of the metabolic disease, familial hyperlipoproteinemia (Type I), which is caused by a deficiency of the protamine-inhibited LPL (La Rosa *et al.*, 1970a).

A. Assay

1. Substrate and Cofactors

Lipoprotein lipases by their original definition (Korn, 1955a,b) hydrolyze the triglycerides of lipoproteins but do not act on the simple artifical triglyceride emulsions that can serve as substrates for pancreatic or microbial lipases. On the other hand, they tolerate concentrations of such interface-blocking substances as phospholipids or albumin that will completely inhibit other lipases, such as pancreatic lipase.

The natural supersubstrate of LPL, chylomicrons, can be isolated from lipemic blood by centrifugation (Bragdon *et al.*, 1956; Korn, 1959; Scanu and Page, 1959). If an artificial emulsion of triglyceride is used, it has first to be activated (Korn, 1955b). This can be achieved by incubation of a 2% emulsion with an equal volume of serum for 30 minutes at 37°C; the substrate is then collected as an oil layer by centrifugation (Korn, 1959). The hepatic lipase is not dependent on substrate activation by serum proteins (Krauss *et al.*, 1973).

In many earlier investigations, commercial oil emulsions such as Ediol, Intralipid, or Lipomul have been employed, but such emulsions are suspect because they may contain partial glycerides that could possibly be hydrolyzed by other enzymes, such as monoglyceride hydrolases. Purified triolein, a reliable substrate, has been used in many recent studies. Emulsification is achieved by sonication in a buffer (usually pH 8.5) containing a detergent such as Triton X-100. A sufficiently high substrate concentration is important; many conflicting results concerning activation, inhibition, and substrate specificity of LPL appear to have resulted from the use of suboptimal substrate concentrations and the resulting measurement of submaximal reaction rates (Assmann *et al.*, 1973b). Krauss *et al.* (1973) report that 7.5 mg of [^{14}C]triolein sonicated with 0.9 ml of 0.05% of Triton X-100 in buffer (containing 2% bovine albumin and 0.10 M NaCl) for 1 minute at 60 W and 4°C produce an emulsion that will saturate the post-heparin LPL of 0.1 ml of plasma.

The optimal ionic strength of the assay solution is usually given by 0.15 M NaCl or 0.05 M $(NH_4)_2SO_4$. For enzyme preparations from tissues, calcium ion may be required as a cofactor (Korn, 1959); in general, calcium ion may serve as acceptor for the liberated fatty acids (Korn, 1955b). An agent accepting fatty acids must be present in any assay because the acids or their anions are strong inhibitors. The role of acceptor is usually given to defatted bovine serum albumin at concentrations of 2–4%. Whayne and Felts (1972) have observed considerable variation in LPL activity dependent on the albumin preparation added to the assay mixture. An albumin depleted of fatty acids resulted in lower activity than the common crystalline product.

Phospholipids are present in chylomicrons as well as in commercial fat emulsions, but if an emulsion of pure triolein is employed for a substrate, phospholipid, added in addition to the activating apolipoprotein, will cause a further increase of the reaction rate (Scanu, 1967; La Rosa *et al.,* 1970b). Egg lecithin activated optimally at 0.6 mg mixed with 10 mg triolein in 1 ml solution; lysolecithin, at the same concentration, activated four times better (Chung *et al.,* 1973), perhaps because of its action as a detergent.

The differentiation between the post-heparin LPL activities of hepatic and extrahepatic origin has been described by Krauss *et al.* (1973). The extrahepatic LPL of the rat was irreversibly inhibited (90%) by incubation of 0.1 ml of plasma with 3 mg of protamine at 27°C for 10 minutes. Inhibition of the hepatic enzyme under these conditions amounted to only 10%. The activities can thus be estimated from the reaction rates obtained with treated and untreated plasma.

Fredrickson and Levy (1972) described a simple qualitative test for postheparin LPL activity in human plasma. Samples of plasma obtained before and after administration of heparin were separated into their constituent lipoproteins by paper electrophoresis. If LPL was deficient, as in a person with Type I hyperlipoproteinemia, the migration of lipoproteins was not affected, while if the enzyme was present, the mobility of the lipoproteins was increased.

2. *Quantitation*

LPL activity can be quantitated by measuring the rate of clearance of lipemic blood or of chylomicron or triglyceride emulsions (Baker, 1957; Korn, 1959) or by the clearance of an oil emulsion stabilized by agar (Scanu and Page, 1959) or agarose gel (Wilson *et al.,* 1973). These methods yield, of course, arbitrary units of activity unless they are standardized by comparison with more direct methods.

The fatty acids liberated by LPL can be titrated after extraction of the reaction mixture (Korn, 1959; Robinson, 1965; Boberg, 1970). Continuous titration of the reaction mixture, the most convenient method for the assay of many other lipases, has not often been employed for LPL because of the low specific activity of the usual enzyme preparations and the inhibitory effect of fatty acid salts (Korn, 1959), but recently, continuous methods using an automatic titrator have been described (Chung and Scanu, 1973; Egelrud and Olivecrona, 1973). Smith (1972) and Miller and Smith (1973) have used the decrease in radioactivity of a monolayer of tri[1-^{14}C] octanoin to measure LPL activity; this method has the common drawback of monolayer assays of not permitting measurement of the maximal reaction rate V (see Chapter III, Section VI).

Because of the low activity of most LPL preparations, radioactive substrates, in particular, glycerol [^{14}C]- or [^{3}H]trioleate, have been used in most studies during recent years. In such assays, the titration of the fatty acids in the final solvent extract is replaced by a count of radioactivity, but since fatty acids, substrate, and partial glycerides are all extracted by the same (acidified) solvent, usually isopropanol–heptane (Dole, 1956), a further alkaline extraction of this solvent is necessary to isolate the fatty acids A fast and convenient method using such a sequence has been described by Schotz *et al.* (1970; Schotz and Garfinkel, 1972b). Alternatively, the fatty acids can be isolated by adsorption on an ion exchanger (Kelley, 1968). Zieve (1973) separated oleic acid from tri-, di-, and monoglyceride with a small disposable column of TEAE-cellulose, a method requiring less than 3 minutes.

An assay method specifically suited for LPL, though not for lipases in general, has been described by Schotz and Garfinkel (1972a). It relies on the fact that most LPL preparations hydrolyze triolein completely to glycerol and fatty acids with little accumulation of partial glycerides. The substrate employed is [2-^{3}H] glycerol trioleate. After the reaction, the remaining substrate, partial glycerides, and fatty acids are precipitated with trichloroacetic acid together with the proteins (mainly albumin) of the mixture, and the free glycerol in the supernate is counted.

B. Sources

LPL activity has been detected in many tissues and fluids: adipose, mammary, muscle, heart, aorta, liver, milk, plasma, lung, spleen, kidney, medulla and diaphragm (Desnuelle, 1972; Frederickson and Levy, 1972). Post-heparin plasma contains lipolytic activity toward both triglycerides and monoglycerides (Shore and Shore, 1961; Biale and Shafrir, 1969). Greten *et al.* (1969) presented evidence that these activities were separate entities. As compared to the triglyceride lipase, monoglyceride hydrolase activity was (a) higher, (b) much less heat sensitive, (c) not affected by the three typical inhibitors of TG tissue lipase: sodium chloride, protamine, and pyrophosphate, (d) not affected by drastic changes in diet, and (e) normal in Type I hyperlipoproteinemia. The last observation especially was considered evidence for a separate monoglyceride hydrolase, since Type I is characterized by a genetic deficiency of LPL (Fredrickson and Levy, 1972). Vogel *et al.* (1971) also noted normal post-heparin hydrolase activity toward monoglyceride in some individuals with hypertriglyceridemia and deficiences in LPL.

To further complicate the problem of enzyme identity, La Rosa *et al.* (1970a) presented data that indicated that post-heparin plasma contained two triglyceride lipases, possibly representing a mixture of enzymes from liver and adipose and other extrahepatic tissues. This suggestion was based on similar responses of liver and plasma LPL to sodium chloride, protamine, and pyrophosphate, i.e., little or no inhibition, as compared to strong inhibition of adipose tissue LPL. Krauss *et al.* (1973) have obtained data that confirm the presence in rat post-heparin plasma of hepatic and extrahepatic LPL. The LPL activity from post-heparin plasma and liver perfusate was not inhibited by sodium chloride and protamine, whereas the LPL from the supradiaphragmatic part of the rat (extrahepatic) was inactivated by both. This differential sensitivity to protamine was the basis for a selective assay of the liver LPL. The existence of two enzymes was confirmed by assays of post-heparin plasma from partially hepatectomized rats: the remaining quantity of liver was related to the protamine-resistant LPL but not to the protamine-sensitive LPL in post-heparin plasma.

Much, if not most, of the post-heparin LPL of the blood seems to originate in liver cell plasma membranes (Assmann *et al.*, 1973a; Krauss *et al.*, 1973). The question of the location of LPL in other tissues, especially adipose tissues, has been reviewed by Robinson (1970). For adipose tissue, it has been postulated (Blanchette-Mackie and Scow, 1971; Scow *et al.*, 1972) that chylomicron triglycerides are hydrolyzed in the capillary endothelial cells and in the subendothelial space near the pericytes, but not in the capillary lumen or near the fat cells.

The lipoprotein lipase of milk (Quigley *et al.*, 1958; Korn, 1962) seems to be involved in the transfer of triglyceride fatty acids from blood to milk fat. About 40–60% of the fatty acids in bovine milk are derived from serum lipoprotein, the principal source being the VLDL (Patton, 1973a,b). The triglycerides are hydrolyzed by LPL in the endothelium lining the capillaries of mammary tissue (Schoefl and French, 1968; Scow *et al.*, 1972). In rats, guinea pigs, and other nonruminants, the chylomicrons provide the fatty acids of the milk. Both LPL activity and transfer of blood triglycerides into mammary tissue are increased by lactation (Barry *et al.*, 1963;

McBride and Korn, 1963, 1964; Robinson, 1963a; Askew *et al.,* 1970). After parturition in the rat, the mammary gland LPL activity increases sharply and the adipose tissue enzyme activity decreases. Cessation of suckling completely inhibited the mammary tissue activity. Although the two enzyme activities have not been rigorously shown to be identical, it seems that milk LPL is plasma enzyme that has been transferred into milk.

C. Purification

Numerous attempts have been made to isolate LPLs. Completely homogeneous enzymes seem not to have been obtained so far, but preparations of an estimated purity around 80% have been described. The starting materials used have been post-heparin plasma, rat liver, and bovine milk; the methods: flotation of enzyme–supersubstrate complexes, selective adsorption to calcium phosphate gel, and, most recently, affinity chromatography on heparin gels.

1. Post-Heparin Plasma

Blood collected 2–15 minutes after injection of heparin (1–5 mg/kg body weight) is oxalated and centrifuged. The LPL of the frozen plasma is indefinitely stable (Korn, 1959). A 600-fold concentration of its activity was obtained by Nikkila (1958) by adsorption of the enzyme to aged calcium phosphate gel and elution with 0.01 M sodium citrate.

Further purification of plasma LPL was achieved by Fielding (1969), who utilized the strong adsorption of the enzyme to its supersubstrate (Anfinsen and Quigley, 1953) and the stabilizing effect of potassium oleate or linoleate (Fielding, 1968). Postheparin rat plasma was flotated with intralipid emulsion, the enzyme–supersubstrate complex was disintegrated in a deoxycholate–oleate–heparin solution, and the soluble enzyme was further concentrated by adsorption and desorption on calcium phosphate gel (Table IV-19). The final product gave one band on disk gel electrophoresis, and its specific activity, 40, was not further increased by chromatography on Sephadex or acetone fractionation. Human post-heparin plasma LPL was purified 16,000-fold by a similar procedure to yield a product of slightly lower specific activity (Fielding, 1970a). Ganesan and Bradford (1971) repeated the procedure and obtained a preparation that still contained admixed apolipoproteins. Since the LPL preparations of Fielding were inactive against triglyceride emulsions in the absence of lipoprotein (Fielding, 1968, 1969, 1970a), it may be supposed, in the light of recent studies (Assmann *et al.,* 1973a; Krauss *et al.,* 1973), that they represented LPL of extrahepatic origin; this is also suggested by the high specific activity.

TABLE IV-19

Purification of Rat Post-heparin Plasma Lipoprotein Lipase[a]

Step	Specific activity	Yield (%)	Purification (fold)
Crude plasma	0.02		
Enzyme–supersubstrate complex	9.3	50.0	499
Soluble enzyme extract	22	41.8	1285
Calcium phosphate eluate	40	30.0	2345

[a] Fielding (1969).

Greten *et al.* (1972) found a lipase in the infranate of post-heparin human plasma that had been adjusted to a density of 1.21 and centrifuged to remove lipoproteins. Repeated affinity chromatography on a Sepharose gel coupled with heparin—a method introduced by Iverius (1971)—yielded an enzyme, homogeneous by disk gel electrophoresis, of a specific activity of 0.06, much lower than that of Fielding's preparations. The enzyme did not require added plasma for the hydrolysis of triglyceride emulsions, and it also hydrolyzed micellar solutions of monoglycerides. Since these characteristics were similar to those of a LPL in liver (La Rosa *et al.,* 1972), Greten *et al.* (1972) postulated that their plasma LPL originated in liver.

2. Liver

A lipase that is released from rat liver by heparin has been purified by Assmann *et al.* (1973a). Plasma membranes were collected by centrifugation. Purification of the enzyme was achieved on a heparin affinity column. This column did not bind the enzyme released from plasma membranes by heparin, but up to 50% of the enzyme could be released by incubation of the plasma membranes in glycine buffer alone, and this portion could be chromatographed (Table IV-20). The specific activity of the most active fractions was 1.12, or twenty times higher than that of the plasma LPL of Greten *et al.* (1972) of presumable hepatic origin; different assay methods may account for this discrepancy. The liver LPL preparation was not homogeneous; it showed two major and several minor bands on gel electrophoresis. Assmann *et al.* (1973a) reported that additional activity was found in liver microsomes and especially cytosol, which contained, in fact, the larger part of the alkaline LPL activity.

3. Milk

LPL was discovered in milk by Quigley *et al.* (1958). Korn (1962) and Olivercrona and Lindahl (1969) achieved partial purifications of the

TABLE IV-20

Partial Purification of Liver Plasma Membrane Triglyceride Lipase[a]

Fraction	Specific activity	Yield (%)	Purification
Liver homogenate	0.003	100	—
Top layer of 15,000 g sediment	0.004	2.96	1.2
Plasma membranes	0.028	3.07	8.9
Released material (pH 9.5, 1 hour incubation at 27°C)	0.059	1.70	18.9
Heparin affinity column	1.12	0.53	363

[a] Assmann *et al.* (1973a).

bovine enzyme; Ribeiro (1971) concentrated the enzyme from guinea pig milk. Olivecrona *et al.* (1971) noticed specific ionic binding between bovine milk LPL and heparin covalently bound to agarose gel, and used the effect for purification. In an improved procedure (Egelrud and Olivecrona, 1972, 1973), the lipase was dissociated from the casein of rennet curd with 1.16 M NaCl. Fractionation of the supernate with ammonium sulfate and acetone led to a stable powder with a specific activity of 0.6, a tenfold purification. Chromatography of this material on the heparin gel yielded preparations with the specific activity of 354 at 25°C (580 at 37°C), and over 80% purity as judged by disk gel electrophoresis.

4. Tissues

Extraction of acetone powder of adipose tissue or heart with 0.025 M ammonia and lyophilization yields a stable powder of crude LPL (Korn and Quigley, 1957; Korn, 1959). An eightyfold purification of the enzyme from hen adipose tissue has been achieved by chromatography on heparin agarose gel (Egelrud, 1973). Garfinkel and Schotz (1973; Schotz and Garfinkel, 1972b) separated extracts of rat epididymal adipose tissue on agarose gel and obtained LPL in two forms, a and b, of different apparent molecular weights.

There are several reports describing the influence of nutritional status on LPL activity (Bagdade *et al.,* 1970; Shafrir and Biale, 1970; dePury and Collins, 1972; Landes, 1972; Reichl, 1972; Schotz and Garfinkel, 1972b; Struijk *et al.,* 1973).

D. Characteristics; Activation and Inhibition

LPL from plasma, milk, and extrahepatic tissues have pH optima between 8 and 9. The enzyme of liver plasma membranes has an optimum of activity at pH 9.5; in the cytosol of rat liver, activity is optimal around pH 8 (Assmann *et al.,* 1973a).

The LPL of rat post-heparin plasma (probably of extrahepatic origin) had an apparent molecular weight of 72,600, determined by sedimentation (Fielding, 1969). The LPL of bovine milk had 62,000 to 66,000 as determined by SDS-gel electrophoresis (Egelrud and Olivecrona, 1972).

1. Heparin

The acidic polymeric carbohydrate heparin effects the release of LPL from tissues into the blood; this is true for both the hepatic and the extrahepatic enzymes (La Rosa *et al.,* 1970a; Assmann *et al.,* 1973a). Heparin also activates soluble enzyme preparations such as those from rat heart (Korn, 1955b) or bovine milk (Iverius *et al.,* 1972). The heparin-induced release of the enzyme could be based, in principle, on either of two mechanisms: the binding of LPL to cationic membrane sites from which it is replaced by heparin; or its binding to negative sites from which it is transferred to the more negative (or more specific) heparin. The second mechanism is more probable, because the enzyme can be shown to bind strongly to the polyanion (Olivecrona *et al.,* 1971; Assmann *et al.,* 1973a; Egelrud, 1973). Heparin could, therefore, be regarded as a prosthetic group of the enzymes, but this interpretation is doubtful. LPL from different sources have responded differently to added heparin. For example, the enzyme from chicken adipose tissue is not stimulated by heparin (Korn and Quigley, 1957), and the liver enzyme binds heparin but does not require it to act on its substrate (Assmann *et al.,* 1973a). The effect of heparin upon both crude and purified milk LPL has been found to vary with the purity of the enzyme, the method of substrate activation, and the ionic strength of the assay medium (Iverius *et al.,* 1972). For example, heparin had almost no stimulating effect on the purified enzyme activated by HDL at ionic strength 0.16, some effect at ionic strength 0.40 with HDL_3 or serum (ionic strength 0.16), and a marked effect with crude enzyme activated by serum at ionic strength 0.16. Apparently heparin can increase the activity but only when conditions are not optimal. Both crude and purified enzyme preparations contained small quantities of endogenous heparin-like substance. Iverius *et al.* (1972) concluded that milk LPL was stabilized rather than stimulated by heparin, possibly by interfering with the effect of inhibitors. The release and activation of LPL in blood can also be triggered with synthetic, heparin-like anions (Frandoli and Spreafico, 1972).

It appears that the effect of heparin is rather unspecific. This cofactor is not likely to be immediately involved in the catalytic mechanism of the enzyme. However, a role in enzyme activation in addition to its role of releasing or stabilizing the enzyme cannot be excluded. An allosteric mechanism of activation has been suggested (Patten and Hollenberg, 1969; Whayne and Felts, 1970), and the possibility that heparin mediates enzyme–supersubstrate binding or induces the formation of a supersubstrate binding site must be considered.

2. *Protamine; Salts*

Protamine, a strongly basic protein, inhibits the LPL of extrahepatic tissues and milk but leaves the hepatic LPL almost unaffected (Krauss *et al.,* 1973). The inhibition cannot be reversed by dialysis but can be prevented by heparin in concentrations high enough to bind the protamine (Korn and Quigley, 1955). Protamine may be presumed to either bind directly to the LPL or to bind the heparin that functions as a cofactor. Other polycations, e.g., polylysine, are equally inhibitory (Korn, 1962).

The inhibition by sodium chloride, also first described by Korn (1955a,b), is irreversible by dialysis and must therefore be described as an inactivation. It is dependent not on the nature of the ions or their molarity but on the ionic strength of the solution (Fielding, 1968). Again, the LPL of liver plasma membranes is little affected (Krauss *et al.,* 1973).

3. *Apolipoproteins; Phospholipids*

Korn (1955b) noticed that LPL from rat heart hydrolyzed coconut oil only if the oil was preincubated with serum, and he located the activator in the HDL fraction. In human serum, the apoproteins of the HDL fraction consist of at least five proteins (Shore and Shore, 1969), two major constituents, apoLP-Gln and apoLP-Thr (as identified by their carboxy-terminal amino acid residue), and three minor ones, apoLP-Ser, apoLP-Ala, and apoLP-Glu, (Fredrickson *et al.,* 1972b). These three are also components of VLDL. The LPL activating activity has been found to reside among these minor proteins (Fielding *et al.,* 1970), in particular, in apoLP-Glu (Havel *et al.,* 1970; La Rosa *et al.,* 1970b). ApoLP-Ala, especially a subfraction, R_1-Ala, can also serve as an activator but is much less potent (Havel *et al.,* 1973). The activation of LPL by the proteins is multiplied by the addition of phospholipids (Scanu, 1966; La Rosa *et al.,* 1970b; Chung *et al.,* 1973).

Miller and Smith (1973) have studied the association of milk LPL and proteins by monolayer techniques. ApoLP-Glu formed a surface film to which the enzyme was bound. This complex dissociated in 2 *M* NaCl. However, other HDL (and VLDL) components, apoLP-Ala, and apoLP-Ser, also formed complexes with LPL, though less effectively than apoLP-Glu, but did not activate the enzyme.

Activation by apolipoprotein is presumed to be obligatory for LPL preparations from extrahepatic tissues, but it is not needed by the LPL from liver plasm membranes (Assmann *et al.,* 1973a). The LPL from bovine milk has been found to need serum, HDL, or VLDL for activation of the substrate (Korn, 1962; Iverius *et al.,* 1972). However, in a recent study, Egelrud and Olivecrona (1973), using a highly purified milk LPL, found that the preparation could hydrolyze emulsified trioleate and tributyrin, micellar monooleate, the carbohydrate fatty acid ester Tween 20, and even the soluble esterase substrate *p*-nitrophenyl acetate without any addition of serum or apolipoproteins. The authors believed that LPL and not a contaminating enzyme hydrolyzed these substrates and arrived at the conclusion that the enzyme has no absolute requirement for serum proteins, but that these "cofactors" merely serve to achieve maximal reaction rates by modifying the substrate.

4. Other Activators and Inhibitors

Calcium or ammonium ion were found to be obligatory activators of rat heart LPL (Korn and Quigley, 1955). Posner and Morales (1972) found that LPL activity could be suppressed by EDTA and restored by Ca^{2+}, which they believed to be bound to the enzyme–substrate complex. There is little evidence, however, that would suggest that LPL is a genuine metalloenzyme; it is more likely that Ca^{2+} and other cations affect in some way the enzyme–supersubstrate association.

LPL of serum is inhibited by the serine reagent DFP (Shore, 1955), DNP inhibits LPL from rat heart (Korn and Quigley, 1957) and serum (Katz, 1957; Fielding, 1972). The cysteine reagent PCMB inhibits the rat heart enzyme (Korn and Quigley, 1955). It can be tentatively concluded that LPLs, like other lipases, are serine enzymes that may also contain sulfhydryl groups whose modification can interfere with enzyme–substrate or enzyme–supersubstrate binding.

F. Substrate Specificity

The substrate requirements of LPL have so far not been rigorously examined becaue a pure enzyme has not been available. Post-heparin plasma or extract of tissues all contain hydrolytic activity against mono- and diglycerides as well as against triglycerides, and it has not yet been possible to separate these activities clearly. LPL appears not to make much distinction between fatty acids of different chain length and saturation (Korn, 1961). Laurell (1968) found that phytanic acid esters resisted hydrolysis by post-heparin human plasma. An LPL preparation from milk which was believed to be over 80% pure hydrolyzed long chain and short chain triglycerides, monoglyceride in bile salt solutions, and even *p*-nitro-phenyl acetate (Egelrud and Olivecrona, 1973). The substrate specificity of the enzyme appears, then, to be very broad, wider than that of pancreatic lipase. Jansen and Hülsmann (1973) suggested that LPL of post-heparin rat plasma can also hydrolyze CoA esters.

The question of the positional specificity of LPL has repeatedly found attention. Korn (1961) found that chylomicron triglycerides were hydro-lyzed to glycerol with little accumulation of diglycerides and monoglycer-ides, and he concluded that the enzyme lacked the positional specificity shown by pancreatic lipase. Korn's findings left the possibility that the fatty acid esters in position 2 of the glycerol were hydrolyzed only after migra-tion to positions 1 and 3. This conclusion, was, in fact, reached be Nils-son-Ehle *et al.* (1971), who found that 2-monoglyceride and 1,2(2,3)-diglyceride accumulated during the reaction. Data obtained by Greten *et al.* (1970), which suggested a preferential hydrolysis of the secondary esters, may have arisen from suboptimal assay conditions (Krauss *et al.,* 1973). Morley and Kuksis (1972) and Assmann *et al.* (1973b) have con-firmed the specificity of LPL from milk, post-heparin plasma, and rat liver for the primary positions of triglycerides. The hydrolysis of secondary esters is not excluded, though it may be slower. Assmann *et al.* (1973b) found that LPL of both hepatic and extrahepatic origin hydrolyzed a 2-acyl diether glycerol at one-tenth of the rate of a 1-acyl analog. If LPL can indeed hydrolyze *p*-nitrophenyl acetate (Egelrud and Olivecrona, 1973), activity against secondary esters must be expected.

The lack of stereospecificity of rat plasma LPL was established by Karnvovsky and Wolff (1960) by a method identical with that employed for pancreatic lipase (Fig. IV-15). Morley and Kuksis (1972), on the other hand, reported a partly stereospecific hydrolysis of triglycerides by bovine milk and rat postheparin plasma LPL. The composition of diglycer-ides recovered after lipolysis of labeled trioleins was found to be: 1,3-isomer, 9–30%; 1,2(2,3) isomer, 70–91%, with the 2,3-isomer making up 73–96% of the mixture. Thus, both enzyme preparations appeared to hydrolyze preferentially the *sn*-1 ester of the triolein. It may be relevant for the explanation of these results that the stereospecific diglycerides iso-lated by Morley and Kuksis (1972) never exceeded a concentration of 2% of the total glycerides; the nonracemic nature of this small, steady-state pool may have been a result of nonrandom degradation of diglycerides rather than triglycerides.

Assmann *et al.* (1973b) have reconfirmed the nonstereospecificity of LPL of both hepatic and extrahepatic origin. They compared the rates of hydrolysis of *sn*-1-acyl and (racemic) *sn*-1(3)-acyl diether glycerols and found them equal.

G. Conclusion

Lipoprotein lipases have acquired their name because they require the presence of a serum protein, later identified as one of the lipoprotein apoproteins, for the hydrolysis of their substrate, triglyceride. Since this requirement is not absolute for all enzyme preparations, it might be advisable to name "lipoprotein lipase," if the name is to be retained, all those lipases that act on the triglycerides in the blood. These enzymes have the twofold task of preparing the dietary triglycerides of the chylomicrons, by hydrolysis, for incorporation into the body cells of animals, and of hydrolyzing the triglycerides of blood lipoproteins in preparation for the oxidation of the fatty acids.

Lipoprotein lipases are released from many tissues by heparin. The mechanism by which the enzyme is released and activated is not understood. Milk LPL, which is believed to be identical with the corresponding serum LPL, is also activated by heparin. Salt and protamine are inhibitors of the LPLs of extrahepatic origin. It is probable that activators and inhibitors, including the apolipoprotein, act by changing the supersubstrate properties of triglyceride emulsions, thereby influencing the attachment and orientation of the enzyme toward the aggregated substrate. However, the heparin-induced release of the enzyme cannot be explained on this basis; heparin seems to bind to the enzyme itself. Salts can also affect the enzyme directly by denaturing it.

Lipoprotein lipases are possibly serine enzymes like other lipases. They are specific for the primary ester bonds of triglycerides, but probably not as exclusively so as pancreatic lipase, since all preparations so far obtained can hydrolyze triglycerides completely. There is evidence that LPL can hydrolyze esterase substrates such as p-nitrophenyl acetate.

Lipoprotein lipases have been highly, but not completely, purified from liver and from milk. The enzyme isolated from liver has a particularly low specific activity (around 1), 100 or more times lower than that of LPL from extrahepatic sources, e.g., milk, but is nevertheless believed to be responsible for most of the post-heparin lipolytic activity in the blood.

VI. TISSUE LIPASES

A. Hormone-Sensitive Lipase

Hollenberg *et al.* (1961) were probably the first to observe the stimulation of lipolytic activity in homogenates of rat epididymal adipose tissue by epinephrine or ACTH. The enzyme, designated as hormone sensitive lipase, is activated by a variety of similar compounds (Vaughan and Steinberg, 1965). Cyclic AMP apparently mediates the activation. Epinephrine, ACTH, glucagon, TSH and LH, all lipolytic hormones, cause an increase in intracellular levels of cyclic AMP in isolated fat cells before any stimulation of lipolytic activity is observed (Galton, 1971). Compounds affecting phosphodiesterase exert an indirect influence upon lipolytic activity. Caffeine, for example, inhibits the phosphodiesterase, thereby raising the level of cyclic AMP and increasing the activity of the lipase. Since the activation of lipolysis by epinephrine is not inhibited by puromycin (Fain, 1964), enzyme activation rather than enzyme induction is believed to occur. The effects of pharmacological agents upon the levels of plasma FFA are discussed in detail by Kupiecki (1971).

Assay of the enzyme is fraught with two major problems: the activity is low, and reesterification of the partial glycerides occurs. Thus the method must be very sensitive, and measurement of the FFA alone may not suffice. Since adipose tissue contains little or no glycerol kinase activity, glycerol will not be phosphorylated (Steinberg and Vaughan, 1965) or subsequently reesterified, and the measurement of glycerol has been utilized as an index of lipolysis. In addition to the determination of glycerol, Vaughan *et al.* (1964) also assayed FFA. The assay media contained albumin, which acted as an FFA acceptor, and phosphate buffer, pH 7.0.

There are several other lipases in adipose tissue (Jensen, 1971; Steinberg, 1972): lipoprotein lipase, other triglyceride lipases, a diglyceridase, a distinct monoglyceride lipase and various ill defined Tween hydrolases and short-chain acid "esterases."

1. Purification

Rizack (1961) homogenized rat epididymal fat pads in 0.25 *M* sucrose, centrifuged the homogenate, discarded the fat cake, and used the supernate for further study. Centrifugation at 105,000 *g* for 12 hours sedimented the lipolytic activity in a small pellet. The extent of purification was not reported.

Vaughan *et al.* (1964) prepared acetone powders of rat epdidymal fat pads and rabbit fat for their investigation of the hormone-sensitive lipase. The powders maintained activity for several months in a desiccator at room temperature. Homogenates of adipose tissue with 0.154 *M* KCl were also employed; partitioning of activity was noted, with a considerable portion associated with the fat layer after centrifugation.

Strand *et al.* (1964) confirmed that hormone-sensitive lipase of rat epididymal fat pads was tighly bound to the fat of centrifuged homogenates. Most of the fat could be removed by extraction with ether, and about 25% of the original activity was recovered in sediments obtained by centrifuga-

TABLE IV-21

**Purification of Hormone-Sensitive Lipase of Rat
Adipose Tissue**[a]

Fraction	Yield (%)	Purification
78,000 g supernate[b]	100	1.0
pH 5.2 precipitate	94	4.5
$d < 1.12$	42	11.5
$d = 1.06-1.12$	23	12.1
Agarose column effluent	19	27.6
87,000 g supernate[c]	15	106.0

[a] Huttunen *et al.* (1970d).

[b] Total activity, 42 units; total protein, 435 mg.

[c] Total activity, 0.66 units; total protein, 0.6 mg; specific activity: 1.1.

tion. Tsai *et al.* (1970) substituted benzene for ether in the fat extraction, then precipitated most of the activity with ammonium sulfate. The specific activity of the precipitate was about three times higher than that in the benzene-extracted supernatant fluid.

Huttunen *et al.* (1970a,b,c) obtained a soluble lipase preparation when rat adipose tissue was homogenized in 0.25 M sucrose containing 1×10^{-3} M EDTA. Preparative ultracentrifugation of this soluble fraction at $d = 1.12$ resulted in further separation. An enrichment of activity occurred in the floating fraction, $d < 1.12$, when the fat pads were exposed to epinephrine prior to homogenization. The enzyme was not lipoprotein lipase because the activity was not inhibited by 1 M NaCl or protamine sulfate and had an optimum pH of 6.8. The lipoprotein lipase activity, present in the infranate, was active at pH 8.4 and was stimulated by heparin and inhibited by 1 M NaCl or fasting; incubation of the rat pads in a medium not containing glucose lowered its activity to nearly zero. Huttunen *et al.* (1970d) achieved further purification of the hormone-sensitive lipase by the steps listed in Table IV-21. The molecular weight of the lipase was estimated as 7.2×10^{6}; it contained about 50% phospholipid. The purified fraction could be activated and phosphorylated by a cyclic AMP-dependent protein kinase (Huttunen *et al.,* 1970e; Huttunen and Steinberg, 1971).

Schwartz and Jungas (1971) studied hormone-sensitive lipase with the help of sucrose gradient centrifugation. Two types of activity were found in the infranate obtained by centrifugation of rat fat pad tissue homogenates at 40,000 *g*. One of the activities was in the 6 S and the other in the 15 S region. Further purification was not attempted, but partial separation of hormone-sensitive lipase from several other lipolytic activities, i.e., lipoprotein and short-chain lipases, was achieved.

Pittman *et al.* (1972) fractionated the supernates obtained by centrifugation (180,000 *g*) of homogenized rat fat pads and found that two peaks of lipase activity emerged from a column of 4% agarose gel, one in the void volume and the other at about twice this volume. The first form of hormone-sensitive lipase is phospholipid-rich and of high molecular weight and apparently identical with the lipase purified earlier by Huttunen *et al.* (1970a,b,c).

2. Characteristics

Although it is well known that many compounds (e.g., catecholamines) stimulate and others (e.g., caffeine) inhibit hormone-sensitive lipase, these effects are indirect and were detected with crude preparations (Vaughan and Steinberg, 1965; Löffler and Weiss, 1970). It seems best, therefore, to limit this discussion to the observations of Huttunen *et al.* and Schwartz and Jungas on their purified lipases.

The Fraction S_{105} lipase and the second enzyme with the lower molecular weight were both stimulated by epinephrine (Huttunen *et al.,* 1970a; Pittman *et al.,* 1972). If the early purification steps with Fraction S_{105} were done at room temperature, there was a loss in activity that could be prevented by the addition of EDTA 10^{-3} *M* to the homogenizing medium. After further purification, the enzyme was stable in the absence of EDTA. The assays were carried out at pH 6.8 in sodium phosphate buffer in the presence of serum albumin and sodium chloride. The S_{105} activity was not affected by protamine sulfate, heparin, or sodium chloride; sodium fluoride caused about 85% inhibition. Of the adenine nucleotides tested, ADP, AMP, cyclic 3',5'-AMP, and ATP, only the latter at 1.5×10^{-2} *M* decreased activity. Both the S_{105} lipase and the second lipase were activated by a cyclic AMP-dependent protein kinase (Pittman *et al.,* 1972), and the second enzyme was not inhibited by 1 *M* NaCl or affected by addition of 10% serum.

The purified lipase of high molecular weight was activated in the presence of purified rabbit muscle protein kinase, cyclic AMP, ATP, and Mg^{2+} (Huttunen *et al.,* 1970e). The γ-phosphate from $AT^{32}P$ was transferred to the lipase in quantities of 2 to 4 moles of phosphorus per 10^6 gm protein. Both the kinase and the cofactors were required. In order to relate activation to hormone stimulation in intact cells, fat pads incubated with and without epinephrine were purified by the procedure in Table IV-21 (Steinberg, 1972). Activation was observed after each step starting from the control tissues, but the epinephrine-treated tissue showed little or no activation. Apparently the lipase of the tissue incubated with epinephrine was

TABLE IV-22

**Proposed Lipolytic Cascade for Activation of Hormone-Sensitive Lipase in
Adipose Tissue[a]**

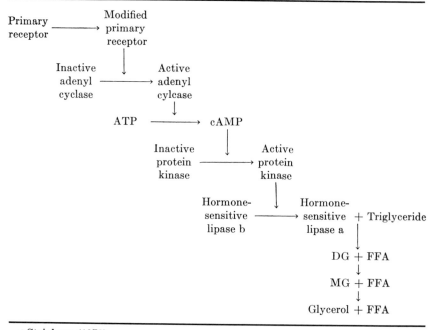

[a] Steinberg (1972).

completely activated prior to homogenization. From these results, Steinberg (1972) proposed the lipolytic cascade depicted in Table IV-22. In this series, epinephrine reacts with an unidentified receptor thought not to be adenyl cyclase. The message may pass through a hypothetical transducer (Braun and Hechter, 1970; Rodbell, 1970) and then on to the adenyl cyclase receptor site. Fat feeding reduces the response of adenyl cyclase to epinephrine, presumably at the transducer (Gorman *et al.*, 1973).

Adenyl cyclase is activated; this is followed by increased synthesis of cyclic AMP, which in turn stimulates the protein kinase. Corbin *et al.* (1973) and Soderling *et al.* (1973) separated inactive protein kinase from the active form on Sephadex G-100 and measured the effect of cyclic AMP. The concentration of cyclic AMP regulates the conversion; an increase in cyclic AMP produces more active protein kinase. The active kinase catalyzes the phosphorylation of the prolipase in an ATP-dependent reaction. The cascade should also have a mechanism for deactivation of the lipase. Crude homogenates have been shown to contain an inhibitory system (Vaughan *et al.,* 1965), and lipase phosphatase activity has been detected in crude adipose tissue fractions (Steinberg, 1972).

Tsai and Vaughan (1972) noted that when supernatant fluid fractions of rat adipose tissue were incubated with ATP and Mg^{2+}, activity decreased. When the fluid fraction was chromatographed on Sephadex G-25, all of the lipase activity emerged in the void volume, but this was not inactivated by ATP and Mg^{2+} unless material from later column fractions was added. Tsai and Vaughan (1972) offered these possible mechanisms: (1) an inactivation of the phosphorylated lipase requires dephosphorylation, the phosphatase being affected by ATP and Mg^{2+} directly or by conversion of the inactive phosphatase to a phosphorylated form, and (2) the lipase is inactive when phosphorylated or adenylated in the presence of ATP and Mg^{2+} and a cyclic AMP-dependent protein kinase activates an enzyme that removes the phosphate or adenylate, thus activating the lipase.

Schwartz and Jungas (1971) identified their "15 S lipase" as the hormone-sensitive lipase by these observations: (a) exposure of the tissue to epinephrine caused the activity in the extracts to increase while exposure to insulin reduced activity; (b) the increased lipolytic activity of fat pads from fasted–refed rats compared to fasted rats was localized in the "15 S" area; and (c) protein moving at 15 S, when incubated with [^{32}P]orthophosphate, picked up radioactivity when the enzyme was incubated with ATP, cyclic AMP, Mg^{2+}, and a protein kinase from rat liver. The 6 S lipase did not exhibit these characteristics, but did coincide with di- and monoglyceride lipase activities. Both of the enzymes were found in washed adipocytes prepared by collagenase digestion; they were probably bound to membranous material. In most respects, the 15 S enzyme is similar to the high molecular weight lipase studied by Huttunen and colleagues; the two activities are probably the same.

Khoo *et al.* (1972a) have demonstrated lipase activation by the protein kinase, ATP, cyclic 3′,5′-AMP system in crude fractions of human adipose tissue. Activation was stopped by addition of a specific protein kinase inhibitor. However, this system gave equivocal results when fat samples from rabbit, dog, sheep, or pig were tested.

In contradiction to the report that hormone-sensitive lipase was associated mostly with the inner surface of the plasma membrane of adipocyte ghosts (Crum and Calvert, 1971), Khoo *et al.* (1972b) found most of the activity in the cytosol.

Practically nothing is known about the chemistry of the hormone-sensitive triglyceride lipase, mainly because purified enzyme preparations have, until recently, been unavailable. Marsh and George (1969), using isolated fat cells, observed that a 10 second exposure to *N*-ethylmaleimide inhibited the lipolytic effects of TSH, theophylline, and dibutyryl cyclic 3′,5′-AMP.

The data were interpreted as indicating that sulfhydryl groups were essential for the activation of the lipase.

3. Conclusion

The preparations of hormone-sensitive lipase so far obtained have very high apparent molecular weights. The S_{105} lipase of Huttunen *et al.* (1970a,b,c) of molecular weight 7.2×10^6 is reminiscent of the fast pancreatic lipase; like this enzyme, it seems to be a phospholipid–protein aggregate. The second lipase fraction described by Pittman *et al.* (1972) appears to have a molecular weight of several hundred thousands, still an improbably large size for a hydrolytic enzyme. Further fractionation and purification of the lipase must be awaited.

The function of hormone-sensitive lipase is the mobilization of fatty acids from the depot fat, and since all three fatty acids esters of the triglyceride molecule must be hydrolyzed, it would seem plausible, from a standpoint of evolutionary purpose and practicality, that the same enzyme can hydrolyze tri-, di-, and monoglyceride with equal facility, the partial glycerides perhaps faster in order to prevent their accumulation at the oil–cytosol interface. The enzyme would then differ from pancreatic lipase in preferring a more open lipid surface and in being able to hydrolyze hydrated fatty acid esters. Such a somewhat less hydrophobic affinity is also likely because the lipase acts on an oil droplet which probably has a surface containing phospholipids.

The final step in the hormonal activation of the lipase seems to be a phosphorylation. The obvious explanation for this mechanism is a conformational change of the proenzyme, which establishes the architecture of the reactive site. However, an alternative explanation in terms of enzyme orientation (Brockerhoff, 1974) might be considered. If the oil droplets are enveloped by phospholipid, they almost certainly have a negative surface charge. The active enzyme should then be facing the surface with a positive, or at least neutral, head. Introduction of the strongly negative and hydrophilic phosphate group, perhaps as a tail, might trigger a necessary reorientation of the lipase. Such a mechanism could, of course, also work in conjunction with conformational activation.

B. Other Lipases

1. Adipose Tissue

Several other rather ill defined lipolytic or esterolytic activities have been detected in adipose tissue; their relationship to the hormone-sensitive lipase is on occasion obscure. We will discuss these enzymes briefly.

Schnatz (1966) found lipolytic activity in human adipose tissue homogenates at pH 7.0 and 37°C in the presence of albumin. Schnatz and Cortner (1967) were able to separate a neutral from an alkaline lipase activity by ammonium sulfate precipitation and gel filtration on Sephadex G-200. The neutral lipase was associated with a large molecular weight, similar to the hormone-sensitive lipase. Schnatz and Cummiskey (1969) determined neutral lipase activity in biopsies of human adipose tissue. Epinephrine infused into three normal individuals at the rate of 6μg/minute did not effect the activity, although whether or not the drug had an opportunity to reach the tissue selected for the test (buttock) may be questioned. The neutral lipase activity could well be identical with the hormone-sensitive lipase. Schnatz and colleagues detected an accompanying activity at pH 8.0 and 47°C, the alkaline lipase, with tributyrin as a substrate. The characteristics of this enzyme strongly suggest that it was an esterase and not a lipase. Boyer (1967) and Boyer *et al.* (1970) assayed neutral lipase activity both in rat and human adipose tissue with a potentiometric and a radioactive method.

Mann and Tove (1966) purified a lipase from a rat adipose tissue particulate fraction approximately 140-fold. The enzyme was quite specific for primary esters in tri- and diglycerides, but did not hydrolyze a series of monoglycerides. The enzyme is one of the few that have been tested for stereospecificity with synthetic enantiomeric triglycerides; no evidence for stereospecific hydrolysis was obtained (Wright and Tove, 1967). The place of this enzyme in the hierarchy of adipose tissue lipases is difficult to assess.

Hydrolytic activities toward substrates with short-chain acids have been studied by several groups. Lynn and Perryman (1960) observed lipolysis of butyryl glycerides and naphthyl esters by a porcine adipose tissue preparation. Similar activities toward nonphysiological substrates were demonstrated by many other workers (Wallach *et al.*, 1962; Schnatz, 1964, 1966; Schnatz and Cortner, 1967; Lech and Calvert, 1968; Crum and Lech, 1969; Schnatz and Cummiskey, 1969; Crum *et al.,* 1970). The pertinence of these activities to the long-chain fatty acid triglyceride lipases in adipose tissue remains obscure.

2. Insects

The fat bodies, eggs, and flight muscle of various insects contain lipases. The fat body of the southern armyworm moth has enzymes capable of hydrolyzing tri-, di-, and monoglycerides (Stevenson, 1972); the resulting FFA is transported in the hemolymph to the flight muscle, where the acids are oxidized for energy. The flight muscle of the southern armyworm moth also contains a monoglyceride lipase that digested monoolein,

monopalmitin, and monolaurin about sixty times more rapidly than di-
or triolein (Stevenson, 1969). All of these lipases were most active around
pH 8.0. The lipase in the eggs from the southern corn rootworm were
inhibited by DNP (10^{-5} M), but only in the presence of the substrate
(Krysan and Guss, 1971). The flight muscles of many insects seem to
contain lipases especially active against diglycerides (Crabtree and Newsholme, 1972).

3. Fibroblasts

The lipases in Littlefield strain L fibroblasts (connective tissue cells)
were investigated by Lengle and Geyer (1973). These authors observed
that when the supply of exogenous fatty acids in the culture medium exceeded the demand for energy, the acids were esterified and taken into
lipid-rich (90%) cytoplasmic droplets. The droplets were apparently contained by the usual membrane. When exogenous FFA were removed, the
triglyceride droplets disappeared at a rate suggesting hydrolysis by an intracellular lipase.

The lipase was partially purified by sonication of the fibroblasts, centrifugation of the sonicate and chromatography on DEAE-cellulose, with 46-fold
purification and 21% recovery with a specific activity of 0.007 in the crude
sonicate and 0.3 in the DEAE-cellulose fraction. This fraction was impure
by disk gel electrophoresis; the estimated molecular weight was 800,000.

Three optimal pHs were observed, 4.5, 6.5, and 8.5, with about 80%
of the activity associated with pH 6.5 lipase. Prior extraction of the crude
cell sonicate with diethyl ester or heptane increased the apparent activity
two to three times. These solvents removed primarily triglycerides. When
more polar solvents were used, those which extract phospholipids reduced
the activity markedly. The pH 6.5 lipase was inhibted by PO_4^{3-}, sodium
taurocholate, and DNP (1 mM), but was not affected by sodium chloride,
heparin, eserine, or cyclic AMP.

4. Liver

The initial detection of lipolytic activity in rat liver homogenate by
Vavrinkova and Mosinger (1965) was soon followed by many corrobora-
tive reports (see Guder *et al.,* 1969; Claycomb and Kilsheimer, 1971; Assmann *et al.,* 1973a, for comprehensive lists of references). There have
been numerous reports on the location of lipolytic activity within the subcellular particulates, with localizations reported as microsomal (Carter,
1967; Guder *et al.,* 1969; Hayase and Tappel, 1970; Muller and Alaupovic, 1970), cytoplasmic (Olson and Alaupovic, 1966; Guder *et al.,* 1969),
lysosomal with alkaline pH optimum (Stoffel and Greten, 1967), lysosomal
with acid pH optimum (Mahadevan and Tappel, 1968; Fowler and de

Duve, 1969; Guder *et al.*, 1969; Hayase and Tappel, 1970; Muller and Alaupovic, 1970), and mitochondrial (Waite and van Deenen, 1967; Claycomb and Kilsheimer, 1971). It is possible that some of the localizations may have been artifactual, i.e., redistributions of the enzymes released from their original site by the fractionation techniques. Heparin-sensitive lipolytic activity (lipoprotein lipase) has been reported in liver (Guder *et al.*, 1969; Assmann *et al.*, 1973a). Liver lipases have been reviewed by Claycomb (1972).

The centrifugation procedures of de Duve *et al.* (1955) and Ragab *et al.* (1967) have been used by most investigators to obtain subcellular particulate fractions with lipolytic activity. Lysosomes and Triton-loaded lysosomes (tritosomes) were isolated by Mahadevan and Tappel (1968) and Guder *et al.* (1969). A 300-fold purification was achieved by the latter group by discontinuous gradient ultracentrifugation. An alkaline lipase (pH 8.5) was associated with the microsomal fraction. Muller and Alaupovic (1970) obtained similar preparations, a lysosomal acid lipase and an alkaline microsomal lipase.

The acid lipase is optimally active in the pH range of 4 to 5 (Mahadevan and Tappel, 1968; Guder *et al.*, 1969). The enzyme was inhibited by PCMB but not by NEM, DFP, or protamine sulfate with both glycerol tridecanoate and *p*-nitrophenyl acetate as substrates (Hayase and Tappel, 1970). The decanoate esters of glycerol were hydrolyzed at the following rates: tri, 100; 1,2-di, 87; and 1-mono, 10. When labeled glycerol trioleate was digested, the results indicated specificity for the primary esters; there was about seven times more 1,2- than 1,3- dioleate formed; monoolein was not found. Hayase and Tappel (1970) suggested that the acid liposomal lipase might be identical with a liposomal esterase that hydrolyzed *p*-nitrophenyl esters.

Kariya and Kaplan (1973) have reported that rat liver lysosomal lipase is inhibited by nucleotides, pyrophosphate, and heparin, and stimulated by acidic phospholipids. Phosphatidylserine could accelerate the reaction rates 35-fold.

Wolman's disease, which is characterized by an accumulation of triglyceride and cholesterol ester in the liver and other organs of the afflicted patients, is associated with a deficiency of liver acid lipase (Patrick and Lake, 1969). The enzyme is lysosomal (Lake and Patrick, 1970) and its deficiency is also observed in cholesterol ester storage disease (Burke and Schubert, 1972). Strangely, the hepatic triglyceride content in the cholesterol ester storage disease is normal. This and other observations led Burke and Schubert to conclude that the two diseases are not completely explained by a deficiency of the acid lipase. However, both lipid accumulation and acid lipase deficiency were detected in fibroblasts from a family with

Wolman's disease (Kyriakides *et al.,* 1972). These diseases have been discussed by Sloan and Frederickson (1972a,b).

Claycomb and Kilsheimer (1971) purified a mitochondrial lipase from rat liver. The mitochondria were solubilized by treatment with Triton X-100 and sonication, and the lipolytic activity was then concentrated 77-fold by the use of Sephadex.

Waite and van Deenen (1967) observed partial inhibition of the liver mitochondrial enzyme with PCMB. The substrate specificity of the enzyme revealed more rapid hydrolysis of the partial glycerides as compared to triglycerides (Claycomb and Kilsheimer, 1971). The positional specificity of the lipase is unknown.

5. Serum

The small lipolytic activity that is found in the serum of normal individuals is much increased in serum from patients with pancreatic disorders, mumps, and nephropathies. The identity of this serum lipase is not known, but in cases of pancreatitis it must be assumed that the activity represents pancreatic lipase that has leaked into the blood stream. The determination of this activity is of obvious value as a diagnostic tool, and there exists a voluminous literature on this subject; relevant references are given by Tietz and Fiereck (1972). The most common assay methods are based on the titration of the fatty acids liberated from emulsified olive oil (Cherry and Crandall, 1932).

VII. MILK LIPASES

Although it has been known for many years that the milk from cows and other mammals contains lipolytic enzymes, intensive study of the enzymes themselves did not commence until about 1950. Large-scale use of pipeline milkers, an improved method for obtaining raw milk, was the technological event which initiated this activity. It was quickly noted that excessive foaming of the warm raw milk in the pipelines, generally caused by air leaks, activated a lipase which upon subsequent storage of the milk at low temperatures partially hydrolyzed the milk fat. Relatively large quantities of butyric and caproic acids are released as both these acids are located in the *sn*-3 position of milk triacylglycerols (Pitas *et al.,* 1967; Breckenridge and Kuksis, 1968) and milk lipase is specific for the 1 and 3 positions (Jensen, 1964). The resulting flavor and odor defect is called rancid in the dairy industry, an unfortunate choice, since with all other food fats rancid means oxidized; a more descriptive term would be "lipolyzed." Milk lipases and lipolysis of milk fat have been reviewed by Jensen (1964, 1971), Chandan and Shahani (1964), Shahani (1966), Groves (1971a,b) and Shahani *et al.* (1973). Milk contains also a lipoprotein lipase (Korn, 1962); this enzyme is discussed in the chapter on lipoprotein lipases. Milk esterases have been reviewed by Shahani *et al.* (1973).

Lipases are presumably present in the milk of all mammals, but no other species has received such intensive study as the enzyme in cow's milk. Activity has also been detected in human milk (Tarassuk *et al.*, 1964; Jubelin and Boyer, 1972). Chandan *et al.* (1968) found the following activities, expressed as micromoles of fatty acid released per minute per 100 ml milk: human, 13; cow, 132; goat, 39; sheep, 9; and sow, 141.

The milk protein of interest is casein. This complex protein is synthesized in the mammary gland, but it is not known if lipase is synthesized in the same organ or transferred to the casein from the blood. There are many types of casein, but those that concern us are the α_s, β and κ fractions. The designation α and β refer to the locations of the caseins in electrophoretic separations; α_s is a calcium-sensitive fraction and κ identifies the micelle stabilizing protein associated with the α_s fraction. These proteins are present in milk as large spherical particles of broad size range which are termed micelles.

The supernate of a solution of whole casein which has been treated with 0.25 M CaCl$_2$ and centrifuged contains κ casein (Machinlay and Wake, 1971). This protein is heterogeneous (genetic variants), gel electrophoresis showing 7 or 8 bands; it contains sialic acid and it is the substrate for rennin, the milk-coagulating enzyme. Rennin converts κ casein to an insoluble protein, para-κ-casein, and a soluble macropeptide.

At least two blood serum proteins, transferrin and albumin, are found in milk, indicating that large protein molecules can be transferred from blood to milk (Groves, 1971a,b).

About 50% of the fatty acids in milk triglycerides are derived from the very low density lipoproteins in the blood via the action of lipoprotein lipase (Patton, 1973a,b). The remainder of the acids are synthesized in the mammary gland. Triglycerides are reassembled by both the monoglyceride and the 3-glycerophosphate pathways in the gland and form into droplets which are expelled from the secretory cell. A portion of the plasma membrane which contains many enzymes accompanies the droplet as the milk fat globule membrane, but lipolytic activity has not been detected in this fraction (Dowben *et al.*, 1967). Milk lipase apparently is not one of the enzymes associated with the milk fat globule membrane upon secretion of the globule. The source and movement of milk lipase during these transfers is not known. Milk secretion has been reviewed by Jenness (1970) and Patton (1973a,b). Observations by Tarassuk and Frankel

(1957) have indicated the presence of two lipases in milk. One is believed to occur in the normal milk from all cows. Activity is associated with the casein fraction, and activation occurs as a result of homogenization and foaming of warm, raw milk. The milk from some cows contains another lipase that is adsorbed to the fat globule as a result of cooling, a curious phenomenon. The relationship between these enzymes and the active fractions to be discussed later has not been elucidated.

A. Assay

Many methods have been applied to the detection of milk lipase activity; at present, the following are in vogue. For rapid screening on a commercial basis, the procedure of Thomas *et al.* (1955) is used, in which milk fat is obtained by detergent separation in a Babcock milk test bottle, then titrated with alcoholic potassium hydroxide to the phenolphthalein end point. The short-chain acids that are completely or partially water soluble are probably not recovered. The silica gel procedure of Harper *et al.* (1956), which is the basis for the extraction procedure described in Chapter IV, Section I, B, recovers all of the free acids, including butyric. The endpoint is much easier to observe if thymol blue is the indicator rather than phenol red. The silica gel technique is more expensive and time consuming than the Thomas method, but it is ideal for research because of the quantitative recovery of all free fatty acids. The results from both procedures are usually reported as acid degree values, i.e., moles of free fatty acid per 100 gm of fat.

Parry *et al.* (1966) adapted the pH-stat method to the determination of free fatty acids resulting from the lipolysis of milk fat, as did Luhtala and Antila (1968). Downey and Andrews (1965a) similarly used the procedure to evaluate their casein fractions for lipase activity. A screening method for lipase activity in milk involving Rhodamine B has been developed by Nakai *et al.* (1970) and improved by Kason *et al.* (1972). Milk fat is extracted and the color resulting from the addition of Rhodamine B is compared to a standard.

B. Purification

It is now firmly established that milk glycerol ester hydrolase is associated with casein, the exceedingly complex major protein mixture of milk (Shahani *et al.*, 1973). Skean and Overcast (1961) were the first to locate lipolytic activity in a specific casein fraction. They applied continuous paper electrophoresis to caseins obtained by both centrifugation and acidification at room temperature and in the absence of added calcium. The casein complexes separated into three fractions which were designated by the authors as α, β, and γ. Lipolytic activity was associated with the α component; the casein obtained by centrifugation yielded twice as much activity as the casein precipitated by acid. Saito (1963) confirmed the association of lipase with casein micelles. Yaguchi *et al.* (1964), using dialyzed skim milk as the starting material, noted that lipolytic activity accompanied several fractions obtained by DEAE-cellulose chromatography and containing κ casein. All fractions also contained relatively large amounts of sialic acid. Negligible activity was detected in the β casein fractions. Fox *et al.* (1967) found that lipase could be removed from κ casein by treatment with dimethylformamide. The enzyme was therefore considered to be a minor component of the casein system and not κ casein itself.

Downey and Andrews (1965a,b) found that enzymatic activity was sedimented with the casein micelles by centrifugation; a further purification with Sephadex revealed that four lipases were present, three with molecular weights ranging from 62,000 to 112,000 and another of smaller molecular weight. In a later study Downey and Andrews (1969) found at least five overlapping active fractions in skim milk. Each fraction was tested for activity with triacetin, tributyrin, triolein, and milk fat emulsions as substrates. The greatest activity was observed with emulsified substrates in all fractions. Most of the fractions were labile and some contained sialic acid.

Downey and Murphy (1970) studied the association of pancreatic lipase with the casein micelles and soluble casein complexes of milk. Gel filtration on Sephadex G-200 and Sepharose 2 B columns equilibrated with synthetic milk serum indicated a molecular weight of $>10^8$ for the micelles, and 2×10^6 for the casein present in a soluble complex which contained α_s, β and κ caseins. When unpurified porcine pancreatic lipase was added to milk, its activity was reduced due to binding of the enzyme to both micellar and colloidal phosphate-free soluble caseins. Pancreatic lipase dissociated from the soluble casein during gel filtration on Sepharose 2 B columns. In contrast to the earlier hypothesis of the authors (Downey and Andrews, 1965a), the binding of the lipase to casein was not dependent on the presence of colloidal phosphate; a complete micellar structure was therefore not required for the binding of pancreatic lipase to casein. The authors suggested that milk lipases are similarly distributed in an equilibrium between micellar and soluble casein. The divergent results as to the type of casein associated with milk lipase activity might be explained by the tendency of the lipase to absorb to and dissociate from the protein during fractionation.

There are indications that milk lipase might associate with milk proteins other than κ casein. Gaffney *et al.* (1966) observed activity in all of eight fractions obtained by DEAE-cellulose chromatography of a water extraction of rennin casein from skim milk. It appears that the milk lipase activity was redistributed among the milk proteins as suggested by Downey and Murphy (1970). Later, Gaffney *et al.* (1968) reported on lipase-rich fractions of milk protein fractionated from skim milk, freeze–thawed skim milk, and a water extract of rennin casein by Sephadex G-25 gel filtration. The first fraction emerging from all sources contained the bulk of the protein with some lipase activity (yields were not reported), while two slower fractions contained considerably more activity, 10- and 250-fold enriched relative to skim milk. Both these fractions had molecular weights of less than 10,000 and contained carbohydrate. Activity was lost when the preparations were lyophilized and retained when concentration was achieved by ultrafiltration. The proteins were heterogeneous when checked by polyacrylamide gel electrophoresis.

Fox and Tarassuk (1968) coagulated fresh skim milk with rennin and separated the curd from the whey by centrifugation. The curd, which contained most of the lipase, was solubilized with 1 M NaCl and centrifugation, and the lipase was concentrated by precipitation with ammonium sulfate. Further purification was achieved by DEAE chromatography, solubilization with dimethyl formamide, precipitation with ammonium sulfate and filtration through Sephadex G-200. The extent of purification and the yields are presented in Table IV-23. The final fraction was homogeneous when tested by starch gel electrophoresis and had an estimated (gel filtration) molecular weight of 210,000. Patel *et al.* (1968) further characterized this preparation, finding that 14.8% nitrogen, 0.16% phosphorus, and 0.6% sialic acid were present. Activity was inhibited by both sulfhydryl- and hydroxyl-binding agents.

Saito and Igarashi (1971) obtained fractions with enhanced lipase activity by Sephadex filtration of skim milk, but the enrichment was not suffi-

TABLE IV-23

Purification, Specific Activity, and Recovery of Milk Lipase[a]

Fraction	Specific activity	Purification	Yield (%)
Skim milk	0.03	1	100
Rennin and sodium chloride extraction	0.15	5	60
First ammonium sulfate	0.18	6	60
DEAE-cellulose	0.9	30	40
Second ammonium sulfate	2.0	100	24
Sephadex	15	500	10–15

[a] Fox and Tarassuk (1968).

TABLE IV-24

Purification, Activity, and Recovery of Milk Lipase Activity from Separator Slime[a]

Fraction	Activity	Purification	Yield (%)
Clarifier slime	0.44	3	100
Aqueous extraction of acetone powder of slime	0.83	6	83
Ammonium sulfate precipitation	7.4	53	28
Sephadex filtration	12.3	88	22

[a] Chandan and Shahani (1963a).

cient to be used for further purification. Chandan and Shahani (1963a,b) obtained a preparation with lipolytic activity from the sediment resulting from the passage of milk through a clarifier. (The purpose of a clarifier is to remove extraneous material by centrifugal force; the process also removes some milk proteins and cellular material stripped from the mammary gland.) The procedure and extent of purification are presented in Table IV-24. The activity was concentrated 88-fold, 250-fold if based on milk solids. The purified enzyme showed a single band by starch gel electrophoresis and had a molecular weight of about 7000; this weight was confirmed by sedimentation–diffusion and osmotic pressure methods (Chandan *et al.,* 1963). Chandan and Shahani's belief that this enzyme and the lipase associated with casein were the same was challenged by Gaffney and Harper (1965) and by Downey and Andrews (1965a). Gaffney and Harper confirmed that the somatic cells in cream separator sediment possessed lipolytic activity, but they had a gel electrophoretic mobility different from that of milk lipase. Downey and Andrews noted an active low molecular weight component in their casein preparations but were unable to obtain a similar peak of activity from clarifier sediment. However, the enzyme studied by Chandan and Shahani has too many similarities to the unpurified milk lipase to be dismissed as an enzyme not originating from milk (Shahani, 1966). Although many of the characteristics—e.g., optimum pH and temperature,—are shared by most lipases, the specificity of both lipase preparations for the primary esters of acyl glycerols is not; most tissue preparations hydrolyze all three ester positions. Although final proof awaits isolation and purification of the casein lipase followed by comparative studies, it appears that the lipase of Chandan and Shahani originates from milk. In support of this conclusion, Richter and Randolph (1971) isolated a low molecular weight (8500) lipase from clarifier sediment using a modification of the procedure of Chandan and

Shahani (1963a). A 110-fold purification was achieved with a recovery
of 2%. Carbohydrate was detected by the nonspecific anthrone reagent.

The observation of Korhonen and Antila (1972) that synthesis of glycerides in fresh milk was enhanced when larger numbers of somatic cells
were present should be considered here. Both hydrolysis and synthesis occurred simultaneously, with hydrolysis predominating.

The conclusion that at least one milk lipase is a low molecular weight
(7000–8000) glycoprotein, adsorbing to caseins, specifically κ casein, but
possibly to others, is very attractive. On the other hand it is not impossible
that some of the higher molecular weight milk lipases are pancreatic or
other lipases which have "leaked" into the mammary gland from blood.
The observations that milk lipase has no apparent function in the mammary
gland and is inactive until the milk has been released from the gland (Jensen, 1964) supports this hypothesis, although the lipase could play some
as yet unknown role in the nutrition of the young.

Another milk lipase preparation to be mentioned is the one of Forster
et al. (1959, 1961). These investigators found three esterases, A, B, and
C, in milk, by utilizing the inhibitory effects of various organophosphates
and differential hydrolysis of esters. B esterase hydrolyzed aliphatic and
aromatic esters but not choline esters, and was resistant to certain organophosphates but not eserine. The enzyme, concentrated on magnesium hydroxide (Montgomery and Forster, 1961), was very labile and possessed
the same specificity for primary esters of triacyl glycerols as unpurified
milk lipase (Jensen *et al.,* 1961a) and the lipase of Chandan and Shahani
(Jensen *et al.,* 1961a, 1964c).

C. Characteristics

The amino acid composition is available for a preparation obtained by
Patel *et al.* (1968). It differs from that of most milk proteins and also
from both porcine and rat pancreatic lipases. The enzyme contained 0.6%
sialic acid, as did several other preparations. It is possible that sialic acid
is an integral part of the milk lipase molecule, although this has been questioned by True *et al.* (1969), who were unable to relate the sialic acid
content of milk to the lipase activity. It should be mentioned, however,
that several milk proteins contain sialic acid (Graham *et al.,* 1970).

Some information is available concerning the active site. PCMB and
N-ethylmaleimide partially inhibit milk lipase (Tarassuk and Yaguchi,
1959; Chandan and Shahani, 1965; Robertson *et al.,* 1966; Patel *et al.,*
1968). This suggests that free sulfhydryl groups are located in the vicinity
of the reactive site but are not involved in the catalytic mechanism, a situation as found for pancreatic lipase (Verger *et al.,* 1971). It must be men-

tioned, however, that one milk lipase preparation (Chandan and Shahani, 1965) was completely blocked by PCMB.

DFP inhibits the enzyme (Robertson *et al.,* 1966); DFP at 1×10^{-3} *M* and DNP at 5×10^{-4} *M* destroy the activity completely (Patel *et al.,* 1968). The enzyme is also inhibited by dinitrofluorobenzene (Robertson *et al.,* 1966), and it is rapidly inactivated by photooxidation (Frankel and Tarassuk, 1959; Patel *et al.,* 1968). These observations indicate that milk lipase is an enzyme of the serine–histidine type. The active site seems to be sterically less hindered than that of pancreatic lipase, since it reacts with DFP whereas the pancreatic enzyme does not.

D. Specificity

Native milk lipase acting on homogenized raw milk yielded predominately 2-monoglycerides (Jensen *et al.,* 1960); specificity for the primary hydroxyls of milk fat was thus indicated. With lyophilized raw skim milk as a source of lipase, 90% of the FFA hydrolyzed from *rac* 18:1–16:0–18:1 and *rac* 16:0–18:1–16:0 were acids that had been esterified to the primary positions (Gander and Jensen, 1960). Later, the milk lipase preparations of Forster *et al.* (1959, 1961) and Chandan and Shahani (1963a,b) were found to be specific for primary esters (Jensen *et al.,* 1961a; 1964c). Data are presented in Tables IV-25 and IV-26.

Milk lipase, like most other lipases, will digest tributyrin more rapidly than long chain acylglycerols (Patel *et al.,* 1968). This has been accepted as evidence of specificity for short chain fatty acids. However, it is inadmissible to compare emulsions of, for example, tributyrin and triolein. Further,

TABLE IV-25

Fatty Acid Composition of the Intact Triglycerides and the Products of Milk Lipase Hydrolysis of an Equimolar Mixture of *rac*-Glycerol 1-Palmitate 2,3-Dibutyrate and Triolein[a]

Substrate and products of lipolysis[b]	Fatty acids (mole %)		
	4:0	16:0	18:1
Intact triglyceride	31.6	15.8	52.6
Residual triglyceride	25.8	12.1	62.2
Free fatty acids	37.0	33.0	30.0
Diacyl glyceride	36.2	12.8	52.0
Monoglyceride	78.4	7.0	14.6

[a] Jensen *et al.* (1964c).

[b] Samples incubated for 15 minutes at 37°C. Lipase preparation of Chandan and Shahani (1963a).

TABLE IV-26

**Fatty Acid Composition of the Free Fatty Acids, Monoglycerides,
and Diglycerides Produced by Milk Lipase Hydrolysis of
rac-18:1–6:0–6:0(A) and *rac*-16:0–4:0–4:0(B)**[a]

Substrate	Fatty acid	Free fatty acid	Diglyceride	Monoglyceride
A	6:0	52.6	78.3	97.9
	18:1	47.4	21.7	2.1
B	4:1	42.9	70.8	100
	16:0	57.1	29.2	—
—	Theoretical[b]	50.0	75.0	100
		50.0	25.0	0

[a] Jensen *et al.* (1962). Milk lipase preparation of Forster *et al.* (1959, 1961).
[b] Values calculated assuming specificity for primary esters and none for fatty acids.

Patel *et al.* and many other investigators did not use equimolar amounts of substrate in their studies. Thus, about four to five times more tributyrin than triolein molecules were available to the enzyme in the investigation by Patel *et al.* (1968). These investigators observed an approximate five-fold difference in rates of digestion between the two substrates, with the tributyrin hydrolyzed more rapidly.

In order to reduce the variables inherent in emulsions of triglycerides differing greatly in molecular weight, *rac*-glycerol 1-palmitate 2,3-dibutyrate and triolein were mixed in equimolar quantities and digested by the concentrated milk lipase of Chandan and Shahani (1963a,b) and Jensen *et al.* (1964c). The results in Table IV-25 reveal comparatively rapid hydrolysis of the butyroyl glyceride with less 4:0 and 16:0 than 18:1 in the residual triglycerides, whereas the contents should be equimolar. The other values reflect this trend; there is more 4:0 and 16:0 than 18:1 in the free fatty acids, etc. The 18:1 in the free fatty acids must be divided by two for purposes of comparison because of the specificity of the lipase for primary esters. If there had been no carbon number specificity, the 18:1 content would have been twice that of either 4:0 or 16:0.

The preferential hydrolysis of the triglycerides with lower carbon number can perhaps be explained by the physical properties of the substrates. The triglyceride species with lower molecular weights are more polar and thus more hydrophilic than the long-chain species and therefore more likely to accumulate at the oil–water interface. They may also be faster to diffuse to the place of reaction. The apparent specificity of the lipase is, then, the result of different concentrations of the substrates at the interface.

The faster digestion of tributyrin by milk and other lipases led investigators to accept the results as short-chain fatty acid specificity. From the discussion above and the research described below, it is now known that milk lipase does not differentiate between 4:0 or 6:0 and long-chain acids attached to the same triglyceride molecule. To further complicate matters with milk lipase and its action on milk fat, the primary ester specificity of the enzyme, coupled with *sn*-3 position location of 4:0 in milk fat (Pitas *et al.,* 1967; Breckenridge and Kuksis, 1968) resulted in relatively large accumulations of this acid in the lipolysate. Many researchers were convinced that here was a true specificity for butyric acid. However, this was not confirmed in experiments by Jensen *et al.* (1961a,b, 1962). The first lipase studied was the B esterase •preparation of Forster *et al.* (1959, 1961), and the substrates were glycerol 1-oleate 2,3-dihexanoate and glycerol 1-palmitate 2,3-dibutyrate. From the results in Table IV-26, it is clear that the enzyme did not differentiate between 4:0 or 6:0 as compared to 18:1 or 16:0 in the primary positions. Similar data were obtained when glycerol 1-palmitate 2,3-dioleate and glycerol 1-oleate 2,3-dihexanoate were hydrolyzed by another purified milk lipase (Chandan and Shahani, 1963a; Jensen *et al.,* 1962).

Further confirmation for the predominance of positional rather than chain-length specificity was obtained when it was found that both butyric and palmitic acids were insignificantly digested when esterified to the 2-positions of 1,3-dichloropropanol and 1,3-diethoxypropanol (Gander *et al.,* 1961). Finally, the rates of sequential digestion were determined with laurate acyl glycerols and Chandan and Shahani's preparation, with the following results: trilaurin, 100; 1,3-dilaurin, 20; 1-monolaurin, 4; and 2-monolaurin, 0 (Jensen *et al.,* 1963). Similar data have been noted with pancreatic lipase.

Although milk lipase has not been tested for stereospecificity, e.g., differentiation between positions *sn*-1 and *sn*-3 of enantiomeric triglycerides, it is unlikely that the enzyme possesses this faculty.

The data in Table IV-27 indicate that milk lipase may produce diacyl glycerols more suitable for stereospecific analysis (Brockerhoff, 1971b) than pancreatic lipase, since with the substrates tested there was no preferential hydrolysis of oleic acid, as it has been noted occasionally with pancreatic lipase. Milk lipase is readily available in the form of lyophilized raw skim milk and hence should be investigated more thoroughly as a tool for structural analysis of triacylglycerols, although none of the preparations enjoy the stability of crude pancreatic lipase.

E. Factors Affecting Activity

In common with most glycerol ester hydrolases, milk lipase has a pH optimum of 8–9 and a temperature optimum of 35–40°C. If we assume

TABLE IV-27

Fatty Acid Composition (Mole%) of Diglycerides Formed from Synthetic Triglycerides by Various Milk Lipase Preparations

Substrate	Diglycerides fatty acids			
	4:0	6:0	16:0	18:1
rac-18:1–16:0–18:1[a]	—	—	51.0	49.0
rac-16:0–18:1–16:0[a]	—	—	48.0	52.0
rac-18:1–6:0–6:0[b]	—	78.3	—	21.7
rac-16:0–4:0–4:0[b]	75.7	—	24.3	—
rac-16:0–18:1–18:1[c]	—	—	25.0	75.0
rac-18:1–6:0–6:0[c]	—	79.0	—	21.0

[a] Gander and Jensen (1960).
[b] Jensen *et al.* (1964a).
[c] Jensen *et al.* (1964c).

that the enzymatic mechanism of the milk lipase is similar to that of pancreatic lipase, there will be no need for calcium. The evidence for an active serine also indicates that milk lipase is not a metalloenzyme. Nevertheless, calcium is probably useful in the lipolytic reaction because the inhibitory long-chain fatty acids must be bound. Similarly, sodium ions may serve to provide counter ions at the interface. These effects are discussed in the chapter on pancreatic lipase.

Activity of the lipase is inhibited by oxygen, light, heat, N-ethylmaleimide, PCMB, iodoacetate, hydrogen peroxide (Shahani, 1966); some antibiotics (Shahani and Chandan, 1962); DFP and 2,4-dinitro-fluorobenzene (Robertson *et al.*, 1966). Glutathione stabilizes (Shahani, 1966). Robertson *et al.* (1966) have studied the effect of inhibitors on the distribution of fatty acids released from milk fat by the lipase, noting little or no effect. Homogenization and foaming of raw milk stimulate lipase activity, the former perhaps because of disruption of the milk fat globule membrane or release of lipase from the casein molecule, and the latter because of increased fat globule surface areas and recoating of the globule with casein (Jensen, 1964).

Campbell *et al.* (1968) noted a diminution in activity when lecithin or phosphatidylethanolamine were added to milk. These polar compounds would locate themselves at the serum-fat globule interfaces and interfere with lipolysis by displacing the substrate from the interface. Since these compounds oxidize rapidly, the oxidation products could have an inhibitory effect. Such an effect has been observed with pancreatic and Geotrichum candidum lipases (R. G. Jensen, unpublished data, 1973; Pokorney *et al.*, 1965).

F. Conclusions

For further significant research on the subject of milk lipase, a method must be developed for the isolation of the enzyme or enzymes. The principal problem has been their instability, which may be due to a lability of the sulfhydryl groups and might be controlled by careful use of protective agents such as dithiothreitol. It should be mentioned, however, that porcine pancreatic lipase is stable in spite of two free sulfhydryl groups.

Does bovine milk contain several lipases or are all the fractions of different molecular weights only monomer and polymers of the same enzyme, or aggregates of the enzyme with other proteins? The existence of a spectrum of isoenzymes cannot be discounted, especially in view of the known genetic variability of milk proteins. On the other hand, if the lipases are glycoproteins, as seems probable, part of the differences in molecular weight may be due to differences in the size of the carbohydrate moiety. We are reminded of the lipases of *Rhizopus arrhizus,* one form of which can be converted to another by the cleavage of a glycosidic portion of the molecule (Sémériva *et al.,* 1969). Such a situation may well exist for milk lipase.

If the existence of the enzyme of a formula weight of 7000–8000 could be established, this milk lipase would be the smallest of all lipolytic enzymes, and considering that only a part of the molecule may be protein, perhaps the smallest of all enzymes. The determination of its structure would probably be a relatively simple task, and milk lipase might replace chymotrypsin as the darling of the enzymologists.

Milk lipase appears to be a serine enzyme like pancreatic lipase; the evidence is stronger, since the enzyme of milk is inhibited by the serine reagent DFP and the pancreatic enzyme is not, probably because of steric hindrance. We can conclude that the reactive site of milk lipase is sterically less hindered. This is in accordance with the finding that milk lipase appears to discriminate less between the chain lengths of its substrates, insofar as this discrimination in the case of pancreatic lipase may be interpreted in terms of steric hindrance.

VIII. PLANT LIPASES

Most of the efforts in this area have been devoted to seed lipases. Seeds are generally rich in triacylglycerols, which serve as a compact source of energy for the newly emerging plant. During germination of the seed, the triacylglycerol stores disappear. Since the fatty acids cannot be oxidized to provide energy until they are released from the triacylglycerol, lipolytic enzymes are probably rate controlling during germination. As germination usually is rapid, lipolytic activity is relatively high.

Crushing or storage generally activates dormant lipases in a seed, and the resulting accumulation of free fatty acids can cause an industrially important oil to become unacceptable or to require additional processing to remove the acids. Nevertheless, investigators have neglected the lipases in most of the important food fat oil seeds, e.g., soybean, cottonseed, corn, safflower, coconut, sesame, peanut, as well as other industrial seeds such as tung. Very few reports on the lipases of these seeds are available. Most attention has been paid to seed lipases that exhibit some exotic property, for example, the acid lipase of castor bean. Discussion will be limited to these lipases.

A. Castor Bean (*Ricinus communis*)

The lipase in this seed was first described by Green (1890) and has received the scrutiny of investigators since then. In recent years, many publications have originated from the Southern Regional Research Laboratory of the United States Department of Agriculture; they have been reviewed by Ory (1969). The dormant seed is known to contain a lipase with a pH optimum of 4.2. The germinating seed has been reported to have a neutral lipase, with an optimum pH of 6.8 (Yamada, 1957). Ory and St. Angelo (1971) could not confirm the presence of this enzyme, but Thanki *et al.* (1970) have reaffirmed the presence of a neutral lipase.

Ory *et al.* (1968) were able to localize the acidic lipase in the spherosomes of the bean, the oil-bearing organelles of the seed. Using an ingenious technique involving lead ions, which did not inhibit the enzyme but did precipitate the fatty acids hydrolyzed by the enzyme, and subsequent electron microscopy, Ory *et al.* determined that the enzyme was associated with the spherosomes, possibly as part of, or near to, the membrane.

1. Purification

The purification method described by Ory *et al.* (1962) is given in Table IV-28. The dehulled beans were ground and centrifuged to yield the precipitated debris, an aqueous supernate and a fatty layer which contained the lipase. Ricin and various allergenic proteins were located in the debris and aqueous supernate, which were discarded. The particulate material remaining after extraction of the fat pad with ether and salt solution was dialyzed and lyophilized to produce the crude lipase. In one experiment, the specific

TABLE IV-28

Purification of Castor Bean Lipase[a]

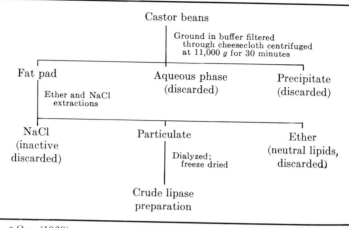

[a] Ory (1969).

activity of the enzyme was 5.1 in the fatty layer and 7.9 for the purified lipase.

The authors emphasized the potency of the allergenic proteins by suggesting the use of rubber gloves and avoidance of dry seed powders during the early stages of purification.

2. Characteristics

Activity was completely inhibited by mercuric ion and PCMB; the inhibition was reversed by cysteine and EDTA (Ory *et al.,* 1962). Ory and Altschul (1964) observed inhibition of castor bean lipolytic activity by the cyclopropene compound, sterculic acid. The inhibition was not reversible with excess cysteine; this suggested to Ory and Altschul an irreversible binding of a sulfhydryl group at the active site. Sterculic acid also inhibits chicken liver stearic acid desaturase; this has similarly been attributed to irreversible binding of an active site sulfhydryl. The possibility must be considered that castor bean lipase is sulfhydryl enzyme.

The enzyme was found to hydrolyze rapidly all of its emulsified endogenous substrate, i.e., castor oil, which contains large quantities of ricinoleic acid (Ory *et al.,* 1960). The partially purified lipase, which contained about 55% protein and 5% fat, completely hydrolyzed emulsified triacylglycerols at pH 4.2. As with many lipases, short-chain, monoacid triacylglycerols were hydrolyzed more rapidly than those composed of long-chain acids.

Castor bean lipase was reported by Savary *et al.* (1958) to hydrolyze esters from triacylglycerols regardless of location; i.e., without positional specificity. Ory *et al.* (1969) obtained contradictory results observing lipolysis of the 2-position of synthetic substrates only after 7–10 minutes of incubation. With shorter periods of incubation, the preparation hydrolyzed the primary acids only. The substrates 2,3-butane dioleate, *n*-hexyl oleate and 2-hexyl oleate were resistant to hydrolysis. The authors postulated that the appearance of acids from the 2-position in the lipolysate could be ascribed to acyl migration from 2 to 1 and subsequent lipolysis, although Borgström and Ory (1970) observed that the acidic condition alone (pH 4.2) did not cause acyl migration.

Borgström and Ory (1970) hydrolyzed [³H]glycerol labeled triolein and *rac*-1,2-diolein with castor bean lipase preparation. The hydrolysis of triolein was rapid and complete, with very little di- and mono-glycerides detected. Free glycerol appeared early. The 1,3-diglyceride isomer made up about 30–40% of the total diacylglycerols and was also present in the reaction mixture when *rac*-1,2-diolein was the substrate. In the absence of enzyme, there was no isomerization. The enzyme released insignificant quantities of labeled oleic acid from the diether substrate. Attempts to separate the hydrolase and isomerase activities were unsuccessful.

The specificity of the enzyme was further investigated by Noma and Borgström (1971) with several labeled substrates. The results revealed that the 1,3-diacylglycerol, which appeared immediately (1 minute) during digestion, had been largely formed by reesterification of 1-monoacylglycerols. Both 1- and 2-monoacylglycerols were hydrolyzed at equivalent rates as was either acyl group from 1,2-di-olein. Isomerization from the 2- to the 1-position appeared to be quantitatively unimportant.

3. Factors Affecting Activity

When *n*-butanol was employed in an attempt to solubilize the enzyme preparation (Ory and Altschul, 1962), an oily material was extracted from the particulate matter, and the residue (apoenzyme) had little or no lipolytic activity. When the butanol extract, a viscous oil, was added back to the apoenzyme, lipase activity was completely restored. The cofactor was identified as a cyclic tetramer of ricinoleic acid with each acid esterified to the 12-hydroxyl group of the adjacent acid (Ory *et al.*, 1964). The exact role of the cofactor remains to be determined, but when the lipid was present, coalescence of the substrate and enzyme was enhanced whether or not the enzyme was inhibited (Ory *et al.*, 1967a).

A heat-stable protein activator was isolated from a suspension of the particulate castor bean lipase (Ory *et al.*, 1967b) that when combined with the apoenzyme greatly increased lipolytic activity. The activator was devoid of lipolytic activity. It was further purified by DEAE-cellulose column chromatography. The activator, while not identified positively as a castor bean allergen by immunoelectrophoresis, was a part of this fraction (Ory, 1969). The amino acid composition of the purified activator was reported by Ory *et al.* (1967b). All three components—apoenzyme, lipid cofactor, and protein activator—were required for maximum rates of lipolysis.

The lipase was inhibited by mercuric ions, PCMB, and sterculic acid. However, inhibition was prevented by the addition of cysteine before the assay (Ory and Altschul, 1964; Ory, 1969). DFP, protamine sulfate, heparin, and lead were not inhibitory. The enzyme did not require calcium ions or albumin as a fatty acid acceptor.

The presence of neutral lipase reported to be present in germinating castor bean seeds was investigated by Ory and St. Angelo (1971). Activity was not detected, and treatment of sterile seed with gibberellic acid or actinomycin D did not cause appearance of the enzyme. Gibberellic acid should promote the synthesis of the lipase and actinomycin D inhibit it. Since there was no effect, the authors suggested that the earlier observation of a neutral lipase (Yamada, 1957) might have resulted from microbial growth. However, Thanki *et al.* (1970) detected neutral lipase activity (pH 6.8) in germinating γ-irradiated seeds grown in a sterile medium. This observation might be ascribed to a difference in varieties or to the effects of irradiation.

B. *Verononia anthelmintica*

The seed of *V. anthelmintica,* an import from India, contains vernolic (*cis*-12,13-epoxy-*cis*-9-octadecenoic) acid. The acid occurs primarily as the simple triglyceride, trivernolin, which can be readily recovered by wet-flaking the seed at low temperature with petroleum ether followed by one crystallization at $-20°C$ of the petroleum ether extract (Krewson and Scott, 1964). Grinding and incubation of the seed prior to extraction resulted in the recovery of 1,3-divernolin and free fatty acid (Krewson *et al.,* 1962). This suggested that an enzyme specific for the 2-position might be present in the seed. Since such a lipase would be valuable for structural analysis of triglycerides, an investigation of enzyme was undertaken (Olney *et al.,* 1968).

An acetone powder extract of the ground seed was partially purified (Table IV-29) by differential centrifugation, with a thirtyfold purification and a yield of about 1% (Olney *et al.,* 1968). Both acetone powder extract and partially purified preparation were chromatographed on columns of Sephadex G-200. With the crude extract, two protein peaks were seen,

TABLE IV-29

Partial Purification of Vernonia Seed Lipase by Differential Centrifugation[a]

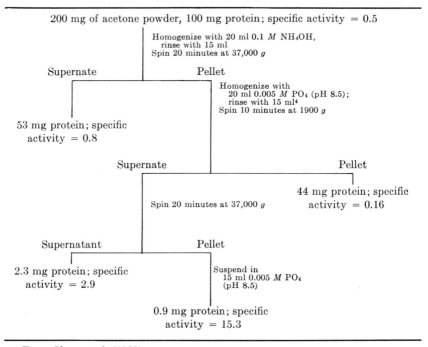

200 mg of acetone powder, 100 mg protein; specific activity = 0.5

Homogenize with 20 ml 0.1 M NH₄OH, rinse with 15 ml. Spin 20 minutes at 37,000 g

Supernate — 53 mg protein; specific activity = 0.8

Pellet — Homogenize with 20 ml 0.005 M PO₄ (pH 8.5); rinse with 15 ml⁴. Spin 10 minutes at 1900 g

Supernate — Spin 20 minutes at 37,000 g

Pellet — 44 mg protein; specific activity = 0.16

Supernatant — 2.3 mg protein; specific activity = 2.9

Pellet — Suspend in 15 ml 0.005 M PO₄ (pH 8.5)

0.9 mg protein; specific activity = 15.3

[a] From Olney et al. (1968).

and lipase activity was associated with the largest peak which eluted immediately after the void volume (phosphate buffer, pH 7.0 or 0.01 M NH₄OH solution). An eightfold increase in activity over the crude extract was observed. When samples partially purified by differential centrifugation where chromatographed, activity coincided with a single rapidly moving peak, and only slight increases in activity were obtained. The elution volumes and the centrifugation pattern indicated that lipolytic activity was associated with a molecule or aggregate of molecular weight over 200,000, or a particulate fraction. Several of the preprations were analyzed by polyacrylamide gel electrophoresis. The only band with high lipolytic activity did not leave the sample slot, another indication of high apparent molecular weight.

Assays of the activity were conducted at pH 7.5 (phosphate buffer) and 43°C (Olney et al., 1968). While activity declined above 48°C, some lipolysis was still evident at 80°C; thus, the enzyme is remarkably thermostable. Calcium ions and bile salts did not affect activity. DNP and DFP both increased the activity of the crude enzyme (acetone powder); this suggests the inhibition of proteolytic enzymes (Carpenter and Jensen, 1971). Nothing further is known about the chemistry of the enzyme.

Preliminary observations with synthetic triglycerides and trivernolin as substrates showed that both 1,2- and 1,3-diglycerides were produced (Olney *et al.,* 1968). The observation of Krewson *et al.* (1962) that 1,3-divernolin was the only species of diglyceride in extracts of crushed seed was thus due to an artifact of isolation. The 1,2 isomer is more soluble in petroleum ether than the 1,3 species and the latter is more readily crystallizable. However, the ester in 2-position was also hydrolyzed. To further delineate the specificity of the partially purified enzyme, four synthetic triglycerides were used as substrates (Table IV-30). To summarize the data: (1) The enzyme did not differentiate between oleic and palmitic acid. (2) Initial hydrolysis must have occurred at both primary and secondary positions to account for the quantities of 1,2- and 1,3-diglycerides and for the proportions of 16:0 and 18:1 in the free fatty acids. (3) The ratio of 1,2 to 1,3-diglycerides was about 2:1. (4) The composition of the 1,2-diglyceride showed a slight predominance of the acid from the primary position, whereas the 1,3 isomer contained considerable amounts of the 2-acid. Isomerization of the 1,2- and 2,3-iosmers to the 1,3-isomer was therefore part of the lypolytic sequence; an enzyme catalyzing acyl migration was possibly present. (5) The monoglyceride usually contained less of the acid from position 2 than either diglyceride. In conclusion, the enzyme preparation hydrolyzed the 2- as well as the 1-position.

C. Wheat Germ

An enzyme from wheat germ hydrolyzing monobutyrin and designated as a lipase was studied by Singer and Hofstee (1948a,b), with the activity being determined manometrically. Although the enzyme preparation also digested tributyrin and long-chain "Tweens" to some extent, there is considerable doubt that the activity was a lipase as we have defined it. The presence of impurities, e.g., monobutyrin, in commercial preparations of tributyrin is documented (Jensen, 1971), and it was probably this glyceride that was the substrate in the apparent hydrolysis of the tributyrin. Lipolysis of the ill defined Tweens is not a valid criterion for the definition of a lipase. It seems likely that Singer and Hofstee studied an esterase rather than a lipase. However, subsequent investigations by Stauffer and Glass (1966) established that wheat germ does contain a true lipolytic enzyme as well as an esterase and a "tributyrinase."

Another hydrolytic enzyme associated with the lipase activity is a nonspecific phosphomonoesterase (Anonymous, 1972). In fact, the com-

TABLE IV-30

Products from the Lipolysis of Purified Synthetic Triglycerides by Purified *Vernonia anthelmintica* Seed Lipase[a]

Triglyceride[b]		Residual Triglyceride	Diglycerides		Mono-glyceride	Free fatty acids	Glycerol[c]
			1,2	1,3			
POP Fatty acid							
Composition (mole %)	16:0	66	54	79	91	70	—
	18:1	34	46	21	9	30	—
Relative moles of glyceride[d]	—	2086	324	160	430	1600	86
Diglycerides	16:0–18:1	—	298	67	—	—	—
	16:0–16:0	—	26	93	—	—	—
OPO Fatty acid							
Composition (mole %)	16:0	33	47	26	16	36	—
	18:1	67	53	74	84	64	—
Relative moles of glyceride[d]	—	1056	320	171	509	1375	−44
Diglycerides	16:0–18:1	—	301	89	—	—	—
	18:1–18:1	—	19	82	—	—	—
OSS Fatty acid							
Composition (mole %)	18:0	64	80	66	72	62	—
	18:1	36	20	34	28	38	—
Relative moles of glyceride[d]	—	2566	389	216	323	1462	72
Diglycerides	18:0–18:1	—	156	147	—	—	—
	18:0–18:0	—	233	69	—	—	—
PSS Fatty acid							
Composition (mole %)	16:0	34	21	31	32	41	—
	18:0	66	79	69	68	58	—
Relative moles of glyceride[d]	—	1635	262	89	614	1683	35
Diglycerides	16:0–18:0	—	110	55	—	—	—
	18:0–18:0	—	152	34	—	—	—

[a] From Olney et al. (1968).

[b] POP = *rac*-glycerol 2-oleate 1,3-dipalmitate; OPO = *rac*-glycerol 2-palmitate 1,3-dioleate; OSS = *rac*-glycerol 1-oleate 2,3-disterate; PSS = *rac*-glycerol 1-palmitate 2,3-disterate.

[c] Determined by difference; $\frac{1}{3}$(free fatty acids − diglyceride − 2 monoglyceride)

[d] On basis of 1000 moles of triglyceride hydrolyzed: diglyceride + monoglyceride + glycerol = 1000.

mercial supplier of wheat germ "lipase" sells a preparation as "acid phosphatase," which is the preparation of Singer (1948), who concentrated the "lipase" from defatted ground whole wheat germ by extracting the powder with water, centrifuging the extract, and adjusting the supernate to pH 5.5. Another centrifugation and adjustment of pH to 6.6 was followed by precipitation with ammonium sulfate. The precipitate was preserved by lyophilization.

Stauffer and Glass (1966) separated wheat germ into three fractions: lipase, tributyrinase, and esterase. The ground hexane-extracted seed was dispersed in water and centrifuged, and the resulting supernate was acidified to pH 5.5 with dilute acetic acid. Centrifugation resulted in a precipitate which contained the lipase. Saturation to 0.4 with ammonium sulfate and centrifugation produced a precipitate with which tributyrinase activity was associated. Finally, an esterase concentration was collected by ammonium sulfate saturation (0.7) and centrifugation.

The lipase precipitate was suspended in phosphate buffer, dialyzed against water, and centrifuged to yield a yellow solution used for characterization of the lipase. The substrate was an emulsified monoolein–olive oil mixture, and liberated glycerol was determined by periodic acid oxidation. No taurocholate or calcium chloride were added; the temperature 37°C. The optimum pH for lipase activity was about 8.2. The enzyme did not hydrolyze tributyrin. It seems unlikely that the lipase ignored the olive oil in which the monoolein, the presumed substrate, was dissolved.

The tributyrinase activity observed by Stauffer and Glass (1966) deserves comment. The pH optimum of this enzyme was about 6.5. Of the substrates studied, tributyrin (100) was hydrolyzed most rapidly; tripropionin, 30; triacetin, 5; triolein, nil; ethyl acetate, 0; and ethyl propionate, 4.

Drapron *et al.* (1969) localized lipolytic activity in the coleptile of the germinating seed. Lipolytic activity was much lower during active photosynthesis than in the dark. Apparently, during the absence of light, hydrolysis yields the free fatty acids necessary to provide energy via oxidation.

Caillat and Drapron (1970), with purified olive oil as a substrate, examined the effect of calcium ions and bile salts upon the activity of the lipase from the germinated wheat gemmule. Bile salts were activating at low concentrations and inhibitory at higher levels, presumably above the critical micellar concentration. The role of calcium ions was more complex; apparently they acted as cofactor in addition to binding free fatty acids. Activity was completely inhibited by addition of EDTA and renewed by a subsequent excess of Ca^{2+} ions.

In endosperm halves from germinating wheat grain, lipase activity was induced in the endosperm starch by glutamine (Tavener and Laidman,

1972). Activity was induced in the brain of the endosperm halves by indole acetic acid in the presence of glutamine. Activity produced by both processes was reduced by inhibitors of energy metabolism, RNA synthesis, and protein synthesis. Azaserine, a glutamine analog, also inhibited induction.

With 4-methylumbelliferone butyrate as the substrate, Pancholy and Lynd (1972) determined that wheat germ lipase was activated by Ca^{2+} ions and inhibited by CN^+, aflatoxin, Cu^{2+}, Fe^{3+}, S^{2+}, and EDTA. A commercial wheat germ lipase was employed. It seems possible that esterase rather than lipase activity was measured. The compound used is a very poor substrate for lipases and the heterogeneous nature of the commerical enzyme preparations used has been discussed. Karnovsky and Wolff (1960) showed that wheat germ lipase did not differentiate between enantiomerically labeled trioleins. The method has been described on p. 57.

D. Oats

A genuine lipase was detected in oat grains (Martin and Peers, 1953) that hydrolyzed both tributyrin and triolein. Di- and monobutyrins were not digested; the authors interpreted this as specificity for the triglyceride. A more likely explanation is that the increased solubility of the di- and monobutyrins in water relative to tributyrin caused them to be unavailable to the lipase. The purity of these substrates was also doubtful.

Berner and Hammond (1970b) further examined the specificity of oat lipase. The enzyme hydrolyzed acids from both 1- and 2-position of lard and cocoa butter, but preferentially hydrolyzed 18:1 (52.0%) from the latter and 18:2 (22.7%) from the former. The original contents were 35.6% and 8.9%, respectively.

E. Rice Bran

Funatsu *et al.* (1971) obtained a lipase concentrate (480-fold) from rice bran—by ammonium sulfate precipitation and chromatography on DEAE-cellulose and Sephadex—with a specific activity of 4.7. The preparation was homogeneous by both polyacrylamide gel electrophoresis and electrofocusing and had a molecular weight of about 40,000 and an optimal pH of 7.5 with tributyrin as the substrate. The possibility of hydrolyzable impurities in the tributyrin must be considered.

Shastry and Rao (1971) similarly obtained a purified lipase (80–90-fold) from rice bran with the additional step of adsorption on calcium phosphate gel. The preparation was homogeneous by polyacrylamide gel electrophoresis; however, both ethyl acetate and triglycerides were hydrolyzed. Two pH optima were observed, 5.5 and 7.4–7.6. Activity was partially inhibited by DFP, EDTA, and PCMB, which indicates the possible involvement of hydroxyl and sulfhydryl groups as well as Ca^{2+} ions. The molecular weight was approximately 41,000.

IX. MICROBIAL LIPASES

In the past, interest in microbial lipases resulted from investigation of food spoilage, especially of dairy products. The short-chain FFA released are directly responsible for flavor defects, while the long-chain fatty acids, themselves innocuous with respect to flavor, could presumably be converted more readily to carbonyls and other volatile compounds as free acids. In contrast, FFA in some dairy products, notably cheese, contribute to desirable flavor. For example, *Penicillium roqueforti* spores are deliberately added during the preparation of Roquefort cheese so that action of the lipase resulting from mold growth can contribute to the flavor. Many other foods utilize microbial lipases, either *in situ* or added, to obtain desired flavors and textures (Seitz, 1973).

More recent attention to the enzymes has arisen because the production of lipases may assist in classification of microorganisms and detection of those that are pathogenic (Lawrence, 1967). Although lipolytic activity has been considered to be taxonomically important, most investigators have carried out their studies within rather than across genera. Both lipolytic and nonlipolytic strains were found, for example, in *Staphylococcus aureus* and *albus* (Davies, 1954). Alford *et al.* (1964), in a cross-generic study, observed hydrolysis of mainly the primary ester of triglycerides by several species of *Pseudomanas* and a variety of molds. *Staphylococcus aureus* and *Aspergillus clarus* attacked all three positions, while the mold *Geotrichum candidum* released oleic acid regardless of position.

Since the triglyceride content of many microorganisms, particularly the true bacteria, is low or even nonexistent, the role of endocellular lipases in the cellular economy is puzzling (O'Leary, 1967). *Actinomycetales,* yeast, and fungi contain more triglycerides, although the contents are usually minor when compared to phospholipids. The function of the exocellular lipases is more obvious: they hydrolyze exogenous triglycerides to provide FFA, probably for energy. Paradoxically, lipase production by some microorganisms, i.e., *G. candidum* (R. G. Jensen, unpublished data), is greatly enhanced when triglycerides are absent from the culture medium. Current research on microbial lipases is directed mostly toward purification and characteristics of the enzymes themselves, rather than toward establishing their functions in cellular metabolism.

Although many microorganisms produce lipases, few have been closely studied and the discussion will be limited to these enzymes. Lawrence (1967) has published a comprehensive review on microbial lipases and esterases.

A. Molds

1. Geotrichum candidum

a. Isolation. G. candidum is a septate mold growing as a white pad on the surfaces of cheese and sour cream (Foster *et al.*, 1957). The exocellular lipase secreted by the mold has been studied because of its ability to lipolyze the fat in dairy products and recently because of its specificity for unsaturated fatty acids, first discovered by Alford and Pierce (1961). Most of the subsequent investigations have dealt with this specificity; they have been reviewed by R. G. Jensen (unpublished data).

As the lipase is an exocellular enzyme, it has been isolated by concentration of the culture media in which the mold was grown. In the first investigation (Alford and Pierce, 1961), the cells and mycelia were removed from the media by centrifugation or filtration and the supernate or filtrate used at a source of enzyme. The mold was grown at 20°C in peptone broth (pH 7, 0.05 *M* phosphate broth). In a later study (Alford *et al.,* 1964), the filtrate was dialyzed against water, concentrated to about 5–10% of its original volume by dialysis against polyethylene glycol (Kohn, 1959), and lyophilized to a dry powder. The powder is very stable at −20°C, retaining most of the original activity for at least 7 years (R. G. Jensen, unpublished data). Alford and Smith (1965) have described further improvements in the production of the enzyme. The specific activities of these preparations were not reported, but for purposes of comparison, the activity with milk fat as a substrate was about 2% of that of a crude porcine pancreatic lipase powder (Jensen *et al.,* 1965). The preparations of Alford were the sole source of the enzyme until December 1972. However, Worthington Biochemicals has recently produced the lipase and a "highly purified, freeze-dried powder" with a specific activity of not less than 300 is available from Miles Laboratories, Kankakee, Illinois. Kroll *et al.* (1973) have concentrated the enzyme by Sephadex filtration to a specific activity of 460 (35°C).

A crystalline and electrophoretically homogeneous lipase with the specific activity 447 has been obtained by Tsujisaka et al. (1973) by chromatography on DEAE-Sephadex and Sephadex G-100 and G-200. The preparation (yield, 20%) contained about 7% carbohydrate, mainly mannose, and a very small amount of lipid. The lipase has an isoelectric point of pH 4.33, and a molecular weight of 54,000, and contains neither cysteine nor methionine.

b. Factors Affecting Activity. Alford *et al.* (1964) observed optimal activity of *G. candidum* lipase at 20°C and pH 7.0. Kroll *et al.* (1973) report an optimum at pH 8.5 and 35°C. The crystalline lipase of Tsujisaka *et al.* (1973) showed activity between pH 4.5 and 9.5 with an optimum of pH 5.6–7.0. A systematic study of factors affecting lipolytic activity of the enzyme during assay (Marks *et al.*, 1968) showed that tris buffer promoted greater activity than phosphate buffer, with a broad pH optimum of 8–9. Calcium chloride did not increase the rate of lipolysis nor did variations in temperature of incubation between 15° and 40°C.

Alford *et al.* (1964) found that DFP (10^{-4} M) reduced the activity of *G. candidum* lipase by 25–35%. Carpenter and Jensen (1971) observed similar reductions when the enzyme was exposed to DNP, Parathion, or oxidation products from polyunsaturated acids (R. G. Jensen, unpublished data). Sodium oleate also inhibited the enzyme.

c. Substrate Specificity. Alford and Pierce (1961) reported that the enzyme released mainly unsaturated acids from several oils. Both specificity for and lack of distinction between 18:1 and 18:2 were found. Some saturated acids were also hydrolyzed. With synthetic triglycerides as substrates, Alford *et al.* (1964) confirmed the specificity for 18:1 noting that the acid was hydrolyzed regardless of location, primary or secondary, from the triglyceride (Table IV-31). The anomalous results observed with 18:1–16:0–16:0 can be ascribed to either use of an impure substrate or to methodological difficulties, since with the same substrate, synthesized and lipolyzed separately, 96.1 mole % 18:1 were later observed in the FFA (Marks *et al.*, 1968).

Additional data on the specificity of the lipase are presented in Tables IV-32 and IV-33. In Table IV-32, it can be seen that *G. candidum* lipase hydrolyzes elaidic, petroselinic (6–18:1), vaccenic (11–18:1), and short- or long-chain saturated esters at a slow rate, about one-tenth of that of *cis* 9–18:1, while this acid, 16:1, and 18:2 were rapidly hydrolyzed. There was no differentiation between 18:1 and 18:2.

Table IV-33 shows the content in unsaturated fatty acid of the FFA and monoglycerides from the hydrolysis of each of fifteen mixed synthetic triglycerides containing 12:0, 14:0, 16:0, 18:2 and double bond isomers of *cis* 18:1 from Δ2 to Δ16 (Jensen *et al.*, 1972). The specificity of *G. candidum* lipase is obvious; the 18:1 isomers other than *cis* Δ9 were poorly digested, as were trans monoenoic acids, including *trans*-Δ9 (elaidic acid). Linoleic acid was preferentially hydrolyzed in the absence of *cis*9–18:1, but there was no differentiation between the two acids in the triglyceride

TABLE IV-31

Lipolysis of Synthetic Triglycerides by
Geotrichum candidum **Lipase**[a]

Triglyceride substrate[b]	Percent of free fatty acids		
	16:0	18:0	18:1
18:1–18:0–18:1	—	1	99
18:0–18:1–18:0	—	2	98
18:1–16:0–18:1	1	—	99
16:0–18:1–16:0	20	—	80
16:0–18:1–18:0	9	1	90
18:0–16:0–18:0	50	50	—
16:0–18:0–16:0	99	1	—
18:1–18:0–18:0	—	2	98
18:0–18:1–18:1	—	1	99
18:1–16:0–16:0	35	—	65
16:0–18:1–18:1	5	—	95

[a] Alford *et al.* (1964).
[b] For the notation see Chapter II, Section I.

TABLE IV-32

Oleic Acid Content of Free Fatty Acids from the Lipolysis of Synthetic
Triglycerides by *Geotrichum candidum* **Lipase**[a]

Substrate	FFA 18:1 (mole %)	Substrate	FFA 18:1 (mole %)
t-18:1–18:1–18:1[b]	95	16:1–18:1–18:1	65
18:1–18:1–16:0	90	11–18:1–18:1–18:1[d]	93
16:0–18:1–18:1	96	16:1–6–18:1–6–18:1[e]	95[f]
16:0–18:1–16:0	94	18:1–6:0–6:0	99
18:0–16:0–18:1	98	18:1–12:0–12:0	86
18:2–18:2–18:2[c] ⎱ 18:1–18:1–18:1 ⎰	62		

[a] Jensen *et al.* (1965); Marks *et al.* (1968).
[b] t-18:1–18:1–18:1 is *rac*-glycerol 1-elaidate 2,3-dioleate. All substrates are racemic.
[c] Derived from a 60:40 mole % mixture of triloein and trilinolein.
[d] 11–18:1 is *cis*-vaccenic acid.
[e] 6–18:1 is *cis*-petroselenic acid.
[f] Mole % 16:1.

TABLE IV-33

Unsaturated Fatty Acid Contents (M%) of the Free Fatty Acids and
Monoglycerides Resulting from Hydrolysis by *Geotrichum candidum*
Lipase of Triglycerides Containing Isomeric *cis*-18:1 Acids[a]

18:1 isomer[b]	FFA[c] (mole %)		Monoglyceride (mole %)	
	18:1	18:2	18:1	18:2
Δ2	0	59	10	4
Δ3	3	65	8	0
Δ4	0	80	33	3
Δ5	2	76	29	4
Δ6	2	67	35	0
Δ7	6	71	19	0
Δ8	4	75	55	2
Δ9	51	15	0	0
Δ10	7	74	22	3
Δ11	4	74	21	3
Δ12	7	44	49	7
Δ13	6	56	18	0
Δ14	5	57	19	0
Δ15	4	75	41	0
Δ16	2	67	0	0

[a] Jensen *et al.* (1972).
[b] Numeral indicates location of double bond.
[c] Both FFA and monoglyceride contained 12:0, 14:0, and 16:0.
Saturated fatty acid contents may be obtained by totaling the
unsaturated fatty acid contents and subtracting from 100.

containing both. The composition of the monoglycerides indicates an enrichment of 18:1 isomers other than Δ9 and suggests an approach to the concentration of these acids from partially hydrogenated food fats. The compositions of the diglycerides and the original and residual triglycerides are available (Jensen, 1973b).

From the data presented above, Jensen *et al.* (1973) hypothesized that *G. candidum* lipase should hydrolyze *cis,trans*-9,13–18:2 but not the trans, cis or trans,trans isomers. Astonishingly, both cis,trans and trans,cis isomers were digested while, as expected, the trans,trans acid was not (Table IV-34). The authors did not offer an explanation for the lack of distinction between the cis,trans and trans,cis isomers.

Results from the lipolysis of several fats and oils confirm the specificity of the lipase. For example, the FFA derived from palm oil and milk fat contained 82 and 62 mole % of 18:1 while the contents of the intact triglyceride were 39 and 20 mole %. (Jensen *et al.*, 1965). A triglyceride

TABLE IV-34

Fatty Acid Composition of Products from the Hydrolysis of Two Triglycerides Containing Geometric Isomers of Linoleate by *Geotrichum candidum* **Lipase**[a]

	9,12–18:2 isomer (wt %)			
	Triglyceride 1		Triglyceride 2	
Product	cis,cis	trans,trans	cis,trans	trans,cis
Original triglyceride	48	52	46	54
Residual triglyceride	35	65	49	51
Free fatty acids	85	15	45	55
Diglycerides	29	71	50	52
Monoglycerides	tr[b]	90+	44	51

[a] Jensen *et al.* (1973).
[b] Peaks were small, but only a trace of cis,cis isomer was observed.

species (two double bonds) from *Crambe abyssinica* oil was hydrolyzed to yield mostly 18:1; the erucic acid (*cis*13–20:1) present was not digested (Gurr *et al.*, 1972). Other compounds or acids lipolyzed by the enzyme are palmitoyl oleate, cholesteryl oleate and *cis*9–14:1 esters, while arachidonic ester, stearolic ester (octadecynoic), *cis*5–14:1, oleyl palmitate and dilinoleoyl phosphatidylcholine are not appreciably hydrolyzed (R. G. Jensen, unpublished data). These specificity data are summarized in Table IV-35.

The enzyme has been applied to the structural analysis of triglycerides. These studies are fully described by Litchfield (1972) and Jensen (1973b)

TABLE IV-35

Substrate Specificity of *Geotrichum candidum* **Lipase**

Esters hydrolyzed rapidly	Esters hydrolyzed slowly
cis-9-18:1[a], 16:1 and 14:1	Double bond positional isomers
cis,cis-9,12-18:1[a]	of *cis*-18:1 other than Δ9.
cis,trans- and *trans,cis*-9,12-18:1	*trans*-9-18:1 and *trans,trans*-9,12-18:1
Palmitoyl oleate	4:0–18:0
Cholesteryl oleate	*cis*-5-14:1 and *cis*-13-22:1[b]
	Arachidonic acid
	Octadecynoic acid
	Oleyl palmitate[c]
	Dilinoleoyl phosphatidylcholine[c]

[a] Hydrolyzed regardless of position within the triglyceride.
[b] Gurr *et al.* (1972).
[c] Not hydrolyzed.

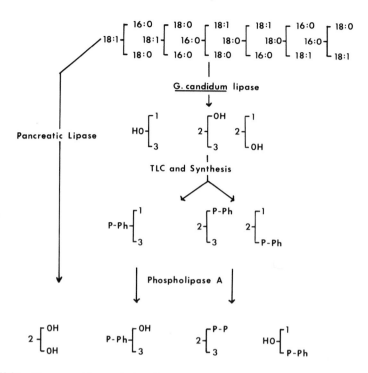

Fig. IV-34 Stereospecific analysis of a triglyceride with the help of *Geotrichum candidum* lipase. From Sampugna and Jensen (1968).

and hence will be only briefly discussed here. In a conventional stereospecific analysis, resolution of individual isomers of a triacid triglyceride mixture is impossible (Brockerhoff, 1966b). In a mixture containing 16:0, 18:1, and 18:0, for example, data from a stereospecific analysis would give the amounts of each acid in each position, but could not assign the quantities to each of the individual enantiomers. The problem can be solved by digestion of the triglyceride with *G. candidum* lipase to form sets of diglycerides which can subsequently be resolved as depicted in Fig. IV-34 (Jensen *et al.*, 1966a, Sampugua and Jensen, 1968). The 1,2 + 2,3 and 1,3 isomers are separated by boric acid–thin-layer chromatography, converted to phosphatidylphenols and digested by phospholipase 2. This enzyme hydrolyzes the *sn*-2 acid from the *sn*-3 substrate and the *sn*-1 acid from the *sn*-2 substrate leaving monoacyl phosphatides and undigested *sn*-1-phenyl phosphatide, which are separable by thin-layer chromatography.

    G. candidum lipase is unique among lipolytic enzymes in its specificity for the remote *cis*-9 double bond. Okuyuma *et al.* (1972) observed a similar specificity with acyltransferases and suggested that π bonds at certain positions were preferred by these enzymes. A binding site for *cis*-9 and *cis*-9,12 π bonds might explain the specificity of the *G. candidum* lipase although enzymatic structures capable of detecting π bonds have not been identified.

2. Rhizopus arrhizus

a. Isolation and Properties. The mold *R. arrhizus* elaborates an exocellular lipase (see Benzonana, 1973, for a review). The enzyme was partially purified by Laboureur and Labrousse (1966). Additional purification has been achieved by Sémériva *et al.* (1967a), who used the lyophilized powder (specific activity, 400) from an extract of the culture medium of Laboureur and Labrousse (1966). An aqueous solution of the powder was eluted from a column of IRF-97 with an ammonium acetate buffer with a yield of 95%; the eluate had a specific activity of 5000. The fractions with high activity were combined, reduced in volume and then passed through a column of Sephadex G-75. The enzyme emerged as a symmetrical peak, with a specific activity of about 8000. The preparation (lipase I) was slowly converted at 2°C to another form (lipase II) of activity similar to the first, but more cationic and more stable. After elution through Sephadex G-100, the two enzymes appeared to have molecular weights of 40,000 ± 2000 and 30,000 ± 1500. The two lipases could also be separated on a column of Amberlite IRC-50 at pH 5.7 (Sémériva *et al.,* 1969). Lipase I could be stabilized with DFP, and lipase II was obtained by allowing lipase I to stand at 20°C for 7 days. These changes were monitored by disk electrophoresis.

Confirming the work of Laboureur and Labrousse (1968), Sémériva *et al.* (1969) found lipase I to be a glycoprotein with a molecular weight of about 43,000. The enzyme contained 13–14 moles of mannose and 2 moles of hexosamine per molecule. The N-terminal amino acids were aspartic acid (or asparagine), serine and alanine.

The conversion, upon long incubation, of lipase I into lipase II produced a glycopeptide portion, "A." The molecular weight of lipase II was estimated to be about 32,000, i.e., 10,000 less than that of lipase I. Lipase II did not contain the sugars found in I and had about twenty fewer amino acids (Table IV-36). In addition, the N-terminal aspartic acid had been replaced by several other residues. Lipase II retained enzymatic activity; the sugars were apparently not required for the catalytic function. Lipase II was regarded as a degraded form of I, but with controlled degradation, since it was homogeneous by disk electrophoresis and retained activity. Fraction A, which appeared during the transformation of I to II upon aging, was a mixture of peptides and glycopeptides. The carbohydrate composition was similar to that in lipase I.

When a solution of the enzyme was (a) treated with 5% trichloroacetic acid at 0°C, (b) heated in a boiling water bath for three minutes, or (c) reduced to pH 1.0 by addition of 6 *M* HCl, a precipitate resulted (Séamériva *et al.*, 1969). However, protein remained in the supernate, which was designated as the glycoprotein fraction. This fraction was purified further by passage through a column of Sephadex G-50. Most of the material containing carbohydrate and protein was found in a single peak eluted with the void volume. The fraction appeared as a single band when analyzed by polyacrylamide gel electrophoresis. The glycoprotein fraction contained 13–14 moles of mannose and 2 moles of hexosamine, similar to the content of lipase I, and had a molecular weight of 7000–8000. The amino acid composition, in Table IV-36, reveals several differences, e.g., less phenylalanine and tryptophan than in lipases I and II.

The amino acid compositions of lipases I, II, the glycoprotein fraction and fraction A are presented in Table IV-36. There was enough similarity

TABLE IV-36

Amino Acid Composition of Lipases from *Rhizopus arrhizus*[a]

Amino acid	Number of residues in			
	Lipase I	Lipase II	Glycoprotein fraction	Fraction A
Ala	23	18	5	5–6
Arg	10	9	1	0–1
Asx	40	27–28	8	11
Cys	6	5	1	—
Glx	29–30	25	5–6	5
Gly	28	23–24	4–5	6
His	8	7	1	0–1
Ile	20	18	2	3
Leu	24–25	18	4–5	7–8
Lys	20	15–16	4	1
Met	3	1	1	1
Phe	15	15	0	1
Pro	24	16–17	5	6
Ser	40	27	2–3	9
Thr	28	24	9	4–5
Trp	6	4	—	—
Tyr	15	12	1	2
Val	29	28–29	1	3
MW	20,281	32,128	6100	

[a] Séamériva *et al.* (1969).

in amino acid composition between the glycoprotein and A fractions to suggest that the slow conversion of I to II occurred in the same manner as the rapid transformation engendered by 5% trichloroacetic acid. The conversion process upon aging may be enzymatic, since DFP stabilized lipase I; this compound inhibits proteolytic enzymes. Sémériva *et al.* (1969) suggested that the glycoprotein portion of the lipase molecule may aid in the transport of this exocellular *Rhizopus* enzyme across the cell membrane.

b. Substrate Specificity. Sémériva *et al.* (1967b), using synthetic triglycerides, established that the specificity of the enzyme for primary ester groups was almost absolute (Table IV–37). The lipase I hydrolyzed the *sn*-1 acids from a variety of phosphoglycerides regardless of the type of the acid (Slotboom *et al.,* 1970a). The racemic phosphatidylcholine (III) (Table IV-38) yielded 100% 18:0 in the free fatty acids when exposed to the lipase (Slotboom *et al.,* 1970a), and the stereoisomers IV and V were digested at equal rates. Therefore, *Rhizopus* lipase is not stereospecific but hydrolyzes either *sn*-1 or *sn*-3 positions.

Several phosphoglyceride analogs were tested as substrates for *Rhizopus* lipase (Slotboom *et al.,* 1970a). Hydrolysis did not occur when the following analogs of phosphatidylcholine were exposed to the enzyme: *sn*-1-benzyl, *sn*-1-vinyl ether (plasmalogen), and *sn*-1-monoacyl (lysophosphatide). The 1-monoacyl-*sn*-3-phosphatidyl ethanolamine and *rac*-1-monoacyl-3-glycerol phosphate were poorly digested as were a group of 1-palmitoyl alkane diolphosphorylcholines. To further elucidate the effects of adjoining groups upon the hydrolysis of *sn*-1 esters, analogs with a long *sn*-2 aliphatic ether or a long *sn*-2 alkyl group were incubated. Neither were hydrolyzed to the same extent as "normal" phosphatidylcholine. The lipase did not hydrolyze the *sn*-2-phosphatidylcholine. The uncharged

TABLE IV-37

Composition of the Fatty Acids Hydrolyzed from Synthetic Triacyl Glycerols by *Rhizopus arrhizus* Lipase[a]

	Acids hydrolyzed (wt %)	
Substrate	18:1	16:0
Glycerol-2-oleolyl 1,3-dipalmitate	1	99
Glycerol-2-palmitoyl 1,3-dioleate	98	2
Glycerol-3-oleoloyl 1,3-dipalmitate	48	52
Glycerol-3-palmitoyl 1,2-dioleate	53	47

[a] Sémériva *et al.* (1967b).

TABLE IV-38

Determination of the Mode of Action of Purified Lipase from the Mold *Rhizopus arrhizus* **by Hydrolysis of Synthetic Mixed-Acid Phosphoglycerides**[a]

			Fatty acid composition of hydrolysis products	
Substrate		Degree of hydrolysis	Monoacyl phosphoglycerides	Liberated fatty acids
III	rac-1-palmitoyl-2-oleoylglycero-3-phosphorylcholine	100	100 % oleic acid 0 % palmitic acid	100 % palmitic acid 0 % oleic acid
IV	1-palmitoyl-2-oleoyl-sn-glycero-3-phosphorylcholine	100	100 % oleic acid 0 % palmitic acid	100 % palmitic acid 0 % oleic acid
V	3-palmitoyl-2-oleoyl-sn-glycero-1-phosphorylcholine	100	100 % oleic acid 0 % palmitic acid	100 % palmitic acid 0 % oleic acid

[a] Slotboom *et al.* (1970a).

dimethylester of 1,3-diacyl *sn*-glycerol-2-phosphoric acid was, however, readily digested, with both fatty acids being hydrolyzed. *Rhizopus arrhizus* and purified pancreatic lipase digested all substrates with identical or similar results.

Slotboom *et al.* (1970a) observed that the results of their hydrolysis experiments described above agreed well with the suggestion (Brockerhoff, 1968) that hydrolysis is stimulated by neighboring electrophilic groups. In the phospholipids with groups other than acyls at the 2-position and thus lacking the activating carbonyl carbon, the ester at C-1 is less susceptible to lipolytic attack. The resistance of the primary acids in the *sn*-2-phosphatidylcholine to lipolysis may be due to the drop in nucleophilicity at the C-1 carbonyl carbons caused by the relative nearness of the electron-donating phosphate group. Removal of the negative charge on the phosphate ester at C-2 by addition of methyl groups restored lability to the lipase. Slotboom *et al.* (1970a) suggested that the nature of the *sn*-2 bond determines the susceptibility of the *sn*-1 ester in phospholipids to lipolysis.

The lipase hydrolyzed micelles of short-chain monoacid triglycerides as well as emulsified globules (Sémériva and Dufour, 1972). The rate of hydrolysis of olive oil triglycerides responded to the presence of sodium chloride, being very low in the absence of the salt, rising to a maximum at 10–15 mM concentration, then dropping sharply and stabilizing at 60 mM.

The enzyme was not activated by bile salts, but competitive inhibition occurred at concentrations above the critical micellar concentration. Calcium ions were not an absolute requirement.

Rhizopus arrhizus and pancreatic lipases have many similar characteristics: pH optima, high catalytic rate, specificity for the primary esters of triglyceride, lack of effect of calcium ions and resistance to DFP. Despite this resistance, both lipases are probably serine–histidine enzymes. *Rhizopus arrhizus* lipase is a glycoprotein three times richer in carbohydrate than the pancreatic enzyme. The carbohydrates are not essential for the catalytic activity; their removal in the form of glycopeptides by autolysis does not reduce activity.

Rhizopus arrhizus lipase can be successfully used for structural analyses of phospholipids and triglycerides, but offers no advantage for this purpose over pancreatic lipase. However, the mold lipase is much more easily obtained in preparations of high specific activity; a lyophilized culture filtrate may be one hundred times more active than pancreatic "steapsin."

4. Aspergillus niger

Aspergillus niger, a ubiquitous mold, secretes a lipase that was purified by Fukumoto *et al.* (1963). The procedure is described in Table IV-39. The mold was grown for 3 days at 30°C and pH 5.6 on a medium containing wheat. The crystalline enzyme was pure as judged by moving boundary electrophoresis and had an optimum at pH 5.6 and 25°C. It did not lose

TABLE IV-39

Purification of *Aspergillus niger* **Lipase**[a]

Procedure	Volume (ml)	Total activity[b]	Specific activity[b]
Aqueous extract of culture medium	12500	267000	5
Ammonium sulfate precipitate	800	218000	25
Acetone	530	180000	70
pH 3.3	650	162000	155
Acrinol[c]	120	97000	250
Duolite A2 column effluent	200	38300	330
Acetone crystallization	—	14000	380
Recrystallization	—	8800	390

[a] Fukumoto *et al.* (1964).

[b] Milliliter of 0.05 N KOH required to titrate the FFA released by the lipase from olive oil. "Specific activity" is the "units" per mg N. Length of incubation was not given.

[c] Acrinol is 2-ethoxy-6,9-diaminoacridinium lactate.

activity for 3 years at 4°C. The enzyme appears to hydrolyze triglyceride without positional specificity.

A "two times crystallized, freeze-dried powder, not less than 95% homogeneous by ultracentrifugal and electrophoretic analysis," with a specific activity not less than 10, is available from Miles Laboratories, Kankakee, Illinois.

5. *Mucor pusillus*

The mold *M. pusillus* secretes a proteinase that has the property of clotting milk, and at least two patents have been issued on enzyme preparations intended for its use in cheese production (Somkuti and Babel, 1968). Lipolytic activity has been noted in these preparations and the resulting FFA could contribute to the natural flavor of cheese made with the crude proteinases, or if excessive could also result in the development of unwanted flavors.

The microorganism was grown in a suspension of 4% wheat bran at pH 6.7 and 35°C. A crude enzyme preparation was obtained by filtration, precipitation with ammonium sulfate, and centrifugation (Somkuti *et al.*, 1969). Both Sephadex and DEAE columns were employed for further purification, although the purity achieved was not reported. The fraction with lipase activity coincided with the one containing activity against methyl dodecanoate.

Information on the chemistry of the enzyme is limited to the observation that lipolytic activity was inhibited by DFP. Cysteine, EDTA, and PCMB had no effect on the activity. The enzyme therefore, appears to belong to the group of serine enzymes, but it differs from most other lipases in that the optimal pH was about 5.5.

Several natural and monoacid triglycerides have been tested as substrates for the lipase. While all were digested, maximum activity was observed with tridecanoin and trioctanoin. Short-chain methyl esters, 4:0, 6:0, and 8:0, were hardly hydrolyzed, while some of the longer esters were. Information on positional specificity and sequence of digestion has apparently not been obtained. The optimal temperature for lipase activity was 50°C.

Few "acid" lipases have been discovered and investigated. A lipase operative at this pH and having serine at the active site should receive attention.

6. *Puccinia graminis tritici*

This fungus, which causes plant rusts, contains relatively large quantities of lipids in the uredospores. It is believed that most of the energy requirements of spore germination are provided by the lipids (Daly *et al.*, 1967).

Knoche and Horner (1970) investigated the lipase of the uredospores and partially purified the enzyme. The spores were disintegrated, the material centrifuged several times in phosphate buffer, and the 5000 g precipitate, representing a 3.8-fold purification, was used for further studies.

Assays were conducted for 30 minutes at 15°C and pH 6.7. With triolein as the substrate, both 1,2- and 1,3-diacylglycerols and a trace of mono-acylglycerol were observed. The enzyme did not appear to differentiate between the 1- or 2-positions of acylglycerols. The isomeric type of mono-acylglycerol was not determined. Several other substrates were exposed to the lipase and the following relative activities were observed: tricaprylin, 0.70: triolein, 0.58; tripalmitin, 0.39; methyl oleate, 0.29; dipalmitin, 0.11; and monopalmitin, 0.02. The results do not yield much information about specificity of the enzyme. Compounds with high melting points such as tri, di-, and monopalmitin would not be expected to be hydrolyzed as rapidly as triolein or trihexanoin. The hydrolysis of methyl oleate is un-expected, although pancreatic lipase will also digest this ester but only very slowly (<0.05). Knoche and Horner (1970) suspected that an esterase might be present, but the addition of an esterase inhibitor, phenylmethyl-sulfonyl fluoride, did not affect the activity toward triolein.

The optimum pH of the enzyme was 6.7 and the optimum temperature about 15°C. The enzyme exhibited considerable activity at 0°C, but was reduced to 20% of its maximal activity at 30°C. PCMB ($10^{-5}\,M$) reduced activity by 20%. Complete inhibition, caused by EDTA ($10^{-2}\,M$), was not reversed by the addition of Mg^{2+}. Calcium was slightly inhibitory. Sodium chloride was not added to the assay mixtures and albumin did not affect activity.

7. Candida cylindracea

This enzyme is commercially available as an impure preparation (Anonymous, a,b), and has been commended for a variety of com-mercial applications. Monsanto has at one time provided the enzyme in bound form, linked to an inert copolymer of ethylene and maleic anhydride.

Tomizuka *et al.* (1966a) purified a commercially prepared powder by extraction with water, centrifugation of the aqueous extract, and precipita-tion with ammonium sulfate. The precipitate was dissolved in water and dialyzed against water. Sodium deoxycholate (0.5%) was added to the solution which was then centrifuged; the supernate was brought to 75% ethyl alcohol and ethyl ether by addition of a 1:1 mixture of the two sol-vents. The resulting precipitate was dissolved in buffer (pH 2.7), the solu-tion was centrifuged, and the supernatant was eluted through SE-Sephadex with a sodium chloride solution. The eluate was concentrated and fractionated with Sephadex G-100. A concentration of 33-fold with a yield of 18% was obtained. The preparation was homogeneous by gel electrophoresis. Data on purification are given in Table IV-40.

Tomizuka *et al.* (1966b) determined the amino acid composition of the purified enzyme. The results, listed in Table IV-41, are given as residues per 1000 residues; a molecular weight of 100,000–120,000 was estimated from physical constants. The enzyme has a large number of hydrophilic residues and considerable leucine. Xylose and mannose were found in the enzyme; the lipase is, therefore, a glycoprotein.

TABLE IV-40

Purification of *Candida cylindracea* **Lipase**[a]

Procedure	Yield (%)	Specific activity
Water extraction	100	34
Ammonium sulfate precipitate	61	47
Sodium deoxycholate treatment	59	51
Ethanol–ether precipitate	45	59
SE-Sephadex	28	628
Sephadex G-100	18	1140

[a] Tomizuka *et al.* (1966a).

TABLE IV-41

Amino Acid Composition of the Lipase from
Candida cylindracea[a]

Amino acid	Residues/ 1000 residues[b]	Amino acid	Residues/ 1000 residues
Ala	89	Lys	34
Amide	68	Met	26
Arg	23	Phe	55
Asx + Glx	123	Pro	60
½Cys	6	Ser	74
Gly	106	Thr	64
His	9	Trp	10
Ile	55	Tyr	36
Leu	109	Val	52

[a] Tomizuka *et al.* (1966b).
[b] Molecular weight from physical constants was 100,000–120,000.

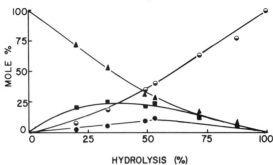

HYDROLYSIS (%)

Fig. IV-35 Hydrolysis of olive oil by the lipase of *Candida cylindracea*. ▲, ■, ●, and ◐ correspond to triglyceride, diglyceride, monoglyceride, and glycerol. From Benzonana and Esposito (1971).

The lipase completely hydrolyzed several natural oils in periods of time ranging from 8 to 20 hours (Anonymous, a,b), but this in itself cannot be taken as proof of random hydrolysis, because acyl migration of 2-monoglycerides to the 1-isomer and of 1,2-diglycerides to the 1,3-isomer would occur during the long incubations. However, Benzonana and Esposito (1971) proved that the enzyme lacked positional specificity. In Fig. IV-35, the time course of lipolysis of olive oil is presented. Glycerol appears almost immediately, in contrast to pancreatic lipolysis. Mono- and diglycerides remained relatively constant; this suggests comparable rates of hydrolysis of all three esters.

Benzonana and Esposito (1971) also studied the specificity of the enzyme with synthetic substrates. The results for the FFA (Table IV-42)

TABLE IV-42

Fatty Acid Composition of FFA, Monoglycerides and Diglycerides from Hydrolysis of Synthetic Substrates by *Candida cylindracea*[a]

Substrate and fatty acid	FFA (mole %)	Monoglyceride (mole %)	Diglyceride (mole %)
16:0–16:0–18:1			
18:1	34	13	8
16:0	67	87	92
18:1–18:1–16:0			
18:1	64	52	61
16:0	36	48	39
18:1–16:0–18:1			
18:1	68	45	53
16:0	32	55	47

[a] Benzonana and Esposito (1971).

confirm the almost random hydrolysis. The di- and monoglycerides, however, do not conform. The theoretical fatty acid compositions for strictly random hydrolysis would be 66.7% for the predominant acid and 33.3% for the minor acid in all cases. Therefore, regardless of position, both mono- and diglycerides were enriched in 16:0. This may indicate selective hydrolysis of diglycerides rather than fatty acid specificity. To confirm the lack of positional specificity, 1,3-dihexadecyl-2-oleoylglycerol was used as a substrate; the oleate was hydrolyzed.

When methyl esters of several fatty acids were hydrolyzed, the relative rates of digestion were: 18:1, 100; 12:0, 85.8; 16:0, 78.1; 14:0, 59.2; and 18:0, 47.0 (Benzonana and Esposito, 1971). If there is some specificity for 18:1, it is slight.

The original assays for activity were done in phosphate buffer at pH 7.0 and 37°C with olive oil as a substrate (Anonymous, a,b). The pH optimum of the purified preparation, with polyvinyl alcohol as an emulsifier, was 7.2; without the stabilizer and with shaking, it was 5.2 (Benzonana and Esposito, 1971); the temperature optimum was 45°C.

8. *Candida paralipolytica*

Ota and Yamada (1966a) grew the mold at 30°C in a medium consisting mainly of soluble starch, defatted soybean meal, and 0.5% olive oil, adjusted to pH 6.2. After clarification of the medium by centrifugation, acetone was added to 60% (v/v) and the precipitate obtained by centrifugation. Further purification was achieved by a series of ammonium sulfate precipitations and centrifugation. Specific activities were not given. Ota *et al.* (1970) obtained a 132-fold purification over the crude acetone precipitate with a yield of 32%, with only one band observable by disk electrophoresis.

Ota and Yamada (1966a) found that bile salts, specifically sodium taurocholate, markedly increased the activity of the enzyme toward an emulsion of olive oil stabilized with polyvinyl alcohol. Several other bile salts were found to be activating. Activity was inhibited by several cationic and nonionic surfactants and slightly by Ca^{2++}. However, when dispersion of the substrate was achieved by shaking without polyvinyl alcohol, Ca^{2+} was required as an activator (Ota and Yamada, 1966b, 1967). The enzyme was assayed using a polyvinyl alcohol-stabilized emulsion at pH 7.0 (phosphate buffer) and 37°C.

In later papers, Ota *et al.* (1968a,b) investigated the lipid requirements for lipase production. When added to the culture medium prior to inoculation, these lipids produced the greatest lipase activity: castor oil, triolein, Span 20, and cholesterol. The last, at a level of 0.1%, was an effective inducer of lipase production. Kalle *et al.* (1972) noted that sorbitan mono-oleate also induced lipase synthesis.

9. Penicillium roqueforti

The lipase of *P. roqueforti* has been extensively investigated, because it is to a large extent responsible for the distinctive flavor of Roquefort and related cheeses (Eitenmeiler *et al.*, 1970). The mold spores are deliberately introduced, developing to blue-green growth following natural fissures and punched holes in the body of the cheese. The mold can be easily isolated from the cheese or the cultures purchased. It grows well on Czapek agar slants. Partial purification of the lipase, about sevenfold, has been achieved by precipitation from the filtered culture medium with ammonium sulfate followed by centrifugation (Eitenmeiler *et al.*, 1970).

Earlier work, with simple triglycerides as substrates, indicated a specificity for substrates with short-chain fatty acids (Shipe, 1951). However, the pitfalls of drawing conclusions concerning specificity from these types of studies have been discussed in Chapter IV, Section I, E. Preferential release of short-chain acids from butterfat was reported by Wilcox *et al.* (1955), but their data do not support this interpretation. Shipe (1951), with an equimolar mixture of tributyrin and tricaproin as the substrate, observed three times more 4:0 than 6:0 in the FFA.

Alford and Pierce (1961) digested corn oil, lard, and coconut oil at different incubation temperatures, 35°C, −7°C, and −18°C, with the lipase. As might be expected, the proportion of saturated acids in the free fatty acid fraction decreased at the lower temperatures. In contradiction to the presumed short-chain fatty acid specificity of the lipase, 8:0 and 10:0 were not preferentially released from coconut oil. Later, Alford *et al.* (1964) reported that the lipase attacked primarily the 1-position. Although defined synthetic substrates were used, data were not presented.

The optimum temperature for lipolysis was 37°C and the optimum pH 8.0 (Eitenmeiler *et al.*, 1970). However, the temperature range of activity was broad, −18°C to 35°C (Alford and Pierce, 1961). Ca^{2+} and sodium chloride were not stimulatory. The enzyme was inactivated within 10 minutes at 50°C (Alford and Pierce, 1961).

10. Torulopsis ernobii

The extracellular lipase of this organism has been purified 230-fold over the culture medium by chromatography on DEAE-cellulose and Sephadex (Motai *et al.*, 1966; Ichishima *et al.*, 1968). The preparation appeared to be homogeneous judging by ultracentrifugation and electrophoretic analysis. Its isoelectric point was at pH 2.95, its optimal activity at pH

6.5, its molecular weight 43,000. The single chain had glutamic acid as N-terminal, threonine as C-terminal residue, three disulfide bridges and no free sulfhydryl groups. It was inactivated by cyanogen bromide, iodoacetic acid, *N*-ethylmaleimide, and photooxidation, and therefore seemed to require methionine and histidine as essential groups; serine reagents were not tested.

B. Bacteria

1. *Leptospirae*

The leptospirae require fatty acids for energy and carbon since they are unable to synthesize long-chain fatty acids (Chorvath and Benzonana, 1971a,b). Consequently, exocellular lipolytic activity is required to provide the necessary free fatty acids from the environment. Patel *et al.* (1964) detected lipolytic activity in all of the fourteen strains of leptospirae studied; the three virulent strains had the highest activity. (The pathogenic leptospirae cause various forms of leptospiroses in man and animals, including Weil's disease, caused by *Leptospira icterohemorragiae.*) Later, Chorvath and Bakoss (1972) could not detect any difference in lipase activity between virulent and avirulent strains.

Patel *et al.* (1964) concentrated the enzyme produced by *L. pomona* LC-12 by centrifugation of the culture medium and subsequent treatment with alcohol and acetone. Berg *et al.* (1969) used cesium chloride gradient centrifugation of dialyzed and lyophilized culture medium of *L. pomona* LC-12 to concentrate the enzyme. The active fractions obtained were pooled and treated with alcohol. Purification of the lyophilized culture medium was also attempted by elution through Sephadex-200.

A 140-fold purification of lipase activity from *L. biflexa* (Patoc I) has been achieved by sequential ethanol precipitation, ultracentrifugation, and Sephadex G-200 chromatography in deoxycholate solution (Chorvath and Fried, 1970a,b). The bile salt was used in the anticipation that it would disaggregate enzyme lipid complexes and improve the purification. Two peaks were observed after Sephadex chromatography; a fast, high molecular weight peak eluted with the void volume and a slow lipase peak with a molecular weight of 40,000 which increased in content after exposure to deoxycholate.

The purified enzyme, as a dry powder, was quite stable at low temperatures (Patel *et al.,* 1964). The pH of optimal activity, in phosphate buffer, was 7.0, the optimal temperature was 30°C. Activity was not affected by sulfhydryl blocking agents. Another lipase preparation exhibited an optimal activity at pH 8.5, in common with most lipolytic enzymes (Chorvath and

Benzonana, 1971a,b). Bile salts were stimulatory below the critical micelle concentration; above, they were inhibitory. Sodium chloride and calcium chloride also stimulated the enzyme.

The lipase from *L. pomona* (Patel *et al.*, 1964) digested tributyrin most rapidly, but long-chain triglycerides were also hydrolyzed. Methyl 18:1 and 18:2 were poor substrates. *Leptospira pomona* DM$_2$H hydrolyzed cocoa butter and lard largely at the primary positions (Berner and Hammond, 1970b), although the positional specificity was not as exclusive as in pancreatic lipase. The free fatty acids obtained from cocoa butter, which consists mostly of *rac*-16:0–18:1–18:0, contained 27% 18:1. Those from lard, in which the secondary ester is occupied almost completely by 16:0 contained 16% of this acid. Lipases elaborated by *L. biflexa* Patoc I, a saprophytic strain, and *L. copenhageni* BP-3, a pathogenic strain, have also been examined for positional specificity. Purified olive oil was hydrolyzed by preparations from both strains and the resultant diglycerides were found to be primarily the 1,2 type as determined thin-layer chromatography (Chorvath and Benzonana, 1971a,b). This pattern of hydrolysis is indicative of specificity for primary esters.

Lipases from *L. biflexa* and *L. copenhageni* hydrolyzed the triglycerides associated with porcine very low density lipoproteins (Chorvath and Fried, 1970a,b; Fried and Chorvath, 1971). Low density lipoproteins were digested to a much lesser extent. Sodium chloride was required for the lipolysis of VLDL.

2. Pseudomonads

Many of these ubiquitous gram-negative bacteria secrete lipases. They have attained economic importance because most of them are psychrophilic (cold loving) and because modern techniques for handling bulk milk and other foods involve long periods of low-temperature storage (Lawrence, 1967). For example, growth of *Pseudomonas fragi* in milk resulted in a distinct off-flavor in cream and butter derived from the milk (Nashif and Nelson, 1953c).

Pseudomonads are readily isolated from water and refrigerated foods (Bjorklund, 1970). Lipolytic activity can be detected by the tributyrin agar double-layer method of Fryer *et al.* (1967). In most of the earlier investigations, cells were removed from the culture medium by centrifugation and the supernate used as the source of enzyme (Nashif and Nelson, 1953a,b). A few attempts have been made at purifications. Alford and Pierce (1963) tested forty-one culture media for their ability to support growth and lipase production. In general, growth was good in all media, but maximum yields of cells did not occur in media that produced high lipolytic activity. The authors then developed a synthetic medium in which

TABLE IV-43

Purification of a *Pseudomonas fragi* **Lipase**[a]

Operation	Total protein (mg)	Total lipase activity (units)	Yield (%)	Specific activity	Purification
Culture Supernatant	320	2890	100	9	1
Ultrafiltration, dialysis, lyophilization	63	1568	54	25	3
Ammonium sulfate fractionation	3.5	850	30	240	27
Chromatography on DEAE-Sephadex	0.55	510	18	925	103

[a] Mencher and Alford (1967).

growth of *P. fragi* produced as much lipolytic activity as in the best commercial medium, Case peptone. The sequence of purification of *P. fragi* lipase (Mencher and Alford, 1967) is shown in Table IV-43. Purification was checked with disk electrophoresis, which detected two bands in the fraction from the DEAE column. Centrifugation of this fraction resulted in two fractions, the first designated H for heavy and second L for light, both with specific activities of about 250. Upon electrophoresis, the H lipase separated into two bands, one corresponding to the location of the L band of activity. The two lipases had the same apparent K_m values for tributyrin (0.8 mM), the same optimum temperature (35°C) and pH (7.5), and the same specificities for synthetic substances. Obviously, neither lipase was completely pure, since the DEAE-cellulose fraction had a specific activity two times higher than either fraction.

Lawrence *et al.* (1967b) passed the ammonium sulfate precipitate of a *P. fragi* (NCDO 752) medium through a column of Sephadex G-100 to remove low molecular weight contaminants, and then eluted the active fractions through Sephadex G-200. Specific activities were not given.

Lu and Liska (1969a,b) also purified an extracellular lipase from a strain of *P. fragi*. After 4–5 days of growth at 20°C in a buffered (pH 7.0) tryptone broth, the cells were removed by centrifugation and a lipase concentrated with ammonium sulfate. The precipitate was dissolved in water, then reprecipitated with acetone. Additional purification to the extent of 100-fold and 1.8% yield was obtained with Sephadex G-200 and DEAE-cellulose. The final preparation contained only one band when examined by disk electrophoresis and had a specific activity of 16.

A lipase concentration of thirty-five times was obtained from a strain of *Pseudomonas aeruginosa* by Finkelstein *et al.* (1970) by passage of the lyophilized culture medium supernate through Sephadex G-200. The microorganism had been found by accidental contamination of an agar plate containing coconut oil.

Neither the lipase of Lawrence *et al.* (1967b) nor the growth of the parent *P. fragi* were inhibited by DNP or DFP. The enzyme studied by Lu and Liska (1969b) was severely inhibited by PCMB, but only moderately by *N*-ethylmaleimide and DFP.

Two studies on substrate specificities have been performed with defined synthetic substrates (Alford *et al.*, 1964; Mencher and Alford, 1967). The data given in Table IV-44 are the compositions of the fatty acids hydrolyzed from the synthetic triglycerides by *P. fragi* 1964. A nearly absolute specificity for the primary esters is apparent. The two observed departures from this specificity may have been due to the relative indigestibility of the intermediate diacyl glycerols. For example, one of the diacyl glycerols resulting from the hydrolysis of 18:1–18:0–18:0 would be OH–18:0–18:0, which because of its high melting point, would not be readily digested. This phenomenon has been observed with pancreatic lipase and the same substrate (Jensen, 1971). *P. fluorescens* and *P. geniculata* also had specificity for primary esters (Alford *et al.*, 1964).

The lipase from another strain of *P. fragi,* purified by Mencher and Alford (1967), was similarly specific for the primary esters of synthetic

TABLE IV-44

Lipolysis of Synthetic Triglycerides by a Lipase from *Pseudomonas fragi*[a]

Substrate	Free fatty acids (wt %)		
	16:0	18:0	18:1
rac-18:1–18:0–18:1	—	2	98
18:0–18:1–18:0	—	97	3
18:1–16:0–18:1	1	—	99
16:0–18:1–16:0	96	—	4
16:0–18:1–18:0	55	45	—
18:0–16:0–18:0	6	94	—
16:0–18:0–16:0	99	1	—
18:1–18:0–18:0	—	33	67
18:0–18:1–18:1	—	15	85
18:1–16:0–16:0	47	—	53
16:0–18:1–18:1	48	—	52

[a] Alford *et al.* (1964).

triacyl glycerols. Again the acids released from the substrates were mostly those in the primary positions. The purified H and L forms of the enzyme both exhibited the same specificity.

The lipase of *P. aeruginosa* studied by Finkelstein *et al.* (1970) hydrolyzed triglycerides, Lipomul, Ediol, monoolein, and dipalmitin. Although the isomers of the last two substrates were not given, they were probably the 1- and 1,3-, respectively, as these are the most stable and easiest to synthesize. The lipase also hydrolyzed the acylglycerols in lipoproteins. As far as can be judged at the present time, the pseudomonal lipases are similar in many respects to pancreatic and *Rhizopus* lipases.

3. Staphylococci

Staphylococci have been known for some time to produce lipases (Lawrence, 1967). Davies (1954) screened one thousand strains of eighteen bacterial genera for exocellular lipases finding the enzyme only among the staphylococci. Of the eighty-eight strains studied, seventy-one had lipolytic activity. Interest in the enzymes has been stimulated by observations that pathogenic staphylococci possess lipolytic activity. The opacity produced in media containing egg yolk by some pathogenic staphylococci was associated with the elaboration of a lipase capable of hydrolyzing unsaturated fatty acids from peanut oil (Gillespie and Alder, 1952). An increase in free fatty acids in human plasma is caused by some coagulase-positive strains of *Staphylococcus aureus* (Weld and O'Leary, 1963). The apparent association between lipolytic activity and pathogenicity was not supported by the work of Rosendal and Bulow (1965) who noted that strains of *S. aureus* which did not produce a Tween-splitting enzyme were more virulent than those which did. The Tween-hydrolyzing enzyme, however, might not have been identical with the lipase.

Shah and Wilson (1963) partially purified *S. aureus* "egg yolk factor" from cell-free culture filtrates by acid precipitation, ethanol fractionation, and zone electrophoresis. The method had been used by Blobel *et al.* (1961) to purify staphylocoagulase, with which tributyrinase activity was associated (Drummond and Tager, 1959a,b). The product resulting from zone electrophoresis was unstable; therefore, the ethanol-fractionated preparation was used in the experiments described later. This fraction had three bands, and the electrophoretic preparation two, when tested by gel diffusion precipitation analysis.

O'Leary and Weld (1964) assayed staphylococcal lipolytic activity with plasma agar plates. If activity was present, surface lipid deposits were seen. Searching for a more readily available substrate than plasma, O'Leary and Weld (1964) tried first egg yolk, then a low density egg yolk fraction, obtaining more satisfactory results with the latter. The supernate obtained by centrifugation (15,000 g) of diluted egg yolk provided the substrate. Although not reported, the supernate must have contained all of the non-polar as well as most of the polar lipids. Shah and Wilson (1963) had also obtained a substrate by centrifugation and proceeded further to prepare the specific substrate, lipovitellenin.

O'Leary and Weld (1964) grew a strain of *S. aureus* capable of lipolyzing plasma lipids on plates of Trypticase soy agar. They removed the single central colony from the agar, minced the remaining agar and added this material to a flask containing phosphate buffer (pH 7.3). At least twenty-five plates were processed in this fashion for each preparation of enzyme. The purified enzyme was then obtained by precipitation with ammonium sulfate, dialysis, precipitation with ethanol, and successive passage through columns of Sephadex G-100 and G-200, then DEAE-cellulose. The material from the latter column contained one major band and one small trailing spot by electrophoresis on cellulose acetate strips. The major band had lipolytic activity and the spot had none. A 225-fold increase in lipolytic activity based on a turbidimetric assay was obtained.

Lawrence *et al.* (1967a,b) purified an extracellular lipase from *Micrococcus freudenreichii* by precipitation with ammonium sulfate and elution through columns of Sephadex G-100 and G-200. A single peak of lipolytic activity was observed in the eluate from the G-100 column, but two were noted after passage through the G-200 column. The extent of purification was not given, but the purified material had one band by electrophoresis, and hydrolyzed short-chain esters of α-naphthol. This may or may not indicate the presence of a lipase.

Vadehra and Harmon (1967a) achieved a factor of purification of 450 (0.7% yield) with another staphylococcal lipase using procedures similar to those above. The addition of alcohol to the cell-free supernate completely precipitated lipolytic activity. A second alcohol precipitation at pH 4.3 yielded a highly active preparation with a specific activity of 200. Two peaks were eluted from columns of Sephadex G-200; most of the activity was concentrated in the first peak.

Renshaw and San Clemente (1967) purified an extracellular lipase from a selected strain of *S. aureus* in a similar manner, with a purification of about 2000-fold (above the cell-free supernate) and a 39% yield. This strain of *S. aureus* also produced coagulase, phosphatase, a hemolysin, fibrinolysin, and deoxyribonuclease. None of these activities were present in the final preparation. However, a trace impurity was detected by both gel electrophoresis and gel immunodiffusion. Attempts to detect isozymes were not successful.

Troller and Bozeman (1970) purified the lipase from a strain of coagulase negative staphylococcus isolated from human skin. Using mainly Sephadex G-200 as the mode of purification, a concentrate with a specific activity of 6.4 resulted, with about twelvefold concentration.

The *S. aureus* egg yolk factor of Shah and Wilson (1963) did not possess any lecithinase, lipoprotein lipase, or "triglyceridase" activity. The substrate from which opacity in egg yolk was produced was identified as lipovitellenin, a low-density fraction. Lipolysis produced mostly unsaturated fatty acids; this suggested a specificity for these acids. The lipase purified by O'Leary and Weld (1964) released largely 18:1 from human plasma lipoproteins and diluted egg yolk. Tri- and diolein were also hydrolyzed, but solvent-extracted egg plasma and several synthetic phosphatidylcholines and phosphatidylethanolamines were not.

Alford and Pierce (1961) found that the unpurified lipase from a strain of *S. au*   ʼʼ hydrolyzed the fatty acids from several fats in about the same proportio   a   present in the intact substrates. Later, Alford *et al.* (1964) lipolyzed several synthetic triglycerides with the lipase. The results (Table IV-45) suggest that all positions of the substrates were attacked although not at equal rates. There was no clear-cut specificity for 18:1.

Vadehra and Harmon (1965) studied the lipolysis of milk fat by a lipase from *S. aureus*. Some of the results indicated a preference for myristic acid, in that the concentration in the free fatty acids was about three times that in the original fat. The method employed to esterify and analyze the fatty acids led to loss of the short-chain fatty acids present in milk fat; these were, therefore, not listed by the authors. Vadehra and Harmon (1967b) digested the synthetic substrate, *rac*-16:0–18:1–18:0, with their purified staphylococcal lipase and found 40%, 29%, and 39% each of 16:0, 18:1, and 18:0 in the free fatty acids. Thus the enzyme appeared

TABLE IV-45

**Composition of the Fatty Acids Hydrolyzed from
Synthetic Triacyl Glycerols by a**
Staphylococcus aureus **Lipase**[a]

Triacyl glycerol	Free fatty acids (weight %)		
	16:0	18:0	18:1
18:1–18:0–18:1	—	25	75
18:0–18:1–18:0	—	62	38
18:1–16:0–18:1	30	—	70
16:0–18:1–16:0	78	—	22
16:0–18:1–18:0	49	32	19

[a] Alford *et al.* (1964).

to hydrolyze all three positions, but the primary esters somewhat faster than the secondary.

Renshaw and San Clemente (1967) found that their highly purified preparation (200 ×) digested the lipids in egg yolk and human plasma as well as a number of natural oils. The enzyme purified by Troller and Bozeman (1970) hydrolyzed several natural oils and synthetic triglycerides but not methyl esters of short or long-chain acids, saturated triglycerides with acyl chains of lengths greater than ten carbons, or cholesterol esters. The concentrations of di- and monoacyl glycerols in the lipolysates were quite small relative to the free fatty acids (Vadehra and Harmon, 1965, 1967b).

All of the enzymes were maximally active in the pH range of 7.5–8.5 and at 37°C. Alford *et al.* (1964) did not observe inhibition of *S. aureus* lipase with DFP (10^{-4} *M*), while Renshaw and San Clemente (1967) noted inhibition with high levels of Parathion and eserine (concentrations not given). Lawrence *et al.* (1967a,b), inhibited the "lipase" (α-naphthol esters) from their strain of *M. freudenreichii* with DFP (2×10^{-3} *M*) and DNP (10^{-4} *M*). The lipase of Troller and Bozeman (1970) was inhibited by PCMB. Some of the enzymes required Ca^{2+} as a fatty acid acceptor (Shah and Wilson, 1965; Renshaw and San Clemente, 1967), and Ca^{2+} was present in most of the assay systems.

4. Corynebacterium acne

Lipases produced by this microorganism growing on skin have been implicated as a major factor in the etiology of acne (Reisner *et al.*, 1968; Freinkel and Shen, 1969; Marples *et al.*, 1971). The free fatty acids resulting from hydrolysis of skin lipids are apparently one of the causes of the lesions (Rosenberg, 1969). The enzyme, which also hydrolyzes tributyrin, tricaprylin, and trilaurin, has been partially purified by ammonium sulfate precipitation (Weaber *et al.*, 1971; Puhvel and Reisner, 1972). Purification factors were not given.

Hassing (1971) has accomplished a more nearly complete purification of the enzyme (175–200-fold) using ammonium sulfate precipitation of the culture supernate followed by Sephadex G-100 chromatography. The enzyme was not homogeneous when tested by gel electrophoresis and isolelectric focusing. Optimal activity was at pH 7.5–9.0 with tributyrin.

With trilaurin as a substrate (Kellum and Strangfeld, 1969; Kellum *et al.*, 1970; Weaber *et al.*, 1971) 1,3- and 1,2-diacyl glycerols as well as free fatty acids were produced, but the length of digestion, 4 hours, was long enough for acyl migration to occur. The 1,3 isomer, therefore, could have arisen by acyl migration from the 1,2 isomer, rather than by digestion of the 2-position. Thus, it is not clear from these data if the enzyme acted at all three positions of trilaurin. The point was settled by Hassing (1971) who obtained almost random hydrolysis of *rac*-16:0–18:1–18:1 (FFA: 16:0, 30 mole %; 18:1, 70 mole %) and of *rac*-18:1–16:0–18:1 (FFA: 16:0, 21 mole %; 18:1, 79 mole %). In addition, butane-2,3-diol dioleate was hydrolyzed. Phospholipids and cholesterol linoleate were not digested.

The less completely purified preparations were stimulated by Ca^{2+} and sodium taurocholate (Weaber *et al.*, 1971; Puhvel and Reisner, 1972). The Ca^{2+} activated lipase was completely inhibited by 10^{-4} *M* tetracycline. This antibiotic, applied topically and systemically, is successfully employed in the treatment of acne.

5. *Propionibacterium shermanii*

This microorganism is one of several used in the manufacture of Swiss cheese, which has a relatively high free fatty acid content that is ascribed mainly to the action of lipases (Oterholm *et al.*, 1970a). From a culture obtained from C. Hansen's Laboratory, Inc., Milwaukee, Wisconsin, the cells, obtained by centrifugation of the culture medium, were disrupted with a French press (Oterholm *et al.*, 1970b). Cell-free extracts were obtained by centrifugation. Activity, detected by pH-stat titration of acids liberated from tributyrin, was nil in the culture supernate, but present to the extent of 203 units/mg protein in the cell-free extract from the disrupted cells. Thus, in contrast to most of the microbial enzymes discussed here, this lipase is intracellular.

The enzyme has been purified 141-fold with 44% yield by ammonium sulfate precipitation and elution through columns of Sephadex G-25 and DEAE-cellulose (Oterholm *et al.*, 1970b). Maximum activity was observed at pH 7.2 and 47°C with tributyrin as a substrate. Simple triacyl glycerols were hydrolyzed at the following relative rates: triacetin, 24; tripropionin, 100; tributyrin, 83; trihexanoin, 68; trioctanoin, 17; tridecanoin, nil; and trilaurin, nil. Neither triolein nor milk fat were tested. Some activity toward short-chain monoesters was also noted.

6. *Achromobacter lipolyticum*

This microorganism produced an extracellular lipase at 21°C and pH 7.0 in casitone broth (Khan *et al.*, 1967). Addition of corn oil, olive oil, or milk fat and slow stirring increased the lipase activity about three times. The enzyme hydrolyzed triolein more rapidly than tributyrin; this suggested a possible specificity for unsaturated fatty acids. In confirmation, with milk fat containing 23.0% 18:1, 2.2% 18:2, and 1.0% 18:3 as a substrate the lipase hydrolyzed 47.4% 18:1, 6.5% 18:2, and 8.1% 18:3. For comparison, *G. candidum* lipase hydrolyzed 61.8% 18:1 from milk fat containing 19.9% of the acid. Apparently *A. lipolyticum* lipase did exhibit specificity for unsaturated fatty acids, and this aspect should be investigated.

7. Mycoplasma

Eight strains of *Mycoplasma* investigated by Rottem and Razin (1964) hydrolyzed tributyrin. The saprophytic strains of *M. laidlawii* had less lipolytic activity than parasitic strains. The lipase from *M. gallisepticum* was partially purified, about 25-fold and with 25% yield, by ammonium sulfate precipitation and chromatography on DEAE-Sephadex A-50. The enzyme hydrolyzed tributyrin more rapidly than trilaurin or triolein, and did not digest cholesterol esters. Optimal temperature and pH for activity were 37°C and 7.5–8.0.

8. Anaerovibrio lipolytica

Anaerovibrio lipolytica is an anaerobic rumen microorganism that produces an extracellular lipase during exponential growth (Henderson, 1971). The lipase was partially purified, 38-fold and with 4.9% yield, by ammonium sulfate precipitation and elution through Sephadex G-200. The purest preparations contained two bands by electrophoresis and considerable amounts of nucleic acids. Activity was highest at pH 7.4 and 20–22°C and was enhanced by Ca^{2+} or Ba^{2+}. Trilaurin was hydrolyzed more rapidly than triolein, while diolein, presumably the 1,3 isomer, was digested faster than either triacylglycerol.

C. Conclusions

Many, perhaps most, bacteria and fungi produce lipases. The enzymes are usually extracellular and their biological function is not obvious. Table IV-46 gives a summary of the properties of the better known microbial lipases. Most of the enzymes appear to be serine enzymes that are inhibited by DFP. The *Rhizopus* enzyme is DFP resistant, but it is exclusively specific for the primary esters of glycerol, just like pancreatic lipase which also resists DFP. Apparently, in both cases steric hindrance prevents the approach of secondary esters and organophosphate. The lipase of *Pseudomonas* species, also specific for primary esters, appears to be at least partially resistant to DFP. These observations support the hypothesis that the degree of steric hindrance around the reactive site of enzymes controls the positional specificity toward triglycerides. It can be assumed, on this basis, tha the *Mucor* lipase will hydrolyze secondary as well as primary esters.

The unique specificity of the *Geotrichum* lipase has been firmly established. There are many reports in the older literature claiming preferential hydrolysis of unsaturated esters. Among the enzymes listed in Table IV-46,

TABLE IV-46
Lipases of Microorganisms

Microorganism	Substrate specificity: Primary (1) or secondary (2) ester	Substrate specificity: Fatty acids	pH optimum	Inhibition[a]	Remarks
Leptospira	1 (2)	All	8	PCMB*	—
Pseudomonas	1	All	6	PCMB†	—
	—	—	—	DFP†	—
Staphylococcus	1,2	All	—	DFP†	—
	—	(Unsaturated)	8	PCMB	—
Geotrichum	1,2	*cis*-Δ9 unsaturated	8–9	DFP	Commercially available
Rhizopus arrhizus	1	All	8	DFP*	Enzyme isolated, a glycoprotein; *R. delemar* commercially available
Aspergillus niger	1,2	—	5.6	—	Commercially available
Mucor	?	Long-chain(?)	5.5	DFP	—
Puccinia	1,2	All	6.7	EDTA	—
Candida	1,2	All	7.2–8	?	*C. cylindracea* commercially available

[a] * = no inhibition; † = inhibition partial, or not in all species investigated.

the staphylococcal lipases may have some such preference; so may the *Achromobacter* lipase. More probably, however, the alleged specificity can, in these cases, be ascribed to solubility or orientation effects at the oil-water interface, and nowhere is the specificity for *cis*-Δ9 acids as pronounced as in *Geotrichum*. This lipolytic enzyme, then, is the only one discovered so far to which a fatty acid specificity can be truly assigned, although even here the specificity is not absolute. Two possibilities could be considered for the catalytic mechanism. The fatty acid could be oriented in the enzyme–substrate complex to fit in a matrix about nine carbons long, and the cis double bond would be needed to bend the rest of the chain away from the enzyme; or, there is actual bonding of the *cis*-Δ-9 double bond to a site on the enzyme, by residues yet to be identified. Both mechanisms might work in conjunction, but mechanism I cannot alone be responsible for the specificity since chain-lengths shorter than C_9 are poor substrates. On the other hand, the hypothesis of a binding of the *cis*-9 double bond, as an interaction of the π electrons of the substrate to the enzyme, would seem very speculative and unprecedented.

The example of substrate–enzyme interaction at C_9–C_{10} does not fit our general hypothesis, presented in the Chapter VIII Synopsis, that lipolytic reactions in general proceed without involvement of the paraffinic or olefinic chains of the substrates; these remain buried in the nonaqueous phase. The reaction of the *Geotrichum* lipase must be declared an exception.

Some, and perhaps all, of the microbial lipases are glycoproteins. The enzyme from *R. arrhizus* contains fifteen to sixteen carbohydrate residues, about 3% of all the protein residues and three times more than found in pancreatic lipase. It has been suggested that the carbohydrate may assist in the passing of the lipase through the cell wall of the organism (Sémériva *et al.*, 1969). A recent hypothesis (Brockerhoff, 1973) assigns to the carbohydrate the role of stabilizing the orientation of the lipases at the oil–water interface.

X. MONOGLYCERIDE LIPASES

Hydrolytic activity against monoglycerides is found in many, perhaps all, tissues. This is not surprising since monoglycerides are partially water soluble and can probably be attacked by nearly any esterolytic enzyme and not only by specifically lipolytic enzymes. Monoglyceride hydrolases are similar, in this regard, to lysophospholipases, and it can be assumed that these two activities, where they are found together, may often be expressions of the same enzyme, namely, an unspecific esterase. The separation of the two activities is, therefore, usually quite artificial and not based on any known biochemical or functional properties of the proteins. It is retained in this book only because it allows the classification of the literature on these enzymes according to the titles of the papers. In general, it can be assumed that an enzyme is described as a "monoglyceride lipase" (or monoglyceride hydrolase) if the authors began their study by using monoglycerides as substrates, as a "lysophospholipase" if they began by using lysophosphatidylcholine.

If it is difficult or impossible to distinguish monoglyceride hydrolase activity from the unspecific esterolytic activities of most tissues, it is also difficult to distinguish the enzyme from triglyceride and diglyceride hydrolase activity in those tissues (e.g., adipose) which contain lipases.

It has not been conclusively demonstrated that such enzymes as hormone-sensitive lipase and lipoprotein lipase can hydrolyze monoglycerides, but it is known that pancreatic lipase can attack 1-monoglycerides, and it is likely that other, less position-specific lipases can attack 2-monoglycerides. There is no reason to suppose that the tissue lipases cannot also digest these glycerides. The existence of specific monoglyceride hydrolases in adipose tissue can, of course, not be *a priori* excluded.

A. Intestinal Tissue

The presence of monoglyceride lipase in intestinal tissue has been demonstrated with long-chain monoglycerides as substrates (Schmidt *et al.,* 1957; DiNella *et al.,* 1960; Tidwell and Johnston, 1960; McPherson *et al.,* 1962; Senior and Isselbacher, 1963). The last three groups detected, in intestinal mucosa, activity against monoglycerides but not against di- or triglycerides. The enzyme has been named monoglyceride lipase; a more correct designation is glycerol monoester hydrolase.

Pope *et al.* (1966) obtained intestinal segments from both rats and chickens. The slit sections were washed and blotted to help remove adhering pancreatic lipase; then the mucosa were removed by scraping and washed with 0.9% NaCl. As a result, hydrolysis of triglycerides by the tissue was reduced to trace levels, but there was little loss of activity against monoglycerides. The washed mucosa from the chicken intestines were homogenized in 0.25 *M* sucrose and were subsequently fractionated as shown in Table IV-47. Assays were made on emulsions of monoolein at pH 7.0 by titration of the liberated and extracted oleic acid.

Both 1- and 2-monoglycerides of identical acids were hydrolyzed at approximately equal rates (Table IV-48). 1-Monosterarin was not digested, presumably because of the high melting point of the compound. Glycerides of short-chain acids were hydrolyzed relatively rapidly. Activity was inhibited by fluoride, DFP, taurocholate, and sodium dodecyl sulfate. The most rapid hydrolysis of monoolein occurred between pH 8.5 and 9.0.

TABLE IV-47

Purification of Monoglyceride Lipase from Chicken Intestinal Mucosa[a]

Purification step	Specific activity	Yield (%)	Purification factor
Mucosal homogenate	0.044	100	1
Subcellular fractionation	0.11	40	2.5
Deoxycholate + calcium chloride	0.27	30	6.1
Ammonium sulfate precipitation	1.1	25	25
DEAE-Sephadex column	4.4	20	100
Carboxymethyl-Sephadex treatment	6.6	15	150
Starch electrophoresis	13.0	7	300

[a] Pope *et al.* (1966).

TABLE IV-48

**Relative Rates of Hydrolysis of Glycerides by
Intestinal Microsomal Lipase**[a]

Substrate	Relative hydrolysis rate[b]	Substrate	Relative hydrolysis rate[b]
1-Monostearin	0.0	2-Monooctanoin	1.7
1-Monopalmitin	0.7	Triolein	0.0
1-Monoolein	1.0	Tripalmitin	0.0
1-Monolaurin	1.1	Tributyrin	40.8
1-Monooctanoin	2.3	Triacetin	9.8
1-Monobutyrin	16.0	Diolein	0.0
1-Monoacetin	3.8	Dipalmitin	0.0
2-Monopalmitin	0.7	Diacetin	7.8
2-Monolaurin	1.4		

[a] Pope *et al.* (1966).
[b] Relative to 1-monoolein, set at unity.

The functional role of the intestinal enzymes is obscure, since the bulk of the 2-monoglycerides derived from dietary fats is not further hydrolyzed but used for the resynthesis of triglycerides in the intestinal mucosa. The monoglyceride hydrolase may hydrolyze some portion of the monoglycerides to glycerol and FFA. It has been suggested that this pathway may provide FFA for biosynthesis of triglycerides or phospholipids when the supply of acid moieties is limited (Brown and Johnston, 1964). If this is true, then the lipase activity should be greatest at the end of a fast.

B. Adipose Tissue

Lipolytic activities for mono- and diglycerides have been observed in crude extracts from rat adipose tissue (Gorin and Shafrir, 1964; Strand *et al.*, 1964; Vaughan *et al.*, 1964). These activities accompanied the hormone-sensitive lipase, but differed in the following respects: they were not affected by incubation of the tissue with epinephrine and by levels of isopropanol that inhibited hormone-sensitive lipase; the latter had a sharper pH optimum than the partial glyceridase; and tris buffer inhibited the hormone sensitive lipase as compared to the partial glyceridase activity (Strand *et al.*, 1964). Whether or not separate enzymes were required for hydrolysis of di- and monoglycerides could not be ascertained. Vaughan *et al.* (1964) and Strand *et al.* (1964) employed acetone and ether extracted homogenates of adipose tissue for their studies.

Kupiecki (1966) achieved partial purification, about sixfold, of monoglyceridase activity from rat adipose tissue, mainly by ammonium sulfate and ethanol precipitation. The activity appeared to be different from both triglyceride and lipoprotein lipases by several criteria, e.g., the substrate specificity for monoglycerides and the lack of effect of protamine sulfate which inhibits lipoprotein lipase. Distearin was poorly digested compared to monostearin. Katocs *et al.* (1971) were able to resolve two monoglyceridase activities in rat adipose tissue, one for monolaurin and the other for monoolein. The first lipase was purified about 48-fold via ammonium sulfate precipitation and DEAE-cellulose column chromatography, while attempts to purify the monoolein lipase beyond the 0–0.5 saturated ammonium sulfate fraction were unsuccessful. However, with use of discontinuous sucrose gradient centrifugation of fractions from ammonium sulfate precipitation, monoolein lipase was separated from both monolaurin and tributyrin activity (Katocs *et al.*, 1972). The monoolein lipase activity was unequally distributed through four layers in the centrifuge tube; the upper layer contained the most and purest activity. Heller and Steinberg (1972) tested a partially purified hormone-sensitive lipase activity against mono- and diolein with the results in Table IV-49. Since the purified (100-fold) hormone-sensitive lipase had a very high apparent molecular weight, 5×10^6, the authors postulated a lipid-rich multienzyme complex. Agarose gel electrophoresis revealed as many as four bands in the preparation, two of which had little activity toward triglycerides.

Monoglyceride lipases digest mono- and in some instances diglycerides more rapidly than triglycerides. The acetone powder preparation of Vaughan *et al.* (1964) hydrolyzed 2-monoolein twice as rapidly as 1-monoolein (pH 7.9) and both at a much faster rate than diolein (mostly the 1,3 isomer) and olive oil (pH 7.0). In a comparison study, Strand *et al.* (1964)

TABLE IV-49

**Lipolytic Activities toward Tri-, Di- and Monooleins by a Partially
Purified Preparation from Rat Adipose Tissue[a]**

Fraction	Enzyme activity (μequiv/ml/hour)		
	Triglyceride	Diglyceride	Monoglyceride
78,000 g supernate	0.10	0.73	1.56
pH 5.2 precipitate	4.90	22.7	35.9
$<d$ = 1.12 fraction	4.42	26.7	39.4
Final preparation	0.29	1.05	1.40

[a] Heller and Steinberg (1972).

observed the following rates of hydrolysis relative to monoolein at 100; monopalmitin, 37; monolaurin, 87; diolein (mainly 1,3 isomer), 73; diolein (mainly 1,2 isomer), 43; dipalmitin, <3; dilaurin, 37; triolein, 1; and tripalmitin, 3. Kupiecki (1966) obtained somewhat similar results with partially purified monoglyceridase: monoolein (= 100) was lipolyzed at a more rapid rate than monostearin, 55; triolein, 5; and distearin, 2. The monolaurin hydrolase of Katocs *et al.* (1971) digested monolaurin at a faster rate than trilaurin, but hydrolyzed methyl laurate to an even greater extent than monolaurin. With the monoolein hydrolase, the specific substrate was favored. The enzyme was highly specific for monoglycerides, largely ignoring both triglycerides and methyl esters (Katocs *et al.,* 1972). Of the monoglycerides tested, 1-(18:2), 1-(16:1), 1-(18:1), and 2-(18:1) were digested at approximately the same rates while 1-monoelaidin, 1-monopetroselinin (Δ6-18:1) and 1-monoglyceride-(18:0) were hydrolyzed more slowly. Mono- and diolein were hydrolyzed at a rate fivefold greater than triolein by the 100-fold purified lipase of rat adipose tissue (Heller and Steinberg, 1972).

C. Other Tissues

Wallach (1968) separated and characterized four "lipases" hydrolyzing monoglycerides from rat skeletal muscle. One activity, found in low speed particulates resulting from centrifugation (2500 g) of muscle homogenate, was termed monopalmitin lipase, although other long-chain monoglycerides were also hydrolyzed. High speed centrifugation (35,000 g) produced particulates with greatest activity toward monomyristin and named monomyristin lipase. The supernate from the first high-speed centrifugation contained a heat-stable monolaurin lipase and a heat-labile trioctanoin lipase which also hydrolyzed monolaurin. This activity was concentrated about sixfold by precipitation with ammonium sulfate. The optimum pH range for the monoglyceride activities was 8.0–8.5, while that of the trioctanoin hydrolase was 6.6. All activities were inhibited by DFP, the particulate enzymes also partially by PCMB and N-ethylmaleimide.

Yamamoto and Drummond (1967) have reported on monoglyceride-hydrolyzing activity, pH 6.8, in rat myocardium. The enzyme was inhibited by DFP and N-ethylmaleimide. Vyvoda and Rowe (1973) have studied tri-, di-, and monoglyceride lipase activities in guinea pig brain, Wang and Meng (1973) those in rat lung. Michell and Coleman (1971) found a lipase in rat erythrocyte membranes than was highly active against diglyceride at pH 7.4 in the presence of 0.75 mM $CaCl_2$.

XI. GLYCOSYLDIGLYCERIDE LIPASES

Glycolipids derived from glycerol are among the major lipids of higher plants and photosynthetic microorganisms; they are mainly associated with chloroplasts. Two galactolipids are known: monogalactosyl diglyceride (GDG), i.e., 1,2-diacyl-[β-D-galactopyranosyl $(1' \rightarrow 3)$]-*sn*-glycerol; and digalactosyl diglyceride (GGDG), i.e., 1,2-diacyl-[α-D-galactopyranosyl-$(1'' \rightarrow 6')$-β-D-galactopyranosyl$(1' \rightarrow 3)$]-*sn*-glycerol; and the "sulfolipid," or sulfoquinovosyl diglyceride, i.e., 1,2-diacyl-[6-sulfo-α-D-quinovopyranosyl$(1' \rightarrow 3)$]-*sn*-glycerol. (Quinovose is 6-deoxyglucose.) Under the name glycosyldiglyceride lipase we group together, provisionally, the hydrolases that deacylate these lipids. The galactosyl–diglyceride acyl hydrolases are often referred to as galactolipases.

A sulfolipid acylhydrolase has been found in the alga *Scenedesmus obliquus* (Yagi and Benson, 1962). A cell-free extract of the organism hydrolyzed sulfolipid to lysosulfolipid and further to the completely deacylated compound, sulfoquinovosyl glycerol. Similar activity was found in *Chlorella,* alfalfa, and corn root extract. The *Scenedesmus* extract also hydrolyzed phosphoglycerides but not galactosyl diglyceride.

More information is available on the galactolipases. These enzymes have been found in several higher plants: in runner-bean leaves (*Phaseolus multiflorus*) and other *Phaseolus* species (Sastry and Kates, 1964), in spinach leaves (Helmsing, 1967), and in potato tubers (Galliard, 1970).

Sastry and Kates (1964, 1969) investigated the enzymes of runner-bean leaves. The substrates, GDG and GGDG, were isolated from the same plant. The hydrolytic activities were measured by following the decrease of ester content with the hydroxamate method, or by quantitating the free fatty acids, after conversion to their methyl esters, by gas–liquid chromatography with an internal standard. Activity was found in the chloroplasts and in the cytoplasm fraction. The cytoplasmic activity was concentrated three-fold by ammonium sulfate precipitation to a specific activity of about 0.01. This preparation did not hydrolyze phospholipids or di- or triglycerides. Saturated GDG and GGDG (prepared by hydrogenation of the natural dilinoleoyl lipids) were not hydrolyzed, not even after the addition of methanol, chloroform, or ether. Ether, in fact, suppressed the enzymatic activity. Calcium ion had no influence. Lysogalactolipids, the obligatory intermediates of the deacylation, could not be detected; obviously they were more rapidly degraded than the diacyl compounds. The enzymatic optimum for GDG was at pH 7.0, for GGDG at pH 5.6; furthermore, the activity toward GGDG decreased markedly on storage at 4°C for several days, while the activity toward GDG was much more stable. This was interpreted as evidence for two distinct enzymes, a monogalactoside lipase and a digalactoside lipase. In spinach leaves, Helmsing (1967) found activity optima at pH 7.5 for GDG hydrolysis, at pH 5.9 for GGDG.

The galactolipase of runner-bean leaves has been isolated by Helmsing (1969). From the 105,000 g supernate of a homogenate, the enzyme was precipitated between 25 and 70% saturation with ammonium sulfate, and separated on Sephadex G-200 and DEAE-cellulose columns, and finally by disk gel electrophoresis at pH 8.3, with a final yield of 1%. The preparation showed one protein band on disk electrophoresis at pH 7.5, and at pH 8.3 in 5 M urea. The activities toward GDG and GGDG stayed parallel during the course of purification. The final specific activity, at 30°C, was 0.18 for GDG, 0.06 for GGDG. The molecular weight of the enzyme, determined by gel filtration, was 110,000. Helmsing could confirm the higher thermostability of the GDG activity and the different pH optima for GDG and GGDG hydrolysis (Sastry and Kates, 1964). To reconcile these findings with the strong evidence for a single enzymatic protein, Helmsing (1969) discussed the possibility of pH-dependent allosteric adaptation of the enzyme to the different substrates; this transformation might be affected by partial denaturation. The enzyme was activated by sodium dithionate and metabisulfite, possibly as a result of a reduction of accompanying inhibitory quinones. PCMB and reduced glutathione were partial inhibitors; cysteine, at 1 mM, inhibited completely, though dithio-erythritol did not.

A galactolipase of potato tubers has been characterized and purified by Galliard (1970, 1971a). The enzymatic activity is found in the particle-free supernatant fraction of the homogenate. Acetone precipitation and Sephadex and DEAE-cellulose chromatography yielded a preparation with specific activities, at 25°C, of 12.0 toward GDG and 6.8 toward GGDG, in the same ratio but much higher than for the runner-bean enzyme. Disk gel electrophoresis showed three protein hands with esterolytic activity, perhaps isoenzymes. The molecular weight, determined by gel filtration, was 107,000. The preparation showed transferase activity; in the presence of methanol and long-chain alcohols, methyl esters and waxes were formed. The enzyme was inhibited by 50% with 5 mM DFP. There was no stimulation by ether, and Ca^{2+}, EDTA, cysteine, and PCMB had no effect on the enzyme. The pH optimum toward GDG was around 6, toward GGDG between 4 and 5.

The enzyme from potato tubers had a rather wide specificity, and the activity against different substrates was greatly influenced by detergents. If Triton X-100 was present, phosphoglycerides, fatty acid methyl and nitrophenyl esters, and diolein were all hydrolyzed with rates similar to those for the galactolipids; lyso-phosphatidylcholine and monoolein were attacked three times faster than GDG; triglycerides and sterol esters were not hydrolyzed. In the absence of Triton X-100, with the substrates emulsified by sonication, the hydrolysis rates for the galactolipids and monoolein were much less affected than those for phosphatidylcholine and phosphatidylethanolamine, which were reduced to a few percent of those observed with the detergent present. Fatty acids had an effect similar to that of Triton X-100 (Galliard, 1971b). The detergent also shifted all activity optima toward higher pH, but this effect was dependent on the nature of the substrate. With Triton X-100 in the assay system, the difference in the pH optima for GDG and GGDG hydrolysis tended to decrease. This makes it seem likely that this difference, which is also observed for the enzyme from *Phaseolus,* is an apparent one which is due to differences between interfacial pH and bulk phase pH.

In its wide substrate specificity, galactolipase resembles the lysophospholipases and monoglyceride lipases of animal tissues. The galactolipase of potato also exhibits strong substrate inhibition with a sharp breaking point in the Michaelis–Menten curve (Galliard, 1971a), an indication that the enzyme may prefer monomolecularly dispersed, or at any rate loosely packed, substrates (Gatt *et al.,* 1972). A further similarity with the lysophospholipases is the apparent lack of positional specificity. This, together with the broad substrate specificity, indicates that the carbohydrate moiety of the galactolipids is not bound to the enzyme. Galactolipases seem to be members of the great family of amphiphilic carboxylester hydrolases.

V

Cholesterol Esterase

Enzymes which hydrolyze long-chain fatty acid esters of cholesterol are found in many mammalian tissues. They have been given the name of sterol-ester hydrolases (EC 3.1.1.13) by the Enzyme Commission; however, since cholesterol esters are their natural, if not only, substrates, and since it is not certain if the enzymes can hydrolyze esters of sterols that are not close relatives of cholesterol, it seems at present preferable to use the name cholesterol-ester hydrolase, or cholesterol esterase. On the other hand, both "cholesterol esterase" and "sterol esterase" may be designations that are much too restrictive, as will be discussed in the conclusion of this chapter.

Cholesterol esterases catalyze not only the hydrolysis of cholesterol esters but also their synthesis from cholesterol and free fatty acids. Both functions are important in the absorption and metabolism of cholesterol. Cholesterol esters cannot be absorbed intact by the mammalian intestine; they must be split to fatty acids and cholesterol by pancreatic cholesterol esterase. In the chylomicrons, however, cholesterol is found in esterified form. The resynthesis is probably catalyzed by the cholesterol esterase of the intestinal mucosa. The esters are hydrolyzed again by the cholesterol esterase of the liver and perhaps other organs. A characteristic feature of cholesterol esterase as a synthesizing enzyme is its independence from energy donors, in particular from ATP. This sets it apart from the acyl-CoA:cholesterol O-acyltransferases (EC 2.3.1.–) which also have been found in mammalian tissues. A third pathway to cholesterol esters involves the transfer of fatty acids from position sn-2 of phosphatidylcholine by the enzyme lecithin:cholesterol O-acyltransferase (EC 2.3.1.–). A review by Treadwell and Vahouny (1968) treats the absorption of cholesterol and the role of the esterase; a second review (Vahouny and Treadwell,

1968), which discusses all three enzymes, is especially valuable for its detailed description of many assay procedures.

The presence of the enzyme in pancreatic juice and extracts of dog pancreas was demonstrated by Mueller (1915, 1916). An intestinal cholesterol esterase was detected by Klein (1938) in the cow. These digestive enzymes have since been found in other mammals. The pancreatic cholesterol esterases of the pig and the rat have been investigated most frequently, and the larger part of this chapter will be devoted to these enzymes. The intracellular cholesterol esterases that have been discovered in many mammalian tissues (Kondo, 1910a,b; Schultz, 1912) will be treated at the end of the chapter.

I. ASSAY

Strictly, only the formation of fatty acids or cholesterol from cholesterol esters should be taken as a measure of cholesterol esterase, but in many studies the reverse reaction, the formation of the ester, has been used for following the activity of the enzyme, with the assumption that both activities belong to the same protein; and this assumption is most probably correct if the possibilities of energy-dependent acylation and lecithin:cholesterol acyltransfer have been excluded.

The most important consideration in the assay of cholesterol esterase is the solubilization of the substrate. Cholesterol as well as cholesterol esters, even unsaturated ones such as the linoleate, are solids at the usual incubation temperatures of 25° or 37°C. It has been found that the catalytic rates of the enzymes are highly dependent on the mode of dispersion of these lipids. For measuring the synthetic activity, cholesterol and fatty acid can be prepared as a mixed emulsion stabilized with taurocholate (Hernandez and Chaikoff, 1957) or with bovine albumin (Vahouny and Treadwell, 1968). It should be noted that, for the digestive cholesterol esterases, cholate or conjugated cholate must in any case be present as a cofactor. Cholesterol esters can be dispersed by adding them dissolved in acetone (Deykin and Goodman, 1962) or alcohol (Murthy and Ganguly, 1962) to the incubation mixture; this procedure probably leads to a microcrystalline suspension. The rates of hydrolysis are greatly increased when the esters are dispersed in micellar form. Bile salts alone are not suitable for this purpose, since they do not easily incorporate cholesterol into their micelles (Small, 1971). The solubility of cholesterol esters is greater in mixed micelles of bile salts and fatty acids (Borgström, 1967). Phospholipids, in particular lecithin, are especially effective solubilizers (Fleischer and Brierley, 1961). Rat pancreatic cholesterol esterase hydrolyzed a micellar solution of cholesterol oleate in lecithin seven times faster than an albumin-stabilized dispersion (Vahouney *et al.,* 1964b). Detailed descriptions of these assay mixtures are given in the review by Vahouny and Treadwell (1968).

The increase in turbidity of a cholesterol–lithium oleate dispersion by the formation of cholesterol oleate has been used as the basis of a nephelometric assay (Hernandez and Chaikoff, 1957). The release of oleic acid from the ester has been measured by a manometric method (Korzenovsky *et al.,* 1960a). The activity of the pancreatic cholesterol esterase is high enough to be measured by continuous titration, with a pH-stat, of the fatty acids released from cholesterol ester emulsified with lithium oleate (Coutts and Stansfield, 1967). Alternatively, the fatty acids can be extracted from the assay mixture by Dole's (1956) solvent (hexane–isopropanol–H_2SO_4 10:40:1) and titrated. In the most generally applicable assay procedure, the lipids are extracted with chloroform–methanol, separated by column or thin-layer chromatography, and quantitated (Vahouney and Treadwell, 1968). Labeled substrates are usually employed. A very rapid and convenient separation of the cholesterol esters from the more polar lipids can be achieved by chromatography on microscope slides coated with silica gel, with hexane:ethyl ether:acetic acid 83:16:1 as the developing solvent. The esters migrate with the front of the solvent, while the hydrolysis products stay near the origin; upper and lower half of the silicic acid coating are collected separately and measured in a scintillation counter (Hyun *et al.,* 1969). A separation of unlabeled reaction products by chromatography on aluminum oxide has been described by Murthy and Ganguly (1962); the cholesterol can be isolated and quantitated by colorimetric methods (Sperry and Webb, 1950). The labeled fatty acid released from a cholesterol ester can be rapidly determined by extraction of the lipid extract with 0.1 *M* KOH (Schotz *et al.,* 1970; Sloan and Fredrickson, 1972b).

In measuring the activity of intracellular cholesterol esterases with the help of radioactive substrates, it must be taken into consideration that the enzyme preparation may contain cholesterol or cholesterol esters that will dilute the labeled substrate and thus lead to a manyfold diminished apparent activity. If the indigenous lipid is removed during further fractionation of the preparation, the result can be an imaginary "purification" of the enzyme.

II. PURIFICATION

A partial purification of the enzyme from pork pancreas has been described by Hernandez and Chaikoff (1957). From an acetone powder extract, the protein precipitating between 30 and 60% ammonium sulfate saturation was collected, dissolved in pH 6.2 phosphate buffer, and dialyzed. A sediment formed, which was redissolved and fractionated with 33% ethanol and 9% ammonium sulfate. The enzyme was precipitated from the supernate on suspended cellulose powder with more ammonium sulfate. The cellulose was collected into a column, and the enzyme was eluted by a decreasing ammonium sulfate gradient. The recovery of the activity was 0.7%, and a 450-fold purification had been achieved; however, ultracentrifugation showed that more than one band of protein was present. The specific activity was given in arbitrary nephelometric units.

A purification of cholesterol esterase from rat pancreatic powder has been reported by Murthy and Ganguly (1962). The soluble proteins were adsorbed on calcium phosphate gel and eluted batchwise with pH 7.4 phosphate buffer of increasing strength. This procedure resulted in a separation of a hydrolytic and a synthesizing cholesterol esterase activity. The hydrolase was obtained 55-fold concentrated in a fraction devoid of synthetic activity; the synthetase could be recovered in 690-fold concentration by a final ammonium sulfate extraction of the gel. However, this extraction succeeded in only two out of eight experiments. This is the only such separation reported, and evidence from other sources suggests strongly that there is only one pancreatic cholesterol esterase with both hydrolytic and synthetic ability.

Hyun *et al.* (1969) have isolated the enzyme from rat pancreatic juice. The active protein was precipitated with 35% acetone, then separated on DEAE-cellulose, and finally on a column of hydroxyapatite by discontinuous elution with pH 6.8 phosphate buffer. Mercaptoethanol was included in the DEAE-cellulose chromatography, and dimethyl sulfoxide, at 10%, was added to the buffer in the last step in order to reduce the loss of activity; even so, the overall yield was only 6%. However, the final preparations were stable for months at 2°C. Their average specific activity against cholesterol oleate in lecithin micelles was 45. Disk electrophoresis showed only one protein band. Table V-1 summarizes the course of a purification.

In a similar procedure, Erlanson (1972) achieved a 200-fold purification in a yield of 12%. The preparation appeared homogeneous on a disk gel electrophoresis but had a specific activity of only 8.0 against micellar cholesterol oleate.

On Sephadex filtration of rat pancreatic juice, cholesterol esterase emerges ahead of the bulk of the proteins in a highly enriched fraction (Morgan *et al.,* 1968; Erlanson and Borgström, 1970b), but no method of purification based on this behavior has yet been published. Morin (1972) has described an incomplete purification of the enzyme from rabbit pancreas. It is to be hoped that a simple method will soon be found, starting from a convenient source, which will make a homogeneous cholesterol esterase available in larger quantities.

TABLE V-1

Purification of Cholesterol Esterase from Rat Pancreatic Juice[a]

Fraction	Protein (mg)	Activity (μmoles · min^{-1})	Specific activity (μmoles · min · mg^{-1})	Recovery (%)
Pancreatic juice	384	46	0.12	100
35% acetone precipitate	14.4	28	1.9	60
DEAE-cellulose	2.4	20	8.0	43
Hydroxyapatite	0.06	2.8	41	6

[a] Hyun *et al.* (1969).

A cholesterol esterase from human placenta has been purified by a procedure following that for rat pancreatic juice (Chen and Morin, 1971). The final preparation, obtained in a 2.7% yield and 350-fold purification, showed one band on disk electrophoresis and had a specific activity of 8×10^{-6}. A partial purification of soluble cholesterol esterase from rat liver has been reported by Deykin and Goodman (1962); the resulting enzyme was very unstable.

III. MOLECULAR PROPERTIES OF PANCREATIC CHOLESTEROL ESTERASES

The enzyme of rat pancreatic juice has a molecular weight of 65,000–69,000 (Morgan *et al.*, 1968; Hyun *et al.*, 1971). Precipitated from the juice with acetone, it forms a dimer of MW 135,000 (Hyun *et al.*, 1972). In the presence of cholate an aggregate of MW 400,000, corresponding to six protein molecules, emerges from Sephadex columns; this complex seems to contain three cholate molecules (Hyun *et al.*, 1972).

Teale *et al.* (1972) reported that a cholesterol esterase complex from pig pancreas powder, with the molecular weight of $>800,000$, can be converted by lipid extraction into enzymatic "subunits" of MW 15,000–20,000 that can still hydrolyze and synthesize cholesterol esters. Since such a low weight for a lipolytic enzyme other than a phospholipase 2 is unexpected, a confirmation and extension of the study of Teale *et al.* would be highly desirable.

Human pancreatic cholesterol esterase had a molecular weight of 300,000 according to gel filtration (Erlanson and Borgström, 1970b). This is an extraordinarily large size for a lipolytic or for a digestive enzyme, and it suggests that the human cholesterol esterase may also be a dissociable aggregate. In the presence of oleic acid the enzyme aggregated to a much larger complex, reminiscent of the "fast" pancreatic lipase. Both cholesterol esterase and lipase complexes dissociated in bile salt solutions.

The pancreatic enzyme is vulnerable to digestion by pancreatic proteinases, much more so than pancreatic lipase (Vahouny *et al.,* 1964c), but can be efficiently protected by bile salts (Nedswedski, 1936; Murthy and Ganguly, 1962). The enzyme also seems to be sensitive to surface denaturation; this is attested by the large losses of activity in the final chromatographic procedures and by the lability of the purified preparations, which can be reduced by the addition of albumin (Vahouny and Treadwell, 1968).

As for all lipolytic enzymes, the exact pH optimum for a cholesterol esterase is to a large degree a function of the physical state of the substrate. The optimal pH most often given for the synthetic reaction of the pancreatic and intestinal enzymes is 6.1–6.2; for the hydrolytic reaction, 6.7–7.0. Values as high as pH 8.6 for the hydrolysis of a substrate prepared by injection of an ethanolic solution into water (Murthy and Ganguly, 1962) and as low as pH 5.5 for a synthetic reaction with butyric acid (Fodor, 1950) and pH 4.7 for acetic acid (Swell and Treadwell, 1955) have been reported. In general, the optimal pH for the synthetic reaction is found to be 0.5–1.0 units lower than that of the hydrolytic reaction of the same enzyme preparation. This is an indication that the substrate for the synthesis is the undissociated fatty acid; the low pH optimum for butyric and acetic acid is thus explained by the lower apparent pK of the short-chain fatty acids. The enzyme can be assumed to catalyze a hydrolysis–esterification reaction which leads to a pH-dependent equilibrium, with esterification prevailing at low, hydrolysis at high pH. Such an equilibrium has, in fact, been demonstrated for the rat pancreatic cholesterol esterase (Filipek-Wender and Borgström, 1971). The situation is analogous to that found for pancreatic lipase (Borgström, 1964a).

The equilibrium between cholesterol ester and its hydrolysis products can experimentally be reached from either side, i.e., starting from the ester or from cholesterol and fatty acid (Filipek-Wender and Borgström, 1971), a convincing argument that the same enzyme catalyzes hydrolysis and synthesis. Further support for this conclusion is supplied by the finding that the ratio of both activities remains unchanged during purification procedures (Hernandez and Chaikoff, 1957; Hyun *et al.,* 1969).

Hyun *et al.* (1972) have reported the amino acid composition of rat pancreatic cholesterol esterase (Table V-2); it is similar to that of porcine lipase (Verger *et al.,* 1969), but tyrosine seems to be nearly absent. The tryptophan content was not determined.

TABLE V-2

**Amino Acid Composition of Cholesterol Esterase from
Rat Pancreatic Juice[a]**

Amino acid	Residues (in nearest integers)[b]	Amino acid	Residues (in nearest integers)[b]
Ala	71	Lys	42
Arg	31	Met	10
As(x)	71	Phe	26
$\frac{1}{2}$Cys	6	Pro	39
Gl(x)	63	Ser	56
Gly	63	Thr	36
His	12	Try	N.D.
Ile	30	Tyr	2
Leu	49	Val	49

[a] Hyun et al. (1972).
[b] Assumed molecular weight, 69,000.

IV. SUBSTRATE SPECIFICITY

A. Fatty Acids

Considerable efforts have been expended on the question of "fatty acid specificity." The results can be summarized, in our opinion, in the statement that no such specificity exists. The studies on the subject can be divided into those that have measured synthesis and those concerned with hydrolysis. When both activities are measured with the same enzyme preparation, the specificities for the individual acids are not the same (Swell and Treadwell, 1955). In synthesis, oleic acid is usually the most active acid (Hernandez and Chaikoff, 1957; Murthy and Ganguly, 1962; Shah et al., 1965; Hyun et al., 1969), though sometimes surpassed by linoleic or linolenic acid (Hernandez and Chaikoff, 1957; Murthy and Ganguly, 1962). Saturated acids are less active, very much less when added to the system in ethanolic solution (Murthy and Ganguly, 1962). Short-chain acids, C_2–C_6, are hardly reactive (Swell and Treadwell, 1955; Murthy et al., 1961), probably because they are soluble and completely dissociated.

The hydrolysis of cholesterol esters in albumin-stabilized dispersions showed completely different "fatty acid specificity" patterns. With hog pancreatic extract, the butyrate was the preferred substrate, followed by hexanoate, acetate, and oleate, while long-chain saturated esters were hydrolyzed more than one hundred times more slowly (Swell and Treadwell, 1955). A similar result was obtained with rat pancreatic juice (Vahouny *et al.,* 1964a); however, when the esters were dissolved in mixed micellar solutions of lecithin and bile salt, these differences disappeared and the esters—saturated from 3:0 to 18:0 as well as the unsaturated 18:1, 18:2, and 18:3—were all attacked at virtually the same rate.

It is clear, then, that the observed differences in reaction rates are not due to enzyme specificity as it is normally understood, i.e., substrate–enzyme recognition at specific binding sites, but to the physicochemical properties of the fatty acid dispersion. The enzyme requires only an accessible ester group or protonated carboxyl group. As with pancreatic lipase, steric hindrance may result from substituents near the carboxyl group; e.g., cholesterol-2-ethyl hexanoate is a very poor substrate (Hyun *et al.,* 1964), and the 2-methyl-2-ethyl hexanoate is completely resistant (Vahouny and Treadwell, 1964).

B. Sterols

In the synthesis of sterol oleates, dihydrocholesterol has been found as effective as cholesterol; substitution in the side chain, as in sitosterol and stigmasterol, reduces the activity somewhat (Swell *et al.,* 1954; Hernandez and Chaikoff, 1957; Korzenovsky *et al.,* 1960a,b; Murthy and Ganguly, 1962). The hydroxy group on carbon 3 must be β, i.e., equatorial (Hernandez and Chaikoff, 1957). Dehydrogenation of ring B (ergosterol) results in appreciably lower activity. The same pattern of specificity is observed for the hydrolysis of both oleate and butyrate esters of different sterols (Swell *et al.,* 1954). It might be deduced that cholesterol esterase requires a certain sterol structure in its substrate for optimal enzyme–substrate fit. However, the possibility must be considered that the observed reaction rates derive from changed accessibilities of the ester and alcohol function caused by different crystalline packing or intramicellar orientation. The only genuine specificity effect might be the one against an α-oriented hydroxyl group, a steric hindrance effect; otherwise, the enzyme may be an unspecific carboxyl esterase that only puts certain physicochemical requirements on its substrate. This view is supported by experiments. Morgan *et al.* (1968) could show that the hydrolytic activities of rat pancreatic juice against cholesterol ester, micellar monoolein, and β-naphthyl acetate remained together during chromatography on Sephadex as well as on DEAE-cellulose. Erlanson and Borgström (1968; 1970a,b) could not separate the cholesterol esterase activity of human pancreatic juice from the activity against *p*-nitrophenyl acetate, and a purified cholesterol esterase from rat pancreatic juice hydrolyzed *p*-nitrophenyl acetate, monoolein, and vitamin A palmitate faster than cholesterol oleate (Erlanson, 1972). Mattson and Volpenhein (1966b, 1968; 1972a) described an enzyme ("nonspecific lipase") of rat pancreatic juice that hydrolyzed various primary and secondary long-chain carboxyl esters in the presence of bile salts and appeared to be identical with cholesterol esterase. Hyun *et al.* (1969) reported that this hydrolytic activity had been separated from a purified preparation of cholesterol esterase; however, they used a triple long-chain secondary ester as a substrate, and it seems likely that this compound was not in micellar solution. Micellar solubility, however, seems to be obligatory for substrates of the enzyme; contrary to micellar monoolein, emulsified monoolein is not hydrolyzed by the cholesterol esterase of rat pancreatic juice, and ultrafiltration studies have indicated that it is the micellar rather than the emulsified fraction of a cholesterol oleate in bile salt dispersion that is hydrolyzed by the enzyme (Morgan *et al.*, 1968). On the other hand, the synthesis of the ester is favored if the micelles of a system stabilized with phospholipids are disrupted by Mg^{2+} (Vahouny *et al.*, 1964a); and hydrolysis is favored over synthesis at taurocholate concentrations exceeding 10 mM (Kelly and Newman, 1971), the region of the critical micellar concentration (Hofmann, 1963). These effects may have their explanation in the removal or addition of the ester from the equilibrium by transition from a micellar to a crystalline phase. The finding that the relative rates of synthesis and hydrolysis depend on the electrolyte concentration (Korzenovsky *et al.*, 1960b) may be similarly explained.

The bulk of the evidence leads to the conclusion that neither fatty acid nor sterol are recognized by the enzyme.

V. INHIBITORS

Cholesterol esterase is independent of cations, but it is inhibited by heavy metal ions such as Zn^{2+}, Cu^{2+}, and Hg^{2+} (Murthy and Ganguly, 1962; Hernandez and Chaikoff, 1957). Fluoride (1 mM) has been reported to inhibit the esterifying activity of both the pancreatic and the intestinal enzyme of the rat (Murthy and Ganguly, 1962). The sulfhydryl reagents, *p*-chloromercuribenzoate (PCMB) and analogous compounds, are strong inhibitors. PCMB at 1 mM inhibited the porcine pancreatic enzyme by 76%, and the inhibition could be prevented by reduced glutathione (Hernandez and Chaikoff, 1957). An inhibition of 35–40% was observed with rat cholesterol esterase and 1 mM PCMB (Murthy and Ganguly, 1962). Taurocholate (5 mM) protected the enzyme almost completely from inactivation by PCMB (Hyun *et al.*, 1969); this suggests that the bile acid may bind across the site of the sensitive sulfhydryl group, and that this group is not immediately involved in the catalytic mechanism.

Several organophosphorus compounds that are typical serine reagents have been found to inhibit the digestive cholesterol esterases. Diisopropyl fluorophosphate (DFP) (10^{-4} M) inhibited both the pancreatic and the intestinal enzyme of the rat by 50% (Murthy and Ganguly, 1962). Diisopropyl-*p*-nitrophenyl phosphate, diethyl-*p*-nitrophenyl phosphate, and tetraethyl pyrophosphate were inhibitory at much lower concentrations, 10^{-7}–10^{-9}. The potential serine reagent phenylmethylsulfonyl fluoride (1 mM) also caused inhibition.

It can be provisionally concluded that cholesterol esterase is a serine enzyme that also contains at least one free sulfhydryl group. This group may, as in pancreatic lipase, be located near the reactive center, or it may be involved in the allosteric conformational transition postulated by Hyun *et al.* (1972).

VI. REQUIREMENT FOR BILE SALTS

The stimulating effect of bile salts on cholesterol esterase was observed by Mueller (1916), and Nedswedski showed in 1936 that the esterification of cholesterol proceeded only in the presence of bile salts, and furthermore, that bile salts protected the enzyme from deactivation during the storage of the pancreatic extract. He suggested that the bile acid combined chemically with the enzyme and that its action was not due to a mere surface tension effect (Nedswedski, 1937). Since then, the activating and the stabilizing effect of bile salts have been confirmed in many investigations. Strong evidence supporting Nedswedski's hypothesis has accumulated, but the question whether bile salts—in particular, cholates—are true coenzymic agents has still not been answered conclusively.

Klein (1938) found that cholate and its conjugates, taurocholate and glycocholate, were much more potent activators than deoxycholate. The cholates have three hydroxy groups, 3α, 7α, and 12α, the first one equatorial, the two others in axial configuration, but all three extending to the same side of the molecule, opposite to the two methyl groups. Vahouny *et al.* (1964b, 1965) reported that deoxycholate ($3\alpha,12\alpha$), its glycine conjugate, and chenodeoxycholate ($3\alpha,7\alpha$) were completely inactive as cofactors, although they seemed to be just as effective as the cholates in solubilizing cholesterol oleate. However, in other studies (Swell *et al.*, 1954; Korzenovsky *et al.*, 1960a; Filipek-Wender and Borgström, 1971), deoxycholate has shown activity in stimulating cholesterol ester hydrolysis. Dehydrocholic acid (3,7,12-triketocholanic acid) is completely inactive (Swell *et al.*, 1954; Murthy and Ganguly, 1962).

The inactivation of cholesterol esterase during storage of pancreatic juice or extract (Nedswedski, 1936; Murthy and Ganguly, 1962) is probably due to tryptic digestion (Vahouny *et al.*, 1964b, 1965, 1967). Bile salts

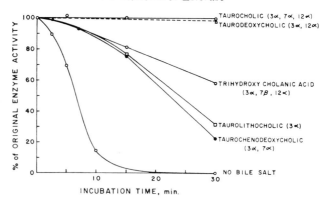

Fig. V-1 Bile salts protecting cholesterol esterase against proteolytic inactivation. Two milliliters rat pancreatic juice, 5 mg bile salt, and 0.5 mg trypsin, pH 6.2. From Hyun *et al.* (1969).

protect the enzyme against trypsin (Fig. V-1), but the mechanism of this protection cannot be entirely identical with that of the activation, since bile salts that did not activate—e.g., taurolithocholate (3α), taurochenodeoxycholate ($3\alpha,7\alpha$) and $3\alpha,7\beta,12$-trihydroxycholanate—nevertheless provided partial protection from proteolysis. The poorly activating taurodeoxycholate ($3\alpha,12\alpha$) was as effective as taurocholate in protecting the enzyme (Fig. V-1) (Vahouny *et al.*, 1967). Taurocholate protected also against deactivation by the sulfhydryl reagent PCMB (Hyun *et al.*, 1969).

Hyun *et al.* (1971, 1972) have reported a bile salt-induced aggregation of rat pancreatic cholesterol esterase. In solutions of 0.9 mM sodium taurocholate, the enzyme emerged from Sephadex columns as a well defined peak with the apparent molecular weight of 400,000, whereas the weight determined without the bile salt or by sodium dodecyl sulfate (SDS)-polyacrylamide electrophoresis is 65,000–69,000. The higher weight corresponds to an aggreggate of six protein molecules. On chromatography on a Sephadex column that had been equilibrated with [^{14}C]cholate, a radioactive enzyme–bile salt complex emerged that appeared to contain three bile salt molecules. The same complex was identified by preincubating the enzyme with labeled cholate and subsequent gel filtration, and by retention by a membrane filter with a withholding limit at molecular weight 100,000. Taurodeoxycholate and taurochenodeoxycholate also induced aggregation of the enzyme, but the aggregates were of even higher molecular weight and did not migrate as well defined peaks during gel filtration. Taurocholate sensitized the enzyme against inhibition by the potential serine reagent phenylmethylsulfonyl fluoride (PMSF). In the absence of taurocholate, 1 mM PMSF caused 16% inhibition; when the enzyme was preincubated with the bile salt, 73% inhibition was obtained. However, a similar degree in inhibition resulted from higher concentrations of PMSF (5 mM), and if the inhibitor was added first and the bile salt later, inhibition was low. These results cannot be easily explained. It must also be kept in mind that sulfonyl fluorides have been found to be very unspecific reagents in studies with pancreatic lipase (Maylié *et al.*, 1972). The more specific organophosphates do not need taurocholate to inhibit cholesterol esterase (Murthy and Ganguly, 1962).

On the basis of their experiments, Hyun *et al.* (1972) have suggested that the aggregate of three trihydroxy bile acid and six enzyme molecules might constitute the active form of rat pancreatic cholesterol esterase. This concept invites further speculations. A bile acid molecule might bind to one enzyme molecule through ionic and hydrogen bonds; this complex might be able to bind a second enzyme molecule, and three such dimers could form a hexamer. Alternatively, two enzyme units might be connected by a bridge of bile acid; again, the dimers would aggregate to hexamers. In any case, the proportion of three bile acid to six enzyme subunits reported by Hyun *et al.* (1972) would require an internal subdivision of the complex in three dimers, because any other arrangement would require that one bile acid molecule binds to three proteins, and this is almost certainly impossible for steric reasons. The sensitive sulfhydryl groups of the enzyme units could be in the interior of the complex, the reactive serine on the periphery. If it is accepted that trihydroxy bile acids induce a conformational change that activates the reactive site of the enzyme, it must be remembered that the bile acid molecule is strongly asymmetrical and could not possibly function as a bridge between two enzyme subunits and at the same time induce the identical conformational change in both of them. This leads to the conclusion that the postulated active six-unit complex cannot have more than three active sites, a conclusion difficult to reconcile with the reported molecular homogeneity of the preparation of the molecular weight 65,000–69,000 (Hyun *et al.*, 1969, 1972).

The absolute dependence of rat pancreatic cholesterol esterase on trihydroxy bile acids has been questioned. In experiments on the hydrolysis–synthesis equilibrium catalyzed by rat cholesterol esterase, Filipek-Wender and Borgström (1971) found that deoxycholate ($3\alpha,12\alpha$) and its conjugates were just as effective as cholate at pH 6, though somewhat less effective at pH 9. They suggested that the role of the bile salts might be to prevent the breakdown of the enzyme by proteolysis, in addition to their role in substrate–bile salt micelle formation.

In summary, some of the effects of trihydroxy bile salts on cholesterol esterase are suggestive of a direct protein–bile salt interaction with allosteric activation of the enzyme, but there is as yet no proof for the existence of such a mechanism, not to mention an explanation. The protection of the enzyme from proteolysis and from sulfhydryl inactivation is better documented.

VII. INTRACELLULAR CHOLESTEROL ESTERASES

A. Liver

Rat liver contains a number of enzymes capable of hydrolyzing cholesterol esters; a "soluble," a microsomal, and a lysosomal cholesterol esterase have so far been described, and further fractionations will very probably be reported. In the following, only those studies will be discussed which allow a distinction of these activities according to the present state of knowledge.

A "soluble" cholesterol esterase was characterized and partially purified by Deykin and Goodman (1962). The activity was precipitated from the 100,000 *g* supernate of rat liver homogenate by 30% saturation with ammonium sulfate. A procedure of stepwise adsorption to calcium phosphate gel led to a final preparation with an apparent specific activity 68 times greater than that of the cell-free homogenate; however, since this activity was measured with labeled substrate and with disregard of the endogenous cholesterol esters, the real degree of purification was much lower. Specific activity units were not supplied, but it appears from earlier studies (Byron *et al.,* 1953) that the activity in liver is at least 1000 times less than in pancreas.

The final enzyme preparation of Deykin and Goodman (1962) was quite unstable and had lost most of its activity after one day's storage at $-18°C$. The enzyme had hydrolytic but no synthetic ability. Its pH optimum was between 6.5 and 7.5. Divalent cations or EDTA had little influence on the activity, but Zn^{2+} and Cu^{2+} inhibited severely. The sulfhydryl reagents *N*-ethylmaleimide (NEM) and PCMB were strong inhibitors; their action could be reversed by reduced glutathione. DFP at 10^{-4} M blocked 81% of the activity. These results indicate a reactive serine, and one or more cysteine residues in the vicinity of the active site or somehow involved in the catalytic mechanism. The enzyme preparation had little activity against *p*-nitrophenyl acetate. Although it hydrolyzed tripalmitin very slowly, it rapidly hydrolyzed ethyl palmitate; the structure of the substrate dispersions may be responsible for this difference. Detergents such as Tween 20 and glycocholate inhibited the reaction.

The "fatty acid specificity" of the soluble liver enzyme has repeatedly been investigated. Deykin and Goodman (1962) found for individual cholesterol esters that had been introduced to the incubation mixture in acetone solution that the oleate and linoleate were most rapidly hydrolyzed, followed by the acetate, then by the palmitate, while the stearate was not attacked at all. In a mixture of esters, however, both stearate and palmitate were hydrolyzed at about half the speed of the unsaturated esters. Sgoutas and co-workers (Sgoutas, 1968; Goller and Sgoutas, 1970; Goller *et al.*, 1970) tested a large number of cholesterol esters of trans and cis unsaturated acids having double bonds in different positions of the chains. It was found that the *cis* Δ9 structures were the best substrates for the soluble enzyme. When the double bond was shifted to either end of the chain or changed to trans, a decrease in activity resulted; however, half of the substrate's activity still remained for *cis* Δ3 or *cis* Δ12 and for *trans* Δ9 acids. The curves presenting the reactivity of the esters as a function of double bond position (Goller *et al.*, 1970) resemble curves that describe the mobility of the corresponding acids in thin-layer and gas–liquid chromatography (Ackman, 1972; Gunstone *et al.*, 1967). Contrary to the deductions of Sgoutas and co-workers, it can be concluded that the "specificities" reflect the structure of the substrate aggregates and that there is no "recognition" of particular fatty acid structures by the enzyme.

A microsomal cholesterol esterase of rat liver, also described by Deykin and Goodman (1962) but not purified, contained 11–32% of the total activity of the homogenate. It had a sharp pH optimum at 6.1 and it was inhibited by Tween 20, taurocholate, albumin, Cu^{2+} and Zn^{2+}, NEM, iodoacetamide, and PCMB. DFP inhibited strongly at 10^{-5} *M*.

The presence of a lysosomal cholesterol esterase in rat liver has been reported by Stoffel and Greten (1967), and an energy-independent cholesterol esterifying activity has been found in rat liver homogenates at pH 4 (Stokke, 1972a). The lysosomal enzyme of human liver has been studied by Stokke (1972a,b). Hydrolytic activity was maximal at pH 5, synthetic activity at pH 3.8. Sulfhydryl reagents inhibited the enzyme, and mercaptoethanol reversed the inhibition. Riddle and Glomset (1973) reported an acidic (pH 4.5) activity associated with plasma membranes. A deficiency of the lysosomal cholesterol esterase of the human liver has been found in cases of cholesterol ester storage disease and the related Wolman disease (Burke and Schubert, 1972; Sloan and Fredrickson, 1972a,b).

B. Other Mammalian Tissues

Adrenals and ovaries have been searched for cholesterol esterase activity because these organs are sites of conversion of free cholesterol into steroids. Dailey *et al.* (1963) found hydrolytic activity in dog adrenals; the synthetic activity seemed to be mainly of the energy-dependent type and not due to cholesterol esterase (Dailey *et al.,* 1962). Brot *et al.* (1963) investigated hog adrenals, and found that an acetone powder extract cata- lyzed both the hydrolysis and the synthesis of cholesterol esters indepen- dent of cofactors. Hydrolysis was much faster at pH 7.4 than at pH 5.7, while the optimum for the esterification was between pH 3 and 5. Shyamala *et al.* (1965) studied the properties and the subcellular distribution of the enzyme of bovine adrenal glands. Crude mitochondrial, microsomal, and supernatant fractions all hydrolyzed cholesterol esters; the synthesizing ac- tivity was concentrated in the supernate. Hydrolysis by the particulate fraction was optimal at pH 7.5; the esterification by the cell sap was optimal at pH 5.0 and did not depend on ATP, coenzyme A, or Mg^{2+}.

Bovine corpus luteum homogenate hydrolyzed cholesterol esters with optima at pH 6.0 and pH 7.5 (Coutts and Stansfield, 1968). DFP and *N,N*-diisopropylphosphodiamidic fluoride inhibited strongly, *p*-hydroxy- mercuribenzoate did not. The hydrolytic activity was suppressed by com- plexing the substrate with lithium oleate and by cholate; this behavior sets the enzyme apart from most other cholesterol esterases. Approximately 90% of the activity was found in the particulate subcellular fractions. In corpora lutea of pregnant rabbits, the cholesterol ester hydrolyzing and synthesizing activity is highest in the mitochondrial fraction (Morin, 1973). In contrast, most of the cholesterol esterase of luteinized ovaries from rats has been found in the supernatant fraction (Behrman and Armstrong, 1969).

A cholesterol esterase from human placenta has been purified by Chen and Morin (1971) by a method modeled after that of Hyun *et al.* (1969). The final preparation was homogeneous according to disk gel electrophore- sis and had a specific activity of 8×10^{-6}.

The cholesterol esterase of the brain is of interest because the developing mammalian brain contains cholesterol esters that during myelination are reduced to the very low levels in adult brain. Pritchard and Nichol (1964) found hydrolytic activity in rat brain with an optimum at pH 6.6–7.6, located mainly in the microsomal fraction. Clarenburg *et al.* (1966) re- ported cholesterol ester hydrolysis in all particulate fractions, and doubling of the specific activity during the period of active myelination.

An extensive study of the cholesterol esterases of rat brain has been undertaken by Eto and Suzuki (1971, 1973a,b). An energy-independent synthesizing activity had a pH optimum around 5.6. It was inhibited by bile salts. Nearly half of the activity was found in the crude mitochondrial fraction and none in purified myelin. The hydrolytic cholesterol esterase activity had two well separated peaks of activity, at pH 4.2 and at pH 6.6. The pH 4.2 enzyme was activated two- to four-fold by cholate, deoxycholate, and taurocholate. It was concentrated in the crude mitochondrial fraction, and further fractionation suggested that the enzyme might be mitochondrial rather than lysosomal. The pH 6.6 enzyme was present in all particulate fractions and also in purified myelin. This activity was depressed by deoxycholate, but stimulated by 4 mM cholate. Taurocholate, at the same concentration, caused fifteen-fold stimulation (Eto and Suzuki, 1971). This finding should be relevant in any discussion of the postulated role of trihydroxy bile acids for the pancreatic cholesterol esterase, insofar as taurocholate certainly cannot be a physiological cofactor for an enzyme of the brain.

Eto and Suzuki (1973a) have been able to separate the pH 6.6 activity of rat brain into two fractions, one with an optimum at 6.0 and associated with microsomes, the other with an optimum at pH 7.2 and located in the myelin. Both activities were activated by taurocholate. The second enzyme could be recovered with a 70–80% yield in the purified myelin fraction. This was the activity that increased greatly at the onset of myelination in brain and spinal cord (Eto and Suzuki, 1973b).

Cholesterol esterases of arterial tissue are of interest in the etiology of atherosclerosis, since plaques are known to contain substantial amounts of cholesterol esters. Hydrolytic activity has been found in normal and atherosclerotic rabbit aortas (Day and Gould-Hurst, 1965), pigeon aortas (St. Clair *et al.,* 1972), and in the 100,000 *g* supernatant fraction of rat and monkey aorta (Howard and Portman, 1966). In the human aorta, optimal cholesterol ester hydrolysis was found at pH 6.6 or 7.4, depending on the nature of the substrate dispersion (Kothari *et al.,* 1970). The enzyme appeared to be associated with lysosomes. It was partially inhibited by heavy metal ions and by PCMB (1 mM). Taurocholate protected the enzyme from inactivation during incubation of the homogenate. Similar enzymes which needed taurocholate for activity were found in acetone powders of rat and rabbit aortas (Kothari *et al.,* 1973). These results recall the protective action of the bile salts on pancreatic cholesterol esterase (Fig. V-1), and since taurocholate is not available *in vivo* to the arterial enzyme, further doubt is thrown on the postulated specificity of the bile salt–cholesterol esterase interaction (Vahouny *et al.,* 1967).

An energy-independent cholesterol ester synthesizing enzyme with a pH optimum of 5.0 has been found in atherosclerotic rabbit intima (Proudlock and Day, 1972).

Peritoneal macrophages contain an energy-independent cholesterol ester synthetase with a pH optimum of 6.3 (Day and Tume, 1969). Similar activity, with an optimum at pH 4.5, has been found in alveolar macrophages (Tume and Day, 1970).

VIII. CONCLUSION

Our knowledge of cholesterol esterase is riddled with uncertainties even in those areas that have been actively investigated for decades; it appears that most of the information so far available is in need of reconfirmation. The main obstacle for such a reinvestigation is the difficulty of preparing sufficient quantities of purified enzyme. Even though methods for the isolation of cholesterol esterases have been described (Hyun *et al.,* 1969; Chen and Morin, 1971), few data have been published on the properties of these preparations. A method to isolate the enzyme from an easily accessible source is urgently needed.

Three areas of contention can be outlined in the presently available literature: the role of bile salts; the substrate specificity; and, finally, the identity of the enzyme; these last two are, of course, parts of the same problem. It seems likely that the principal function of the pancreatic enzyme cholesterol esterase is indeed the hydrolysis of dietary cholesterol esters, and that it is correctly named according to this criterion. About the intracellular "cholesterol esterases" we can be less certain. It is possible that many of the enzymes studied were identical with "lipases" or "lysophospholipases" or similar carboxyl esterases. Even for the pancreatic enzyme, however, the assignation as a "cholesterol esterase" becomes less convincing if the substrate specificity is considered.

There are many reports on "fatty acid specificity" and "sterol specificity" of cholesterol esterases. All of them can be summed up, in our opinion, by the statement that no such specificities exist. Neither sterol nor fatty acid are recognized by the enzyme or bound to it with their ring or chain structures. This is true for the digestive as well as for the intracellular cholesterol esterases. The different rates observed with substrates of different structure are due to different orientation and accessibility of the ester group (or the carboxyl and hydroxyl group) in the aggregated substrates, or in some cases (α-substituted acid, 3α-sterol) to steric hindrance. This interpretention explains the large changes in reaction rates that are observed on transition of cholesterol esters to the micellar state (Vahouny *et al.,* 1964a), and it is in agreement with the conclusion of Borgström and his co-workers (Morgan *et al.,* 1968; Erlanson and Borgström, 1970b; Filipek-Wender and Borgström, 1971; Erlanson, 1972) that pancreatic cholesterol esterase and monoglyceride hydrolase are identical, and that the sole requirement for the enzyme is a carboxyl ester in micellar solution.

Nedswedski, in 1936, postulated that the relationship of bile salt to cholesterol esterase was that of a true coenzyme rather than a substrate activator, and much apparently confirmatory evidence has been produced since then; but the question is by no means settled. Trihydroxy bile acids, i.e., cholate and its conjugates, are especially potent activators (Swell *et al.,* 1953). Tri- and dihydroxy bile acids, and monohydroxy bile acids to a lesser degree, protect cholesterol esterase against tryptic digestion and inactivation by PCMB (Vahouny *et al.,* 1964b; Hyun *et al.,* 1969) and form multimolecular aggregates with the enzyme, of which the cholesterol esterase–cholate complex is particularly clearly defined (Hyun *et al.,* 1972). It has been suggested that this complex is the active form of the enzyme; however, as discussed, the stoichiometry of the complex is puzzling. In other studies, the dihydroxy bile salts have been found almost as effective as activators (Swell *et al.,* 1953; Filipek-Wender and Borgström, 1971). Furthermore, taurocholate is a potent and quite specific activator of a cholesterol esterase from brain (Eto and Suzuki, 1971), and it protects cholesterol esterase from human aorta from inactivation (Kothari *et al.,* 1970); in these cases, the bile salt obviously cannot be the physiological coenzyme. These observations lend weight to the hypothesis (Filipek-Wender and Borgström, 1971) that the bile salts merely protect the enzyme from proteolytic destruction, especially at neutral or alkaline pH. The bile salts might also prevent surface denaturation of the enzyme at the oil–water interface, to which cholesterol esterase seems to be particularly sensitive. Finally, the possibility of substrate activation cannot be entirely discounted; in mixed micelles of trihydroxy bile salts and cholesterol esters, the ester groups are perhaps especially favorably oriented for attack by the enzyme, or the supersubstrate is fluidized by the bile salts.

Cholesterol esterases, digestive and intracellular, seem to be serine enzymes, but they also contain sensitive sulfhydryl groups. In this respect they resemble pancreatic lipase. Their active site is sterically less restrictive than that of the lipase and can accommodate secondary esters. On the other hand, cholesterol esterase (at least, that of the pancreas) cannot attack an ester in a closely packed interfacial structure, such as that formed by long-chain triglycerides, but needs a substrate in micellar solution, perhaps with an isolated, hydrated ester group.

VI

Phospholipases: Carboxyl Esterases

Phospholipases hydrolyze phospholipids, the class of bioorganic compounds that contain long-chain fatty acids and phosphoric acid bound to a common backbone, either glycerol in the glycerophospholipids or phosphoglycerides (A), or sphingosine in sphingomyelin (B).

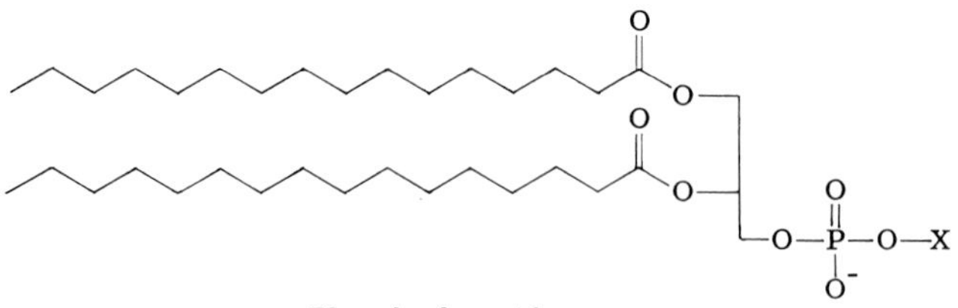

Phosphoglyceride

(A)

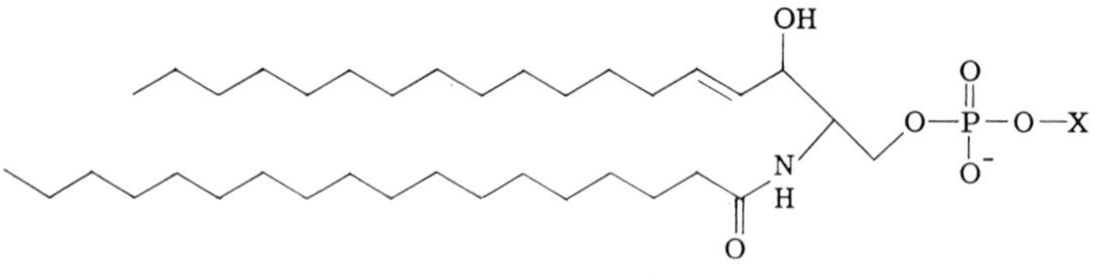

Sphingomyelin

(B)

As the formulas show, there are in these compounds various linkages that might undergo hydrolytic cleavage: each of the fatty acid ester bonds in the phosphoglycerides; the phosphate diester bonds, one carrying a hydrophobic, double chain, primary alcohol, the other a hydrophilic substituent X; and the fatty acid amide bond in sphingomyelin. Phospholipases are known for all these bonds except for the last one. (There exists, however, an enzyme that splits the amide bond of the parent compound of sphingomyelin, *N*-acylsphingosine or ceramide).

Contardi and Ercoli, in 1933, were the first to realize that different enzymes might hydrolyze the different bonds of the phosphoglycerides, and they proposed a scheme of classification for these enzymes. They called them "lecithases" because lecithin (in which the substituent X is choline) was the best known phospholipid. The enzymes are still often called "lecithinases" if the substrate in lecithin, i.e., phosphatidylcholine; "phosphatases" was often used before the term "phosphatides" for the substrates was replaced by "phospholipid." Contardi and Ercoli proposed the name "lecithase A" for enzymes able to split one fatty acid ester bond, "lecithase B" for enzymes splitting both, or the remaining ester in the lysophospholipid. "Lecithase C" would split the choline–phosphate linkage, "lecithase D" the diglyceride–phosphate. The present-day classification is based on this scheme, but some modifications have been made. "Lecithase C" has become phospholipase D, and "lecithase D" phospholipase C, because it seemed more logical and easier to memorize if the enzymes were lettered in a continuous order starting from the ester bond split from the primary ester bond of the glycerol.

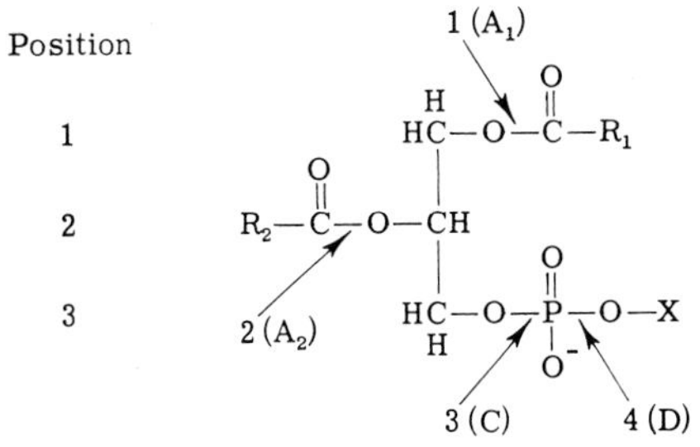

Phospholipases

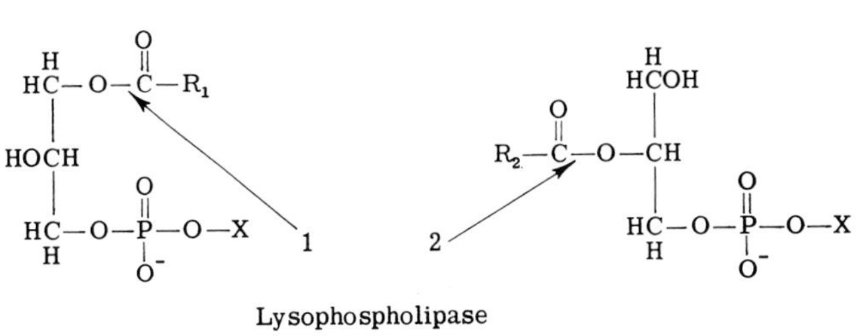

Lysophospholipase

It was also discovered that the phospholipases "A" of venoms and pancreas would hydrolyze the ester bond in one position of the glycerol only, and it became obvious that there must be two types of phospholipases A, one for esters at position 1, another for esters at 2. Van Deenen and de Haas (1966) proposed the names phospholipase A_1 and A_2; this proposal has been generally accepted.

The term "phospholipase B" seems to be slowly falling in disuse, mainly for the reason that there has not yet been any entirely conclusive evidence for the existence of enzymes that fulfill Contardi's and Ercoli's first definition of removing both fatty acids of a phospholipid. Enzymes which remove the remaining fatty acid of a lysophospholipid are known. According to the recommended enzyme nomenclature (Florkin and Stotz, 1965), they may be called phospholipases B, but another trivial name, also recommended, "lysophospholipase," is preferable and commonly used. Theoretically, lysophospholipases 1 and lysophospholipases 2 might exist, but it seems that all the enzymes so far investigated have been of the 1,2-type that hydrolyzes either ester bond. There is reason to believe that all lysophospholipases are of this nonspecific type, in fact, that they are not "phospholipases"—insofar as they do not require a phosphate group in their substrates—but rather they are nonspecific amphiphilic ester hydrolases.

With the disappearance of "phospholipase B" there remains no reason to retain the A in phospholipase A_1 and A_2. A simple and logical nomenclature might use the trivial names phospholipase 1, 2, and 3, according to hydrolysis at the corresponding positions of the glycerol, and phospholipase 4 for the enzyme splitting the phosphate–X bond. The somewhat hypothetical "phospholipase B, first definition" could become phospholipase 1,2. We realize that the naming of these enzymes by letters is historically entrenched, but there is real danger in the present trivial nomenclature of confusing A and B, B and lysophospholipase, and especially C and D. In this book we are therefore using a nomenclature of numbers only.

Our presentation of phospholipases will start with enzyme 2 rather than 1, for the reason that phospholipase 2 is by far the best known of all phospholipases and can serve in many respects as a prototype and reference for the other members of the group.

There are a good number of reviews on phospholipases; for the early history of research, Wittcoff's book "The Phosphatides" (1951); for subsequent developments, Hanahan (1957), Slotta (1960), Kates (1960), van Deenen (1964), van Deenen and de Haas (1966), Hill and Lands (1970), Hanahan (1971), van den Bosch *et al.* (1972), and Gatt and Barenholz (1973).

I. PHOSPHOLIPASE 2

A. Occurrence and Function

By definition, phospholipase 2 is an enzyme that hydrolyzes the ester linkage in position 2 of a phosphoglyceride. Historically, however, it was realized only around 1960 that the well known phospholipases of snake venom and pancreas possessed positional specificity. In the literature before 1967, these enzymes are referred to as phospholipase A or lecithinase A, with the question of specificity undecided. The older literature can nonetheless be interpreted without ambiguity, since it is now known that it was almost exclusively the 2 enzymes that were described. Phospholipases 1 have only been discovered in the last few years.

Phospholipase 2 was first detected in snake venom. In 1903, Kyes noticed that red blood cells in isotonic suspension were hemolyzed by cobra venom only in the presence of lecithin. Willstätter and Lüdecke (Lüdecke, 1905) suggested that an enzyme in the venom acted on the lecithin with removal of oleic acid, and Delezenne and Ledebt (1912) and Delezenne and Fourneau (1914) confirmed this suggestion by isolating the other reaction product, "lysolecithin," which contained mainly palmitic acid. This distribution of fatty acids was due to the circumstance, then unsuspected, that in natural phospholipids the position 1 tends to be occupied by saturated, position 2 by unsaturated acids.

The book on phospholipids by Wittcoff (1951) should be consulted for an account on the history of phospholipase research until the year 1950.

Phospholipases 2 have proven to be of ubiquitous distribution in nature. At present, the enzymes may be conveniently grouped in three classes according to the sources in which they are found: (1) as enzymes of venoms, (2) as digestive enzymes, (3) as intracellular enzymes occurring in the tissues of animals or in microorganisms.

1. Phospholipase 2 in Venoms

The enzymes occur in all reptilian venoms (Mebs, 1970) and also in the venoms of invertebrates, especially arthropods such as bee (Habermann, 1972), scorpion (Kurup, 1965), and ant (Lewis *et al.*, 1968), but also in coelenterates (Stillway and Lane, 1971). Many investigations have been carried out, mainly with snake venoms, concerning the connection between phospholipase activity and toxicity; a review on the subject has been provided by Condrea and de Vries (1965). There have repeatedly been reports, some very recent (Currie *et al.*, 1968; Delori, 1971; Ruebsamen *et al.*, 1971), of toxic phospholipase fractions or of toxins with phospholipase activity; but ever since it was shown that a toxic preparation, the crotoxin from cobra venom, could be separated into a neurotoxin and a nontoxic phospholipase (Neumann and Habermann, 1956) and that bee venom could be similarly fractionated (Neumann and Habermann, 1954; Habermann, 1972), it has been accepted that, in general, the phospholipases are not identical with the toxic principle. It is believed that they may assist the toxin to spread within the victim by hydrolyzing phospholipids and thus breaking down the membranes of cells. In order to attack intact membranes, the enzymes of some snakes require the help of a small basic protein, the direct lytic factor (Neumann and Habermann, 1954), which can be replaced by lipophilic synthetic polypeptides (Klibansky *et al.*, 1968) or even by calcium (Condrea *et al.*, 1970). Some snake phospholipases, for instance those of cobras, will attack membranes only indirectly through the formation of lysocompounds from phospholipids that are not part of the membrane but have been added to the system (Condrea and de Vries, 1965); it is not clear how exactly the lysophospholipids destroy the membrane. The entire problem of the synergistic action of the phospholipases, and indeed of their function in the venom, is still incompletely understood. It seems probable that the enzymes are also active in the digestion of the snake's victim, but it is debatable if this assistance is of much practical importance for the reptile.

The phospholipases 2 of venoms have in common a low molecular weight, usually less than 15,000 for the monomers; they are thus among the smallest of all enzymes. They are also distinguished by their resistance against destruction by heat. Below pH 7 they will survive heating to 95°C for many minutes.

2. Phospholipases 2 as Digestive Enzymes

Among the digestive phospholipases 2, that of the pig has been most thoroughly investigated; those of man and rat have also received attention. Presumably, all mammals possess the corresponding enzyme; it is not known if it also occurs in other animals. Like the enzyme from snakes, the porcine enzyme is heat-resistant and has a low molecular weight (14,000), but while the reptilian enzymes are released from the poison glands in an active form (though often, or perhaps always, together with inhibitory protein), the mammalian digestive enzymes are produced by the pancreas as zymogens or proenzymes which require activation by trypsin. The purpose of this mechanism is obviously the prevention of pancreatic autodigestion, as in the case of the proteinases and their zymogens. The existence of a zymogen of the phospholipase in the pancreas of the pig was first reported by de Haas in 1967 (de Haas *et al.*, 1968b). Arnesjö *et al.* (1967) showed that the phospholipase activity of rat pancreatic juice increased tenfold on treatment with trypsin, and Belleville and Clément (1968a,b) demonstrated a similar effect on porcine, rat, and human pancreatic juice.

3. *Phospholipases 2 as Intracellular Enzymes*

In comparison with the efforts expended in the research on the phospholipases from exocrine glands, the corresponding intracellular enzymes have long been neglected. The last years have seen a burst of activity in the field, in part as a consequence of the work of W. E. M. Lands, who showed that enzymes exist in animal tissues that can acylate lysophospholipids (Lands, 1960), and who postulated (Lands, 1958, 1965) that a metabolic turnover and exchange takes place among the fatty acids of intracellular phospholipids, an exchange in position 1 as well as in position 2 of the glycerol. The phospholipases 1 and 2 that are required for such an exchange have since been found in many animal organs as well as in microorganisms. The enzyme 2 has been found in mitochondria. In lysosomes and "microsomes," phospholipase 2 is accompanied by 1. Microorganisms may contain both enzymes. A 2 zymogen seems to occur in mammalian blood.

The phospholipases 2 of animal tissues seem to fall into two groups: one containing enzymes that are active at neutral or basic pH, are heat-resistant, and are probably of low molecular weight; they are apparently related to the 2 enzymes of exocrine glands. The other group comprises the "acidic" phospholipases 2 of lysosomes, which are active at pH 4–5, and about which little is known to date except their existence.

B. Assay and Kinetics

1. *Assays*

Historically, the first method of assaying phospholipase activity was the one that led to the detection of the enzymes (Kyes, 1903), the lysis of red blood cells in the presence of lecithin. Hemolysis in an agar support may be used with advantage for the scanning of microorganisms for phospholipase activity (Bernard and Denis, 1969), and the hemolytic reaction is still important in research on the action of the enzyme on membranes, but as an assay of the enzyme, it has fallen in disuse. Another indirect method has employed the clearing effect of lysolecithin on lecithin emulsions (Doizaki and Zieve, 1964, 1966; Marinetti, 1965; Habermann and Hardt, 1972). However, the direct method of measuring the release of fatty acids from the substrate has been universally adopted. A circuitous variant that measures the decrease of ester bonds with the hydroxamate reaction has often been used but is now rarely employed (Shapiro, 1953; Augustyn and Elliott, 1969); the direct titration of fatty acids is generally preferred.

The three principal assay procedures now in use differ with respect to the medium in which the reaction takes place: either a solution of the phospholipid in ether, or an aqueous dispersion, or a true solution of a water-soluble phospholipid. The first method is derived from the discovery of Hanahan (1952) that the phospholipases 2 of pancreas and snake venom form ether-soluble aggregates with their substrates; hydrolysis then proceeds in the organic solvent. In a recently described procedure (Saito and Hanahan, 1962; Wells and Hanahan, 1969), the substrate solution consists of phosphatidylcholine, 15 μmoles/ml, in 95% ether and 5% methanol. The enzyme preparation is dissolved in aqueous 0.22 M NaCl, 0.02 M CaCl$_2$, and 0.001 M EDTA (to bind contaminating metals) at pH 7.5. Of this solution, 25 μl are added to 2 ml of the substrate solution, mixing is achieved by shaking, and the reaction is allowed to proceed for 10 minutes. The amount of water in the reaction mixture, 25 μl, is critical, and the enzyme concentration should be such that at most only 25% of the substrate is hydrolyzed. Otherwise, a lysophosphatidyl enzyme aggregate will precipitate and the rate of the reaction changes. The enzymatic hydrolysis is stopped by the addition of 3 ml of 95% ethanol, a drop of indicator is added, and the fatty acid is titrated with 0.02 N NaOH in 90% ethanol.

The method of enzymatic hydrolysis in ether is especially valuable for the analysis of phospholipids, since it allows the convenient recovery of substrates and products. It is also useful in monitoring the degree of enzyme purification during the course of an isolation procedure. For the kinetic analysis of phospholipase 2 action, however, it is less convenient, because it does not permit continuous titration during the reaction. This can be accomplished in the assay systems using aqueous dispersions or solutions of the substrate. For the assay of the phospholipase 2 in pancreatic juice, a procedure by Figarella and Ribeiro (1971), modified from de Haas et al. (1968a), uses an emulsion of egg yolk in water with CaCl$_2$ at 0.001 M and sodium deoxycholate at 0.006 M, at pH 9. The release of fatty acid is then continuously followed and recorded with an automatic titrator. A similar procedure is described by Nieuwenhuizen et al. (1974).

While the egg-yolk method is practical and convenient for routine assays, it is still unsatisfactory from the viewpoint of the enzymologist, since it is biphasic and does not allow the determination of a true substrate concentration and, consequently, of binding constants. The route toward a monophasic system was opened by Roholt and Schlamowitz (1961) with the use of the soluble synthetic phospholipid dihexanoyllecithin. It was shown by these authors that this substrate was dissolved in monomolecular form at concentrations below 11 m*M*. The hydrolysis by the enzyme can be followed by automatic titration. De Haas *et al.* (1971) have recently studied the action of pig pancreatic phospholipase 2 on this substrate and its homologs, diheptanoyl- and dioctanoyllecithin, and shown that the least soluble lipid, the dioctanoyl compound, is the best substrate, and that the enzyme preferentially hydrolyzes substrates in micellar form; this is probably a manifestation of micelle–enzyme aggregation. A further step toward monomolecularly dissolved substrates has recently been taken by Wells (1972) with the use of dibutyryllecithin, which had been shown by van Deenen and de Haas (1963) to be hydrolyzable by phospholipase 2. This substrate is water soluble as a monomer up to a concentration of 75 m*M*.

2. Kinetics

The study of the kinetics of phospholipid hydrolysis is plagued with the same difficulties encountered in the studies of lipase. Most phospholipids when dispersed in water form multimolecular aggregates to which the phospholipases are adsorbed. The Michaelis constant K_m determined under such circumstances will include the enzyme–supersubstrate dissociation constant, and it will have little value as a constant measuring enzyme–substrate affinity. Even the maximal enzymatic hydrolysis rate, V, cannot be determined as a definitive parameter, and turnover numbers can therefore not be calculated, because the substrate may be present in several aggregate forms (Tinker and Pinteric, 1971) that react at different rates. The principal factors that determine these differences are the closeness of the molecular packing and the size of the substrate micelles. The transition between three micellar forms of lecithin–bile salt micelles has been shown to result in three different reaction rates (Olive and Dervichian, 1968).

Following the pioneer study of Hughes (1935) the monomolecular film technique has been applied in countless studies of phospholipase and phospholipids. Much valuable information has been obtained on the nature of lipid layers and membranes; for an introduction in this field, articles by Shah and Schulman (1967), Dawson (1968, 1969), and Colacicco (1969) may be consulted. However, less has been learned from monolayer experiments on the action and mechanism of phospholipases. A study by Zografi *et al.* (1971) elaborates this point. It was shown by these authors that maximal enzymatic hydrolysis of diheptanoyllecithin by phospholipase 2 occurred at a surface pressure of 8 dynes/cm, corresponding to approximately 10^{14} lecithin molecules per square centimeter, the same value as found by Hughes (1935) for the optimal hydrolysis of egg lecithin. The rate then measured had to be considered as a minimal specific activity, because the degree of adsorption of the enzyme to the interface could not be measured. For the same reason, inhibition of the enzyme by competitive inhibitors present in the film could not be studied.

Kinetic studies of phospholipase action in ether might be thought to promise more success because the system is, ostensibly, monophasic. However, this advantage is illusory. In ether–methanol–water mixtures, lecithin also forms aggregates, probable of an inverted pattern with the fatty acid chains on the outside (Misiorowski and Wells, 1971), and the enzyme is bound in these aggregates.

At the present stage of the science of interfacial kinetics, a system that lends itself better to the kinetic analysis of phospholipase 2 action is that of short-chain phosphatidylcholines dissolved in water, introduced by Roholt and Schlamowitz (1958, 1961). These authors first investigated the calcium–phospholipase affinity; from their experiments a meaningful (if less than accurate) figure for a binding constant, $K_{Ca} < 5 \times 10^{-4}$ for the enzyme of *Crotalus durissus terrificus,* can be extracted. Shipolini *et al.* (1971a) have used the same substrate, dihexanoyllecithin, in studies of the enzyme of bee venom. Wells (1972) has extended the method by studying the action of the phospholipase 2 of *Crotalus adamanteus* on monomolecular solutions of dibutyryllecithin.

C. Isolation

The last few years have seen a burst of publications describing the isolation of phospholipases 2, a natural consequence of the introduction of gel permeation and ion-exchange chromatography. The sequence of two such operations often achieves a good purification of the enzyme and even the separation of the different molecular species that are often present in the same source. For this special purpose, preparative electrophoresis and electrofocusing have recently been applied as powerful tools.

Phospholipases 2 have been prepared from the venoms of over a dozen different species of snakes: from *Naja naja* (Salach *et al.,* 1968, 1971a; Braganca *et al.,* 1969); *Naja oxiana* (Sakhibov *et al.,* 1970); *Naja nigricollis* (Wahlström, 1971); *Vipera russellii* (Salach *et al.,* 1971a); *Crotalus atrox* (Wu and Tinker, 1969; Hachimori *et al.,* 1971); *Crotalus adamanteus* (Saito and Hanahan, 1962; Wells and Hanahan, 1969); *Agkistrodon halys blomhoffii* (Kawauchi *et al.,* 1971a); *Agkistrodon piscivorus* (Augustyn and Elliott, 1970); *Bothrops neuwiedii, B. jaracara, B. jaracarussu,* and *B. atrox* (Vidal and Stoppani, 1966, 1971a,b); *Laticaudia semifasciata* (Uwatoko-Setoguchi and Ohbo, 1969; Tu *et al.,* 1970); and *Vipera palestinae* (Shiloah *et al.,* 1973).

The isolation of bee venom phospholipase has been described by Habermann and Reiz (1965), Munjal and Elliott (1971), Shipolini *et al.* (1971a,b), and Jentsch (1972).

The pancreatic enzyme has been isolated from the pig (de Haas *et al.,* 1968a) and from the rat (Arnesjö and Grubb, 1971); from the pig, the zymogen has also been prepared (de Haas *et al.,* 1968b).

Attempts to purify and isolate phospholipases 2 other than those from exocrine glands have not been equally successful. Highly concentrated preparations have been obtained from the blood plasma of the rat (Paysant *et al.,* 1969a), but the homogeneity of the enzyme has not been established. The intracellular lipases are, in most cases, bound to membranous structures; this, of course, complicates their purification. However, since they are of such obvious importance in the metabolism of membranes and since there is brisk activity in this field of biochemical research, the isolation of some of these enzymes will certainly be reported before long. Partial purifications have already been achieved (Waite and Sisson, 1971).

In the following we shall describe the isolation of two phospholipase 2 isoenzymes from the venom of a snake according to one of the best documented procedures (Wells and Hanahan, 1969), the isolation of the major molecular species of the enzyme from bee venom (Shipolini *et al.,* 1971a), and the isolation of the porcine phospholipase and its proenzyme (de Haas *et al.,* 1968a,b).

1. Isolation of Two Phospholipases 2 from a Snake Venom

The purification of the enzyme from the rattlesnake *C. adamanteus* (Wells and Hanahan, 1969) starts with the extraction of the commercially available dry venom (with aqueous 0.1 M NaCl, 0.05 M tris HCl pH 8, 0.001 M EDTA) and centrifugation of the extract, which is then fractionated on Sephadex G-100, then, after concentration of the active fractions, on Bio-Rex 10. The fractions containing the enzymes are located with the assay procedure using the hydrolysis of phosphatidylcholine in ether. They are then applied to a cationic ion-exchange column, DEAE-cellulose (DE-52) from which an increasing NaCl gradient, from 0 to 0.2 M at pH 8, elutes the phospholipases in two well separated peaks (Fig. VI-1). Each fraction is then further purified by chromatography on the anionic ion exchanger SE-Sephadex (SP-Sephadex) in citrate buffer at pH 5.2, and finally by crystallizing the enzymes, α and β, from an aqueous ammonium sulfate solution at pH 5.0–5.5. The final yield for both enzymes is 39%.

Table VI-1 shows the course of the purification. The purity of the enzyme was tested by disk gel electrophoresis and by high-speed centrifugation, and on this basis the preparations were judged to be at least 95% pure. A recrystallization did not increase the specific activity.

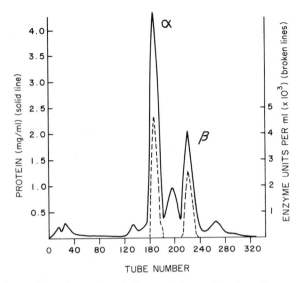

Fig. VI-1   Separation of two phospholipases 2 from *Crotalus adamanteus* venom on DEAE-cellulose. Gradient, 0.05 *M* tris HCl (pH 8) → 0.2 *M* NaCl. From Wells and Hanahan (1969).

The phospholipases of other snake venoms have been isolated in similar fashion. Some procedures include heat treatment to denature and remove contaminating proteins. With the modern method of isoelectric focusing, which separates proteins according to their isoelectric points, it has been

TABLE VI-1

Purification of Phospholipase 2 from *Crotalus adamanteus* **Venom**[a]

Step	Total protein (mg)	Protein (%)	Total units × 10⁻³	Activity (%)	Specific activity	Purifi- cation
Centrifuged solution[b]	22,125	100	2236	100	102	
Sephadex G-100[b]	9,130	41	2072	93	229	2.3
Bio-Rex 10[b]	2,289	11.2	1797	73	722	7.1
DE-52 α	765	3.5	790	35	1030	10.3
DE-52 β	487	2.2	564	25	1160	11.6
SE-Sephadex α	532	2.4	733	33	1375	13.8
SE-Sephadex β	287	1.3	394	18	1370	13.7
α-Crystals	412	1.85	565	25	1415	14.5
β-Crystals	210	0.95	299	13	1425	14.3

[a] Wells and Hanahan (1969).
[b] Mixture of both isomers of the enzyme.

possible to separate nine different molecular species of the enzyme from *Naja naja* and seven from *Vipera russellii* and isolate some of them in a pure state (Salach *et al.*, 1971a).

2. Bee Venom Phospholipase 2

Several different forms of the enzyme are present in bee venom. Shipolini *et al.* (1971a) have described the isolation of the major enzyme species. Lyophilized venom was dialyzed against water to remove small peptides, and then separated, at pH 4.6, on a column of Sephadex G-50. The active fractions (Fig. VI-2) were applied to the anionic exchanger SE-Sephadex in three successive runs. In the third one, at pH 9.1, one peak contained a phospholipase 2 that crystallized on concentration by forced dialysis. The enzyme appeared pure as judged by the standard methods of electrophoresis on starch gel and polyacrylamide gel, isoelectric focusing, and sedimentation and diffusion analysis in the ultracentrifuge, and N-terminal analysis by two methods revealed only one single amino acid, isoleucine.

3. Isolation of Phospholipase 2 from Pancrease

In the first attempt to isolate this enzyme from ox pancreas, Rimon and Shapiro (1959) noticed that the enzyme could be detected in aged pancreas but not in the fresh gland. Furthermore, all activity was bound to

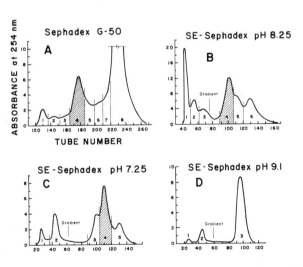

Fig. VI-2 Isolation of the major phospholipase 2 of bee venom by gel filtration and consecutive ion exchange chromatography at different pH. In B, C, and D, a tris citrate gradient 0.05 $M \rightarrow$ 0.03 M was applied. Each shaded peak was applied to the next column. From Shipolini *et al.* (1971a).

insoluble particles. The reasons for these results are made clear in the work of de Haas *et al.* (1968a) on the enzyme of pig pancreas. These authors showed that the pancreas contains a zymogen of phospholipase 2 which is activated by trypsin, and that the enzyme has a strong affinity to lipids, with which it combines above pH 4 to form multimolecular aggregates reminiscent of the "fast" pancreatic lipase.

The isolation of the enzyme according to de Haas *et al.* (1968a) starts with the storing of a homogenate of defatted pig pancreas for 12–24 hours at 22°C. The enzymatic activity is measured by the titration the fatty acids released from an egg yolk emulsion at pH 8. When the maximal activity is reached, the homogenate is brought to pH 4 and heated at 70°C for 3 minutes. Insoluble material and a fatty acid layer are removed after centrifugation. The solution is now dialyzed; then, the proteins precipitating at pH 7 between 0.4 and 0.6 saturation with ammonium sulfate are collected and chromatographed by gel filtration on Sephadex G-50. The proteins of the active fraction are separated on an anionic DEAE-cellulose column at pH 8, with a salt gradient increasing to 0.3 M NaCl, and the final purification is achieved on a column of carboxymethylcellulose in acetate buffer of pH 6 and with a salt gradient between 0 and 0.4 M NaCl. Table VI-2 summarizes the course of the isolation procedure. The yield of enzyme is 31%. For the routine preparation of the enzyme it is preferable to isolate the proenzyme and activate it with trypsin (Nieuwenhuizen *et al.*, 1974).

TABLE VI-2

Purification of Porcine Pancreatic Phospholipase 2 from 500 g Defatted Pancreas[a]

Step	Total activity (10^3 units)	Specific activity	Yield (%)
1. Crude homogenate after storage	256	4	100
2. Heat treatment at pH 4.0	205	—	80
3. Removal of precipitated material and floating fat by centrifugation and filtration	175	7	69
4. Dialysis	193	47	75
5. 0.40–0.60-Saturated ammonium sulfate precipitate	150	158	59
6. Sephadex G-50 percolation	135	350	53
7. DEAE-cellulose	88	550	34
8. Carboxymethylcellulose	79	850	31

[a] De Haas *et al.* (1968a).

The enzyme of the pancreatic juice of the rat has been isolated by Arnesjö and Grubb (1971). The activation of the enzyme could not be left to auto-digestion, because this led to large losses of activity. Therefore, the juice was heated to 75°C in order to destroy the native proteolytic enzymes and zymogens that otherwise degraded the phospholipase, and the prophospho-lipase was then quantitatively activated by trypsin. The proteins were fractionated on Sephadex G-50 gel, then on carboxymethyl-cellulose at pH 7 with an NaCl gradient. The active fractions were desalted on Sephadex G-25, and a lyophilized 200-fold purified enzyme was obtained in an overall yield of 41%.

The isolation of the phospholipase 2 from human pancreatic juice, by analogous methods, has been reported by Wittich and Schmidt (1969). Figarella *et al.* (1971) purified the zymogen of human pancreatic juice.

4. Pancreatic Prophospholipase 2

The porcine prophospholipase has been isolated by de Haas *et al.* (1968b) by a procedure very much similar to that for the active enzyme, with one major difference. The endogenous pancreatic proteinases, in par-ticular trypsin, rapidly convert the zymogen to phospholipase unless pre-cautionary steps are taken, for instance, carrying out the chromatographic separations at low temperature and high speed, and especially inhibiting the proteolytic enzymes. In the first steps of the purification, the endoge-nous trypsin inhibitors are still sufficiently active, but at the later stages the irreversible trypsin inhibitor diisopropyl fluorophosphate (DFP) must be added in rather high concentration, 0.01 molar. Some active phospholipase is nevertheless formed, but since the isoelectric points of enzyme (pH 7.4) and zymogen (pH 6.5) are different, a separation can be achieved on DEAE- and carboxymethyl-cellulose chromatography. The proenzyme is isolated in an overall yield of 36%. The large-scale preparation of the zymogen and its activiation have been described by Nieuwenhuizen *et al.* (1974).

D. Molecular Properties

1. Molecular Weights

The phospholipases 2 of exocrine glands are distinguished by low molec-ular weights, almost always below 15,000 (Table VI-3). (For comparison, the molecular weights of pancreatic proteinases are around 25,000; lipase, 50,000; amylase, 50,000; only ribonuclease is similarly low). Table VI-3 lists some exceptions to this rule that require explanations. The monomeric enzymes of the *Crotalus* species have the regular low weight, but the stable

TABLE VI-3

Molecular Properties of Phospholipases 2 of Exocrine Glands

Source	Isoenzymes	Molecular weight		Disulfide bridges in monomer	Isoelectric pH	Miscellaneous properties	Reference
		Monomer	Dimer				
Snake venoms							
Crotalus adamanteus	2	15,000	29,800[a]	7	—	70% α-helix	Wells and Hanahan (1971)
		15,000	29,800[a]	7	—		
Crotalus atrox	1	14,700	29,500[a]	7	—	—	Hachimoro et al. (1971); Wu and Tinker (1969)
Agkistrodon halys	2	13,900	—	7	10.0	20% α-helix, 27% α-helix; C-terminal cysteine; N-terminal pyroglutamic acid	Kawauchi et al. (1971a)
		13,700	—	7	4.0		
Agkistrodon piscivorus	>1	14,000	—	—	>9	—	Augustyn and Elliott (1970)
Laticauda semifasciata	(1)	11,000	—	6	6.6	—	Tu et al. (1970)
Laticauda semifasciata	4	9,800–11,000	—	—	5.6; 6.6; 7.8; 8.9	—	Uwatoko-Setoguchi and Ohbo (1969)
Naja naja	9–11	8,500–20,200	—	(6)	4.6–5.7	—	Salach et al. (1971a)
Naja nigricollis		13,000	—	6–7	7.8	—	Wahlström (1971)
Vipera russelli	7	14,600	—	7	5.5	—	Salach et al. (1971a)
Vipera palestinae	2	<15,000–23,000	—	6	4.6–9.9	0.1% lysophospholipase activity	Shiloah et al. (1973)
		16,000	—				
Bee venom							
Apis mellifera	>5	19,000[a] (14,629)[b]	40,000	6	10.5	Amino acid sequence known; glycoprotein; C-terminal tyrosine; N-terminal isoleucine	Shipolini et al. (1971a,b)
Pancreas							
Pig	1	13,700	—	6	7.4	Sequence known; 55% α-helix; C-terminal cysteine; N-terminal alanine	De Haas et al. (1968a)
Pig, zymogen	1	14,500	—	6	6.5	Heptapeptide attached to end of chain; N-terminal pyroglutamic acid; 50% α-helix	De Haas et al. (1968b)

[a] Stable form.
[b] Without carbohydrate moiety.

form of the isolated enzymes is the dimer (Wells, 1971b). In bee venom, on the other hand, the dimer exists only in concentrated solutions at high pH, and the still relatively large molecular weight of the monomer is the consequence of a large carbohydrate moiety in the enzyme (Shipolini *et al.*, 1971a); the carbohydate-free protein has a weight of 14,600. Munjal and Elliott (1972) report a molecular weight of 18,500 (without carbohydrate), Jentsch and Dielenberg (1972) 10,200 for a phospholipase from bee venom. The large weights of some of the enzyme species from *Naja naja* and *Vipera russellii* are indeed exceptions. In *N. naja,* only one out of nine or more isoenzymes has a weight of 20,000, all others are below 16,000. In *V. russellii,* the bulk of the enzymatic activity is found associated with molecular weights lower than 16,000 (Salach *et al.*, 1971a). A molceular weight of 9400 has been reported for an isoenzyme from *Bothrops neuwiedii* (Vidal *et al.*, 1972a).

2. Disulfide Bridges

A second characteristic of the exocrine phospholipases is their high content of cysteine, together with the absence of free sulfhydryl groups (Table VI-3). On the average, one out of ten amino acid residues is a cystein connected by a disulfide bridge to another part of the molecule. For comparison, in the pancreatic proteinase chymotrypsin only one out of thirty-three residues carries a disulfide bridge.

In combination, the low molecular weight and the abundance of disulfide bonds must make the molecules compact and resistant against unfolding. These features are also a sufficient explanation for the heat resistance of the enzymes. The adaptive purpose of such a structure is probably the protection of the enzyme against denaturation at the phospholipid–water interface, which is its natural place of activity.

3. Isoenzymes

The existence of a family of phospholipases 2 in the venom of snake species was first reported by Doery and Pearson (1961). In bee venom and in most snake venoms, phospholipase 2 occurs in several molecular forms (Table VI-3). So far, only in *Crotalus atrox* has a single enzyme only been found. The enzyme from *Laticauda semifasciata* has been isolated as a single species (Tu *et al.*, 1970), but other authors (Uwatoko-Setoguchi and Ohbo, 1969; Uwatoko-Setoguchi, 1970) have found four enzymatically active proteins in the venom of this snake. The venoms of four *Bothrops* species (not listed in Table VI-3) also contain two isoenzymes each (Vidal and Stoppani, 1971a; Vidal *et al.*, 1972a,b). From pancreatic sources, only single enzymes have been isolated. As for intracellular phospholipases 2, it would be surprising if the occurrence of isoenzymes should not soon be reported.

There can be very much or very little difference between the phospholipases of the same venom. The α and β enzymes of *Crotalus adamanteus* have the same molecular weight and an identical amino acid composition (Wells and Hanahan, 1969); they must differ in the number of their amide groups, since their isoelectric points are different. The two enzymes of *Agkistrodon halys blomhoffii* also have the identical number of amino acids but differ in their amino acid composition. The acidic enzyme, with an isoelectric pH of 4.0, has 13 glutamic acid and 8 lysine residues; the basic enzyme, isoelectric at pH 10.0, has 6 glutamic acid and 17 lysine residues (Kawauchi *et al.*, 1971b). Numerous enzymatic species with very different molecular weights have been found in the venoms of *Naja naja* and *Vipera russellii* (Table VI-3).

A number of phospholipase 2 species are reported to be present in bee venom (Fig. VI-2) (Shipolini *et al.*, 1971a). Since the major molecular species in bee venom has been shown to contain a large carbohydrate moiety, it is tempting to speculate that differences in this part of the molecule may be at the base of some or all of the isoenzymes. Snake venoms, on the other hand, where they have been analyzed (Wells and Hanahan, 1969; Kawauchi *et al.*, 1971b), have been found to contain only insignificant quantities of carbohydrate.

4. Dimers

The major phospholipase 2 of bee venom associates to form a dimer at high pH in concentrated solutions (Shipolini *et al.*, 1971a); normally, the enzyme seems to occur in monomeric form. The two phospholipases 2, α and β, of the snake *Crotalus adamanteus* (Wells, 1971b), however, and the single enzyme species of *Crotalus atrox* (Hachimori *et al.*, 1971) are stable as dimers. A preparation of *C. atrox* enzyme has been described that had a molecular weight of 14,500 (Wu and Tinker, 1969), but a preparation by Hachimori *et al.* (1971) had a molecular weight of 29,500. It could be broken down to a monomer of MW 15,000 by electrophoresis in the presence of sodium dodecyl sulfate and mercaptoethanol. Of the two isoenzymes of *Bothrops neuwiedii,* one forms a dimer at pH 7.5 but the other does not (Vidal *et al.*, 1972a). The α and β forms of the *C. adamanteus* enzyme, MW 30,000, can both be converted to monomers of MW 15,000 at pH 2 or with concentrated urea solutions (Wells, 1971b). The dissociation is fully reversible. The dimeric proteins are composed of identical subunits, and hybrid formation does not take place during the reassociation of the α and β monomers. On dimerization, one lysine residue per dimer becomes exposed (Wells, 1973). The correlation of the empirical molecular weights in urea solutions and the activity of the enzyme against dibutyryllecithin suggested that it is the dimer that is the active form of the enzyme.

5. Amino Acid Sequence

Once purified, phospholipases 2 are good subjects for amino acid sequence studies because of their relatively small molecular weight. Techniques of such sequence studies are described in the original papers and in numerous monographs, e.g., the collection in Volume XXV of "Methods in Enzymology" (Hirs and Timasheff, 1972). The complete sequence of porcine pancreatic phospholipase and its proenzyme and the sequence in the major bee venom enzyme have been worked out. Among the snake venom enzymes, only the partial sequence of one enzyme has so far been reported.

The amino acid sequence of the major porcine pancreatic prophospholipase (Maroux *et al.,* 1969; de Haas *et al.,* 1970a,b) is shown in Fig. VI-3. The N-terminal acid (1) is pyroglutamic acid in which the amino group is blocked by β-lactam formation. In a minor (5%) prophospholipase, residues 1 to 4 are missing (Nieuwenhuizen and de Haas, 1972). Tryptic hydrolysis between arginine (7) and alanine (8) leads to the formation of the active enzyme with alanine (8) as the N-terminal amino acid. The location of the six disulfide bridges has also been established. The three-dimensional structure of the enzyme is not yet known, but Fig. VI-3 already reveals some interesting features. In particular, it gives a good picture of the compactness of the protein. The C-terminal amino acid, cysteine (130), is linked to the remainder of the molecule by a disulfide bridge. The basic amino acids are accumulated in the C-terminal part of the chain; of the nine lysines, five are located between position 114 and 128. Recent hydrogen titration studies (Janssen *et al.,* 1972) indicate the presence of three more acidic groups than given by Fig. VI-3.

The transformation of the pancreatic proenzyme to the enzyme by tryptic hydrolysis has been studied by Abita *et al.* (1972). Activation produces only very limited changes in the folding of the protein. Tryptophan (10) passes from a polar to an apolar environment. The α-amino group of alanine-8, which becomes alanine-1 in the enzyme, is essential for activity; Abita *et al.* proposed that it forms a salt bridge that stabilizes the geometry of the active site. In the most recent investigations (Pieterson *et al.,* 1974b), it was found that proenzyme and enzyme alike are capable of hydrolyzing monomeric lecithins in true solution, but that only the activated enzyme can attack micellar substrates. It appears, then, that the re-

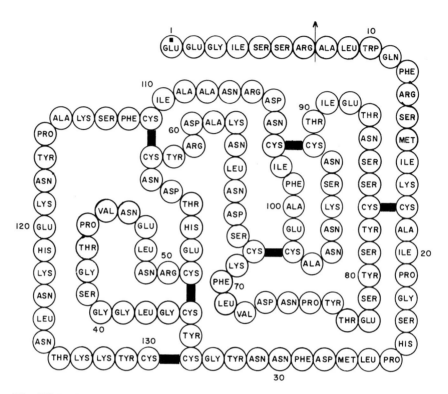

Fig. VI-3 Amino acid sequence and disulfide bridges of porcine pancreatic pro-phospholipase 2. From de Haas *et al.* (1970b).

active site of the enzyme is already functional in the proenzyme. Tryptic activation leads to the formation of a new site that binds the enzyme to the supersubstrate, the micelle. It has been suggested that the binding of this "anchoring site" to the micelle may also induce further conformational changes of the reactive site (Pieterson *et al.,* 1974b). A schematic presentation of the creation of the supersubstrate binding site by tryptic activation is shown in Fig. VI-4.

The primary sequence of the principal bee venom enzyme (Shipolini *et al.,* 1971b) is shown in Table VI-4. The location of the six disulfide bridges is not yet known. Compared with the pancreatic enzyme, the sequence shows interesting differences. The charge distribution appears more even than in the pancreatic enzyme. The C-terminal end is not rejoined to the middle of the chain by a disulfide bridge; it is left dangling from cysteine (108) onward. Perhaps the charged amino acids may form salt linkages to the rest of the molecule. The formula weight of the protein

PRECURSOR   PHOSPHOLIPASE A

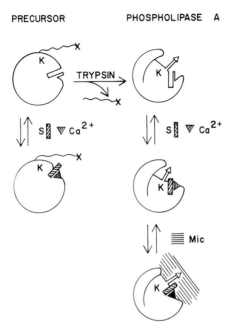

Fig. VI-4 Schematic representation of pancreatic phospholipase 2 and its zymogen according to Pieterson *et al.* (1974b). The monomer (S, substrate) binding site and calcium binding site together constitute the catalytic region (K). During tryptic activation, the zymogen loses its activation heptapeptide (X) and undergoes a conformational change in which an ion pair is formed between the —NH$_3^+$ group of the N-terminal alanine and a buried carboxylate. Linked to the formation of this salt bridge is the creation of a supersubstrate binding site (arrow) required for interaction with micelles. From Pieterson *et al.* (1974b).

of Table VI-4 is 14,629, whereas the molecular weight determined by ultracentrifugation is near 19,000. The difference is made up by covalently bound carbohydrate. For instance, the end of the amino acid chain from isoleucine (1) to lysine (14) contains glucosamine, mannose, galactose, and fucose.

The phospholipase of the snake *Agkistrodon halys blomhoffii* has a C-terminal cysteine like the pancreatic enzyme (Samejina *et al.,* 1970; Kawauchi *et al.,* 1971b) and an N-terminal pyroglutamic acid like the pancreatic zymogen. The C-terminal sequence in the last eight positions is acidic because of the presence of two glutamic acid residues. In speculative anticipation, we might suggest that the enzymes from snakes and from pancreas are of similar construction but vary in those charge distribution patterns that determine the interfacial orientation of the enzyme while it is

TABLE VI-4

Proposed Amino Acid Sequence in Phospholipase 2 from Bee Venom[a]

H₂N–Ile–Ile–Tyr–Pro–Gly–Thr–Leu–Trp–Cys–Gly–His–Gly–Asn–Lys–Ser–Ser–
 1   2   3   4   5   6   7   8   9   10 11 12 13 14 15 16

Gly–Pro–Asn–Glu–Leu–Gly–Arg–Phe–Lys–His–Thr–Asp–Ala–Cys–Cys–Arg–
 17 18 19 20 21 22 23 24 25 26 27 28 29 30 31 32

Thr–His–Asp–Meth–Cys–Pro–Asn–Val–Meth–Ser–Ala–Gly–Glu–Ser–Lys–His–
 33 34 35 36 37 38 39 40 41 42 43 44 45 46 47 48

Gly–Leu–Thr–Asp–Thr–Ala–Ser–Arg–Leu–Ser–Cys–Asn–Asp–Asn–Asp–Leu–
 49 50 51 52 53 54 55 56 57 58 59 60 61 62 63 64

Phe–Tyr–Lys–Asp–Ser–Ala–Asp–Thr–Ile–Ser–Ser–Tyr| |Phe–Val–Gly–Lys–
 65 66 67 68 69 70 71 72 73 74 75 76   77 78 79 80

Meth–Tyr–Phe–Asn–Leu–Ile–Asn–Thr–Lys–Cys–Tyr–Lys–Leu–Glu–His–Pro–
 81 82 83 84 85 86 87 88 89 90 91 92 93 94 95 96

Val–Thr–Gly–Cys–Gly–Glu–Arg–Thr–Glu–Gly–Arg–Cys–Leu–His–Tyr–Thr–
 97 98 99 100 101 102 103 104 105 106 107 108 109 110 111 112

Val–Asp–Lys–Ser–Lys–Pro–Lys–Val–Tyr–Gln–Trp–Phe–Asp–Leu–Arg–Lys–
 113 114 115 116 117 118 119 120 121 122 123 124 125 126 127 128

Tyr–OH
 129

[a] Shipolini *et al.* (1971b).

in action (Brockerhoff, 1973); in the enzyme of the bee, this orientation may be achieved by a hydrophilic carbohydrate tail.

Amino acid compositions are available for a preparation from *Naja naja* (Currie *et al.,* 1968), for the two enzymes of *Naja nigricollis* (Wahlström, 1971), for the enzyme of *Crotalus atrox* (Hachimori *et al.,* 1971), and for the two isoenzymes of *Crotalus adamanteus* (Wells and Hanahan, 1969). The last authors could not detect any *N*-terminal amino acid in either enzyme.

6. Tertiary Structure

The secondary structure, i.e., the location of the disulfide bridges, is known for only one phospholipase, that of the porcine pancreas (de Haas *et al.,* 1970b). X-ray analyses, which would show the shape of the molecule in the crystalline enzyme, are not yet available for any of the enzymes. However, some information on the tertiary structure of several of the enzymes has been obtained from optical data and sedimentation experiments.

Optical rotatory dispersion and circular dichroism of the prophospholipase 2 from porcine pancreas allowed an estimate of its α-helix content as 50%; for the active enzyme it was 55% (Scanu *et al.,* 1969). These high values are in keeping with the stability of the enzyme. The spectra of the zymogen and the enzyme were very similar: little conformational change seems to take place during the activation. High concentrations of urea or guanidine HCl also cause little change.

For the isoenzymes of *Agkistrodon halys blomhoffii,* Kawauchi *et al.* (1971a) estimated, from rotatory dispersion data, a right-handed α-helix content of 27% for the acidic, 20% for the basic species. Circular dichroism spectra suggested that the isoenzymes differ in the conformation of the side-chain chromophores, probably tyrosine and tryptophan residues.

Sedimentation studies of bee venom phospholipase in the dimeric form at high concentrations have led to the calculation of an exceptionally high frictional ratio (f/f_0) of 1.55, from which an axial ratio of 1:7 has been estimated (Shipolini *et al.,* 1971a). The double molecule would therefore have an unusual, elongated shape; but since the carbohydrate moiety of the enzyme, its shape, and its role in the dimerization were not considered in the investigation, no conclusions can be drawn concerning the shape of the monomer. Frictional ratios of similar magnitude have not been found in the phospholipases from other sources (Scanu *et al.,* 1969; Kawauchi *et al.,* 1971a,b).

Wells (1971a) has studied solvent perturbation difference spectra of the phospholipases 2 of *Crotalus adamanteus,* and the changes in circular dichroism and fluorescence during the interconversion of dimers and monomers. He concluded that changes take place during the dimerization that involve two protons per subunits and conformational changes involving tryptophan residues. The circular dichroism spectrum of the enzyme indicated an α-helix content of almost 70%, higher even than that found in the pancreatic enzyme.

E. Specificity and Mode of Action

1. Steric and Positional Specificity

Since the fatty acids split from natural lecithins by snake venoms were always unsaturated, it was assumed that the phospholipases were specific for unsaturated fatty acids until it was found (Zeller, 1952; Hanahan, 1954a,b) that from hydrogenated lecithins saturated acids were released. This indicated that the enzyme was specific for a position in the phospholipids rather than for a certain type of fatty acid. It was then shown in three laboratories that the secondary ester in β-position or position 2 was the only one attacked. To prove this, both Tattrie (1959) and Hanahan *et al.* (1960) employed the known specificity of pancreatic lipase for the primary ester bonds of glycerides (see Scheme VI-1).

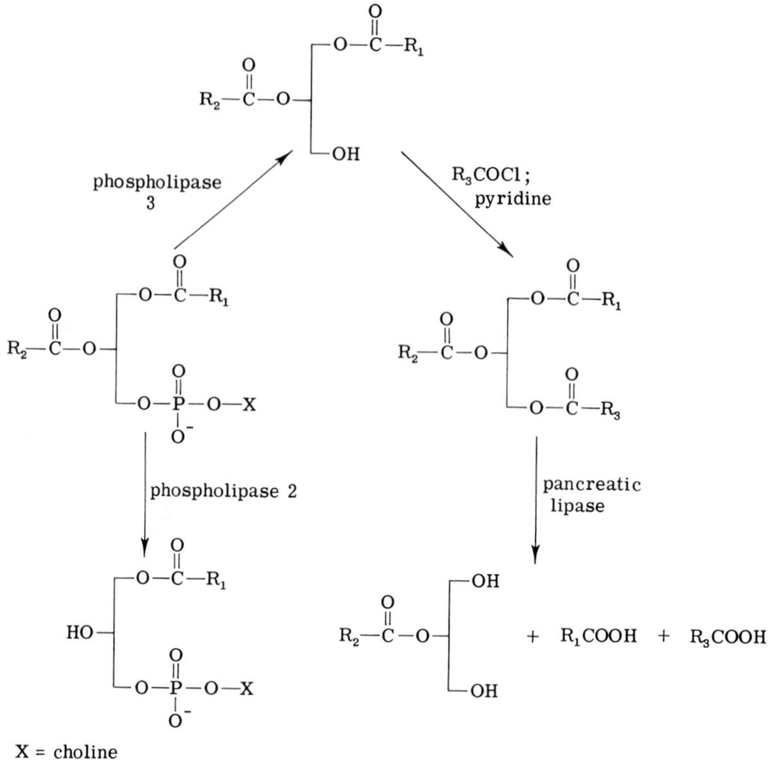

X = choline

SCHEME VI-1

First, the phosphorylcholine group was removed by treatment with the phospholipase 3 from *Clostridium welchii*; then, the resulting diglyceride was acylated with myristoyl chloride (R_3COCl)—the original lecithin did not contain myristic acid. Hydrolysis of the resulting triglyceride gave equal amounts of myristic acid and those acids found in the lysolecithin after the action of the snake phospholipase. The 2-monoglyceride contained the same acids that were split from the lecithin. Action of pancreatic lipase on the unmodified diglyceride (Hanahan *et al.*, 1960) also yielded the same monoglyceride.

De Haas and van Deenen (1961a; de Haas *et al.*, 1960) proved the specificity of phospholipase 2 in a more direct way by its action on synthetic phosphatidylcholines of known fatty acid distribution. They prepared 1-stearoyl-2-oleoyllecithin, from which only oleic acid was released by snake venom enzyme, and also 1-oleoyl-2-stearoyllecithin, from which only stearic acid was released. Subsequently, de Haas and van Deenen and their collaborators studied the enzymatic hydrolysis of a large number of phosphoglycerides, for instance, phosphatidylethanolamines (de Haas and van Deenen, 1961b), phosphatidylethanolamines with a polyenoic fatty acid, linoleic acid, in position 1 or 2 (de Haas *et al.,* 1962a,b), and phosphatidylcholines containing elaidic (*trans*-9-octadecenoic) acid (Dauvillier *et al.,* 1964). In all cases, the ester in position 2 was hydrolyzed, regardless of the nature of the fatty acid. Pancreatic phospholipase showed the same specificity (van Deenen *et al.,* 1963).

The stereospecificity of snake venom phospholipase was first indicated by Long and Penny (1957), who found that only half of a racemic mixture of D- and L-lecithin was attacked. They reasoned that the enzyme was specific for phospholipids with the natural L-α-glycerophosphate (today: *sn*-glycerol 3-phosphate) configuration. This conclusion was confirmed when it was shown that the D-isomer (2,3-diacyl-*sn*-glycerol-1-phosphorylcholine) was completely resistant (de Haas *et al.,* 1960; de Haas and van Deenen, 1961a).

Radioactive synthetic phospholipids are today often used to detect and classify intracellular phospholipases, in pursuit of the lead of Robertson and Lands (1962), who employed lecithins that had been biosynthetically labeled with radioactive fatty acids in the different positions. With these substrates, they confirmed the 2-specificity in several snake venoms and pancreas and then went on to investigate the lipolytic activities of several tissues of the rat.

The phospholipases 2 of venoms and pancreas seem to attack all natural phosphoglycerides, notwithstanding earlier reports to the contrary. The apparent resistance of a phospholipid can always be overcome by more efficient micellar dispersion or by a change of the micellar surface potential. However, not all substrates are hydrolyzed at the same rate, and not all enzymes show comparable rates for comparable substrates. Often, interfacial charge effects are responsible for such differences; sometimes, inductive or steric effects in the substrate molecule seem to play a role. For instance, plasmalogen (1-alk-1'-enyl-2-acyl-*sn*-glycerol-3-phosphorylcholine) is only slowly attacked, in comparison to its diacyl analog, by the enzyme from *Crotalus atrox* (Gottfried and Rapport, 1962; Colacicco and Rapport, 1966; Woelk and Debuch, 1971); this difference can be exploited to isolate plasmalogens. The enzyme from *Naja naja,* on the other hand, hardly makes a distinction between the two phospholipids. The amide linkage of sphingolipids is not hydrolyzed by any of the phospholipases 2.

It is possible that none of the attributes of substrate specificity for the enzymes is absolute. A purified phospholipase 2 from *Vipera palestinae,* for example, has been found to have lysophospholipase activity, although at a rate of only 0.1% (Shiloah *et al.,* 1973), and the pancreatic phospholipase will also hydrolyze *sn*-1 phospholipid at a very slow rate (0.4%) (G. H. de Haas, personal communication).

2. Substrate Requirements

The steric and positional specificity of phospholipase 2 having been established, de Haas, van Deenen, and their co-workers embarked on the synthesis of a great number of phospholipids and phospholipid analogs in order to ascertain the minimal substrate requirements of the enzyme. They first showed that a lecithin containing lauric acid (C_{12}, not normally found in lecithins) was a good substrate (de Haas *et al.,* 1960), later, that the short-chain lecithins diheptanoyl-, dibutyroyl-, and diacetylglycerylphosphorylcholine are also attacked, although at sharply diminishing rates (van Deenen and de Haas, 1963). The hydrolysis of the water-soluble dihexanoyllecithin had already been reported by Roholt and Schlamowitz (1958). Lecithins with one short and one long fatty acid were also attacked (Bird *et al.,* 1965).

The further investigations by van Deenen and de Haas (1963) are best discussed with reference to the structural formulas in Table VI-5. Formula I shows the structure of a natural substrate. The D-isomer, or *sn*-glycerol 1-phosphate derivative (II) is not hydrolyzed. The group X in I does not have to carry an amino or alkylammonium function; in fact, X can be H, as in phosphatidic acid, but also any other group, as, for instance, in III. However, the phosphate must have at least one free acidic hydroxyl: the benzyl-substituted triester phosphoglyceride (IV) is not a substrate; it must be noted, however, that this resistance might also be the result of steric hindrance, and that this point has not yet been conclusively settled. An ester group at position 1 of the glycerol is not essential; it can be replaced by a benzyl ether group (V) or by a free hydroxyl (VII).

It should be noticed that VII is one of the two isomeric lysophospholipids; by definition, therefore, phospholipase 2 is also a lysophospholipase 2.

In any substrate, the ester function in position 2 must of course be present: VI and VIII are not substrates. Finally, the entire grouping 1 of the glycerol can be dispensed with: ethylene glycol analogs of phospholipids (IX) are hydrolyzed by phospholipase 2, though trimethyleneglycol phospholipids (X) (Kaneko and Hara, 1967) are not. The minimal sub-

TABLE VI-5

Substrate Specificity of Phospholipase A (*Crotalus adamanteus*)[a]

Hydrolysis	No hydrolysis

Structure (I):
$$H_2C-O-\overset{\overset{O}{\|}}{C}-R_1$$
$$R_2-\overset{\overset{O}{\|}}{C}-O-CH$$
$$H_2C-O-\overset{\overset{O}{\|}}{\underset{OH}{P}}-O-X$$
(I)

Structure (II):
$$H_2C-O-\overset{\overset{O}{\|}}{C}-R_1$$
$$HC-O-\overset{\overset{O}{\|}}{C}-R_2$$
$$H_2C-O-\overset{\overset{O}{\|}}{\underset{OH}{P}}-O-X$$
(II)

Structure (III):
$$H_2C-O-\overset{\overset{O}{\|}}{C}-R_1$$
$$R_2-\overset{\overset{O}{\|}}{C}-O-C$$
$$H_2C-O-\overset{\overset{O}{\|}}{\underset{OH}{P}}-O-CH_2$$
$$CH_2$$
$$C_6H_4-(CO)_2-N$$
(III)

Structure (IV):
$$H_2C-O-\overset{\overset{O}{\|}}{C}-R_1$$
$$R_2-\overset{\overset{O}{\|}}{C}-O-C$$
$$H_2C-O-\overset{\overset{O}{\|}}{\underset{O-CH_2-C_6H_5}{P}}-O-CH_2-CH_2-N$$
$$(CO)_2$$
$$C_6H_4$$
(IV)

Structure (V):
$$H_2C-O-CH_2-C_6H_5$$
$$R-\overset{\overset{O}{\|}}{C}-O-CH$$
$$H_2C-O-\overset{\overset{O}{\|}}{\underset{OH}{P}}-O-CH_2-CH_2-NH_2$$
(V)

Structure (VI):
$$H_2C-O-\overset{\overset{O}{\|}}{C}-R_1$$
$$C_6H_5-CH_2-O-CH$$
$$H_2C-O-\overset{\overset{O}{\|}}{\underset{OH}{P}}-O-CH_2-CH_2-NH_2$$
(VI)

Structure (VII):
$$H_2C-OH$$
$$R-\overset{\overset{O}{\|}}{C}-O-CH$$
$$H_2C-O-\overset{\overset{O}{\|}}{\underset{OH}{P}}-O-CH_2-CH_2-NH_2$$
(VII)

Structure (VIII):
$$H_2C-O-\overset{\overset{O}{\|}}{C}-R$$
$$HO-CH$$
$$H_2C-O-\overset{\overset{O}{\|}}{\underset{OH}{P}}-O-CH_2-CH_2-NH_2$$
(VIII)

Structure (IX):
$$H_2C-O-\overset{\overset{O}{\|}}{C}-R$$
$$H_2C-O-\overset{\overset{O}{\|}}{\underset{OH}{P}}-O-R'$$
(IX)

Structure (X):
$$H_2C-O-\overset{\overset{O}{\|}}{C}-R$$
$$H-\overset{}{C}-H$$
$$H_2C-O-\overset{\overset{O}{\|}}{\underset{OH}{P}}-O-CH_2-CH_2-NH_2$$
(X)[b]

[a] Van Deenan and de Haas (1963).   [b] Kaneko and Hara (1967).

strate requirement of phospholipase 2 can now be illustrated as follows (van Deenen and de Haas, 1963):

$$\begin{array}{c}
\text{O}\qquad\qquad\ |\ \\
\ \ \ \|\qquad\quad\ \ \ |\ \\
\text{R—C—O—C—H}\\
\qquad\qquad\ |\qquad\text{O}\\
\qquad\qquad\qquad\qquad\ \|\\
\ \text{H—C—O—P—O—}\\
\qquad\quad|\qquad\ |_\ominus\\
\qquad\qquad\qquad\text{O}^\ominus
\end{array}$$

Recently, it was reported by Bonsen *et al.* (1972b) that the phosphoric acid group may be replaced by a sulfonic acid. The resulting sulfolipid, 1,2-diacyl-3-deoxy-*sn*-glycerol-3-sulfonic acid, is a very good substrate for phospholipase 2. The corresponding phosphonolipid, in which the C—O—P linkage is replaced by a C—P bond, is also attacked. It appears, then, that the minimal requirement of the enzyme is a substrate with an ester group and a strong anionic function five or six atoms away from the carboxyl carbon, and in addition a steric configuration analogous to the one above.

In the phospholipid with two susceptible ester bonds, diphosphatidyl-glycerol (cardiolipin), both bonds are hydrolyzed (van Deenen and de Haas, 1963; Marinetti, 1964). An intermediate, monodeacylated cardio-lipin, can be detected (Okuyama and Nojima, 1965).

Most experiments of van Deenen and de Haas were carried out with the enzyme from *Crotalus adamanteus* venom; the enzymes of other snakes and of pancreas have comparable specificities, although some variations can be observed. The enzyme from *Naja naja*, for instance, can reportedly hydrolyze the ethyleneglycol analog of phosphatidylcholine but not that of phosphatidylethanolamine (Kozhukhov *et al.*, 1969), whereas the enzyme from *C. adamanteus* can hydrolyze both (van Deenen and de Haas, 1963).

The discovery that diol lipids could be enzymatically hydrolyzed prompted a reinvestigation of the action of phospholipase 2 on the so-called β-lecithins (1,3-diacyl-*sn*-glycerol-2-phosphorylcholine), which had been reported as resistant (Long and Penny, 1957; de Haas *et al.*, 1960). It was found that they could indeed be hydrolyzed, if rather slowly, by the *C. adamanteus* enzyme (de Haas and van Deenen, 1963a,b). Only one equivalent of fatty acid was released. This indicated that stereospecific attack had occurred and the lysophosphatidylcholine formed was in fact optically active. Proof of a strict stereospecific degradation was obtained when the lysolecithin was reacylated with a different fatty acid and again degraded by phospholipase 2. Only the new acid was released, and quantitatively. De Haas and van Deenen proposed an enzymatic attack on the 1-position of the β-lecithin (see Scheme VI-2).

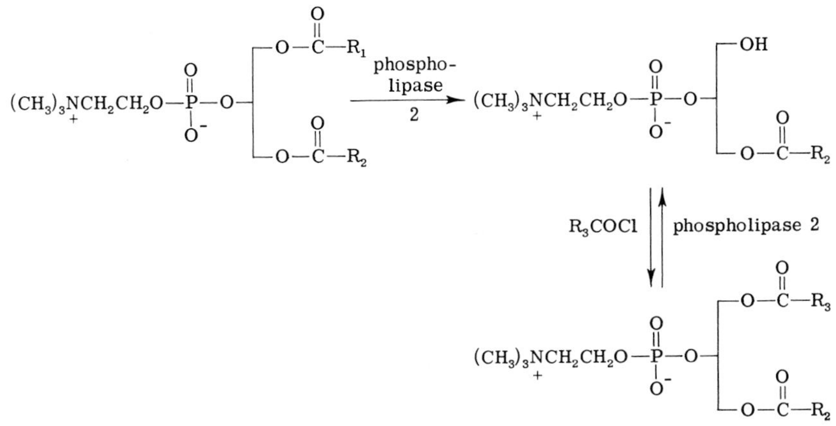

$$(CH_3)_3NCH_2CH_2O\overset{+}{\underset{O^-}{\overset{O}{\underset{|}{P}}}}O \begin{bmatrix} O-\overset{O}{\overset{\|}{C}}-R_1 \\ \\ O-\overset{O}{\overset{\|}{C}}-R_2 \end{bmatrix} \xrightarrow[\text{2}]{\text{phospho-}\atop\text{lipase}} (CH_3)_3NCH_2CH_2O\overset{+}{\underset{O^-}{\overset{O}{\underset{|}{P}}}}O \begin{bmatrix} OH \\ \\ O-\overset{O}{\overset{\|}{C}}-R_2 \end{bmatrix}$$

$$R_3COCl \Big\updownarrow \text{ phospholipase 2}$$

$$(CH_3)_3NCH_2CH_2O\overset{+}{\underset{O^-}{\overset{O}{\underset{|}{P}}}}O \begin{bmatrix} O-\overset{O}{\overset{\|}{C}}-R_3 \\ \\ O-\overset{O}{\overset{\|}{C}}-R_2 \end{bmatrix}$$

SCHEME VI-2

The attack of the enzyme on the *sn*-1 positioned ester was then proven by enzymatic hydrolysis of a stereospecifically labeled β-lecithin, 1-oleoyl-3-palmitoyl-*sn*-glycerol-2-phosphorylcholine (de Haas and van Deenen, 1964); only oleic acid was released. The specificity for the 1-position was also demonstrated by Scheme VI-3 (Brockerhoff, 1967).

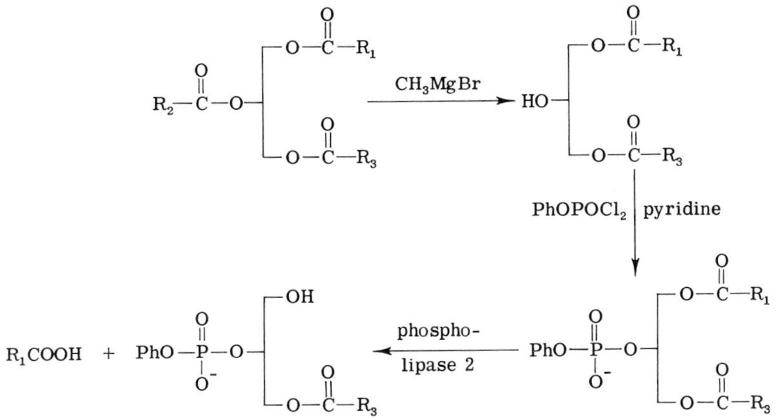

$$R_2-\overset{O}{\overset{\|}{C}}-O \begin{bmatrix} O-\overset{O}{\overset{\|}{C}}-R_1 \\ \\ O-\overset{O}{\overset{\|}{C}}-R_3 \end{bmatrix} \xrightarrow{CH_3MgBr} HO \begin{bmatrix} O-\overset{O}{\overset{\|}{C}}-R_1 \\ \\ O-\overset{O}{\overset{\|}{C}}-R_3 \end{bmatrix}$$

$$PhOPOCl_2 \Big\downarrow \text{ pyridine}$$

$$R_1COOH + PhO\overset{O}{\underset{O^-}{\overset{\|}{\underset{|}{P}}}}O \begin{bmatrix} OH \\ \\ O-\overset{O}{\overset{\|}{C}}-R_3 \end{bmatrix} \xleftarrow[\text{lipase 2}]{\text{phospho-}} PhO\overset{O}{\underset{O^-}{\overset{\|}{\underset{|}{P}}}}O \begin{bmatrix} O-\overset{O}{\overset{\|}{C}}-R_1 \\ \\ O-\overset{O}{\overset{\|}{C}}-R_3 \end{bmatrix}$$

SCHEME VI-3

A semisynthetic triglyceride of known fatty acid composition in positions 1, 2, and 3 was partially deacylated with a Grignard reagent. From the reaction products, the 1,3-diglyceride was isolated and converted into a 2-phosphatidylphenol (1,3-diacyl-*sn*-glycerol-2-phenylphosphate), from which phospholipase 2 split only the fatty acids in position 1.

The β-lecithins are symmetric compounds but yield, on reaction with phospholipase 2, an asymmetric product. Such an enzymatic reaction must be taken as a proof for a three-point attachment of the substrate to the enzyme (Ogston, 1948), or at any rate for a fixed three-dimensional orientation of the substrate in the enzyme–substrate complex. For the phospholipase 2, it is likely that one point of fixation is the negatively charged phosphate group of the substrate; this might be bound by a cationic group on the enzyme. A second point of fixation is, by necessity, the ester group that is to be split. The third coordinate of steric fixation does not necessarily include a third point of bonding between substrate and enzyme. Formula I in Fig. VI-5 shows the binding at the cationic site, B, and the fixation at the reactive ester group of the substrate at site A. The grouping —C(=O)—O—C at position 2, it should be remembered, is planar because of the partial double bond character of the carboxyl C—O linkage. Quite likely, nucleophilic attack occurs on the carbonyl C in a perpendicular direction. If we imagine a plane laid through A and B, parallel to and in front of the page, it can be seen that the nonreactive ester group and C—1 of the glycerol point away from this plane. In the optical antipode of the substrate, the resistant *sn*-1-phosphoglyceride, this group would point toward the plane. We can speculate that the enzyme is situated above the plane of the page, and that the unreactive substituent has to point away from the enzyme surface because in the other configuration it would overlap with amino acid residues of the enzyme. Only a C—2 hydrogen atom can be accommodated on the side of the substrate facing the enzyme.

Of course, we must wait for a steric picture of the enzyme to confirm such speculations. Our present argument is this, that it is not necessary to postulate an additional binding site on the enzyme to account for its

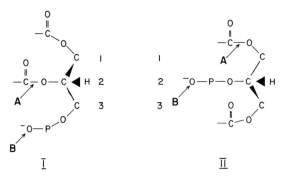

Fig. VI-5 Hypothetical fixation of *sn*-3 (I) and *sn*-2 (II) phosphoglycerides to phospholipase 2. (A) cleavage site; (B) cationic binding site. Notice that the unreactive third group of the glycerol (position 1 in I, position 3 in II) extends behind the plane through A, B, and the central carbon in both models.

stereospecificity. In particular, binding at the second carboxyl ester group of the substrate is not required.

Our hypothesis—that the unreactive end of the substrate must extend away from the enzyme—allows a rationalization of the enzymatic specificity toward the ester-1 in the *sn*-2-phosphoglycerides (Fig. VI-5, II). The unreactive group of the substrate, 3, is again directed away from the plane through A and B. If the ester in 3 should be attacked at site A of the enzyme, the substrate would have to rotate around the P—O—C(2) axis, and it can be seen that the now nonreacting glycerol-1 group would point toward the imaginary surface of the enzyme.

The hypothetical two-point enzyme substrate fixation, made stereospecific by the restricting location of the enzyme surface, has also this to recommend it: that the substrates lack any functional groups in the chain of atoms between A and B that could be expected to bind to the enzyme. From the models it can be predicted that a substitution of the H bound to the C—2 of the glycerol will give totally unreactive compounds.

F. Use of Phospholipase 2 in Analysis and Synthesis of Lipids

The steric and positional specificity of the phospholipases 2 has made them valuable tools in lipid chemistry and biochemistry. The most obvious application is, of course, the determination of the fatty acid distribution, or "positional distribution of fatty acids," in phosphoglycerides. For such an analysis, the reaction is carried out either in ether (Hanahan, 1952) or, more generally, in aqueous buffer saturated with ether, or a buffer–ether mixture (Magee and Thompson, 1960). The progress of the reaction is followed by thin-layer chromatography on silicic acid, and the products, fatty acids from position 2 and lysophosphoglyceride with the fatty acid in position 1, are separated in the same manner and analyzed by gas–liquid chromatography. Today, the description of a newly isolated phosphoglyceride is no longer considered complete without such an analysis of its fatty acid distribution. On the other hand, the multitude of such analyses found in the literature makes it possible, in many cases, to predict the general distribution pattern from the total fatty acid composition of a lipid. Among the rules that can be applied, the most general one states that if the phospholipid contains comparable amounts of straight-chain saturated and of mono- and polyunsaturated acids—and this type of composition is prevalent in phospholipids of animals and common in plants—then the saturated acids will occupy position 1, the polyenoic acids position 2, and the monoenoic acids (usually mostly oleic acid) also position 2 unless they have to make up for a deficit of saturated acids in 1.

The specificity of phospholipase 2 is also employed in the stereospecific analysis of triglycerides, i.e., the determination of the fatty acid composition in positions *sn*-1, 2, and 3 of a triglyceride (Brockerhoff, 1971b). One such method (Brockerhoff, 1967) has been described in the previous chapter: the triglyceride is partially deacylated with a Grignard reagent, and the isolated 1,3-diglyceride is converted to a β-phosphoglyceride from which the enzyme removes fatty acid 1. Another method (Brockerhoff, 1965b), more often used, starts with the mixture of 1,2- and 2,3-diglycerides that results from the action of pancreatic lipase on a triglyceride (Scheme VI-4).

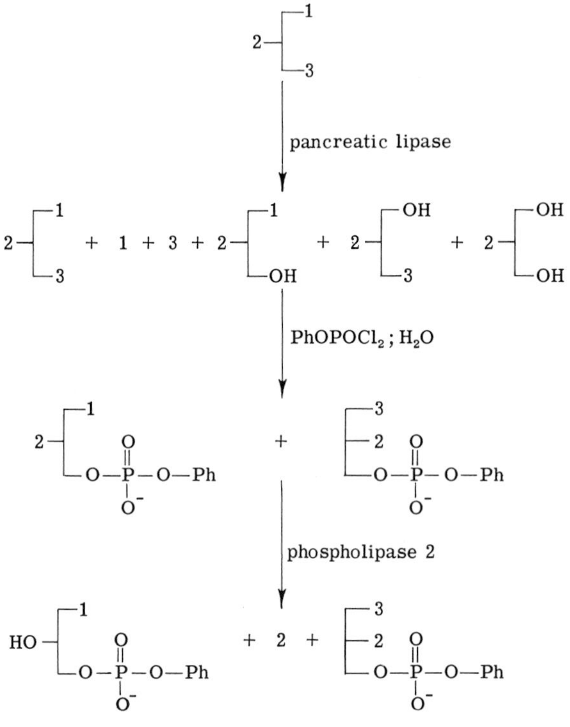

SCHEME VI-4.   1, 2 and 3 stand for the fatty acids in these positions.

Treatment with phenyldichlorophosphate gives a mixture of *sn*-3 and *sn*-1 phosphoglycerides from which only the *sn*-3-phosphate derivative is hydrolyzed by phospholipase 2. The resulting lysophospholipid contains the fatty acids in position 1; position 2 is analyzed in the monoglyceride found among the products of pancreatic lipolysis; the composition in 3 is calculated by difference from the original triglyceride or from the *sn*-1

phospholipid. A variant of the method employs the diglycerides obtained
with the lipase from *Geotrichum candidum,* which is specific for *cis*-9 unsaturated acids (Sampugna and Jensen, 1968).

In the synthesis of lipids, phospholipase 2 is useful for the preparation
of "mixed-acid" phosphoglycerides. For example, the synthesis of a chemically defined lecithin may lead from a natural lecithin (which contains a
large number of different molecular species) over *sn*-glycerol-3-phospho-
rylcholine, obtained by mild deacylation, to a "single-acid" lecithin, which
is then enzymatically deacylated in position 2, and reacylated with a different fatty acid (Hanahan and Brockerhoff, 1960) (see Scheme VI-5).

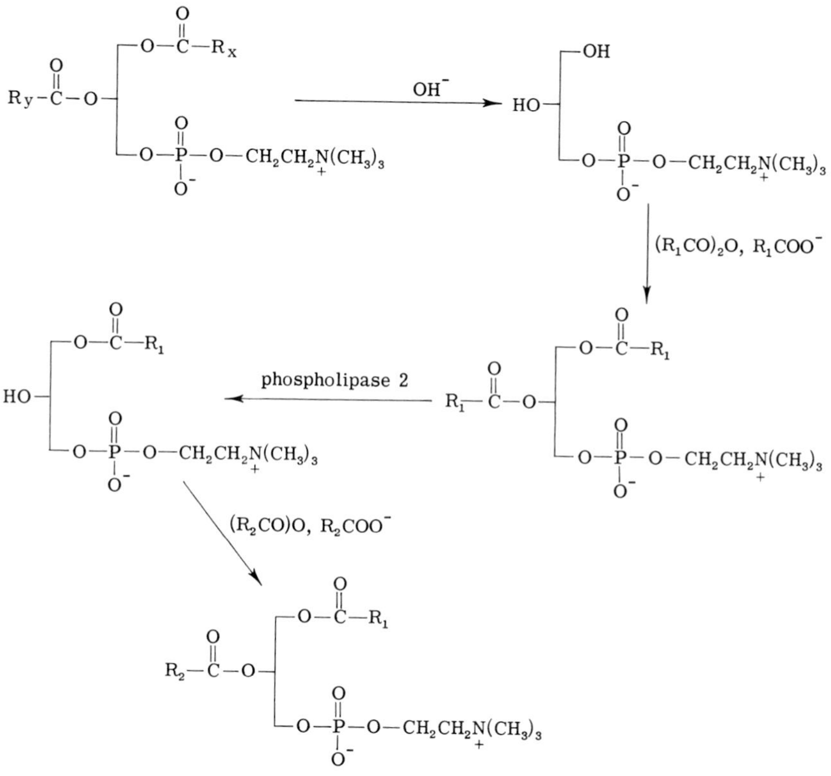

SCHEME VI-5

Removal of the phosphorylcholine with the help of phospholipase 3
yields a mixed-acid diglyceride, 1,2-diacyl-*sn*-glycerol, which can be acylated with a third fatty acid to give a triglyceride of defined stereospecific

fatty acid distribution (Hanahan and Brockerhoff, 1960; Brockerhoff, 1965b) (see Scheme VI-6).

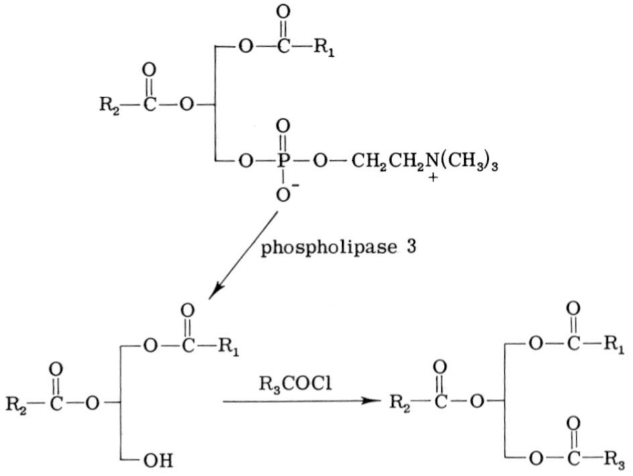

SCHEME VI-6

G. The Physical State of the Substrate: Dispersion and Surface Potential

Phospholipids dispersed in water form large lamellar structures that are not usually attacked by phospholipases. Enzymatic hydrolysis can be induced by a variety of means: sonication (Dawson, 1963b; Beare, 1967); use of a lipophilic buffer such as collidine (Magee and Thompson, 1960; Shah and Schulman, 1967); addition of an organic solvent, e.g., chloroform (Kyes, 1903) and especially ether (Magee and Thompson, 1960); addition of an anionic or cationic surfactant (Dawson, 1964) or a neutral surfactant (Dennis, 1973). Especially bile salts are often very effective or even necessary cofactors (de Haas et al., 1968a; Olive and Dervichian, 1968; Ibrahim, 1970). Presenting the substrates in association with protein, in the form of lipoproteins, also often accelerates the enzymatic hydrolysis (Condrea *et al.,* 1962). Albumin at low concentration also stimulates (Smith *et al.,* 1972). Even the adsorption of phosphoglycerides on silicic acid has been found to stimulate their breakdown by phospholipase 2 (Goerke *et al.,* 1971).

It appears that two factors have to be considered to explain these effects: first, the spatial arrangement of the supersubstrate, i.e., the surface of the substrate aggregates; second, the electrostatic conditions at this surface. The first of these factors is well documented, if not well understood; the nature and importance of the second are more controversial, mainly because it is difficult to disentangle it from the first, since the introduction of an electrical charge into a micellar surface will, of course, also alter the geometric structure of this surface. The spatial factor can be exemplified with the action of the phospholipase 2 of the cobra (*Naja naja*) on lecithin, and the electrostatic factor with the phospholipase 2 from pancreas and with phospholipase 3 and lysophospholipase.

Prolonged sonication of an aqueous dispersion of egg yolk lecithin breaks down its multilayered aggregates to single shell vesicles containing a few thousand molecules (Finer *et al.*, 1972). In this state, the lecithin is attacked by cobra venom phospholipase 2 (Dawson, 1963b). Evidently, a change in the surface structure of the substrate aggregates has taken place, perhaps leading to wider spacing or to increased mobility of the single molecules. The effect of ether (Magee and Thompson, 1960) probably rests on the same principle: the ether is incorporated into the lipid phase and makes for less rigid spacing, or else breaks the aggregates into smaller micelles (Dawson, 1963b). The incorporation of deoxycholate in lecithin aggregates also changes the micellar structure. Transitions from one micellar form to another are correlated with large differences in the rates of enzymatic hydrolysis (Olive and Dervichian, 1968). However, here the effect is possibly compounded, since bile salts also change the surface potential of the micelles.

The enzymatic hydrolysis of phosphoglycerides in mixtures of aqueous buffer and ether must be distinguished from the reaction of the substrate–enzyme–water complex in a homogeneous ether solution according to Hanahan (1952). This system probably contains inverted phospholipid aggregates with the fatty acid chains outside and an aqueous enzyme solution inside (Wells, 1971c).

An electrostatic effect seems to be operative in the action of pancreatic phospholipase. The porcine enzyme cannot even attack the lecithin of egg yolk lipoprotein unless deoxycholate or an acidic phospholipid is added to the system (Ibrahim *et al.*, 1964; de Haas *et al.*, 1968a). The pancreatic phospholipases 2 of rat and man also respond to the addition of deoxycholate (Belleville and Clément, 1966). The lysophospholipase (or phospholipase 1,2) of *Penicillium notatum* requires a negative surface potential of the supersubstrate; the phospholipase 3 of *Clostridium perfringens* needs a positive surface charge (Dawson, 1965).

Electrostatic attraction and repulsion of enzyme and supersubstrate may be partly responsible for these effects; but an enzyme such as pancreatic phospholipase 2 is activated by a negative surface potential at a pH where its own overall charge is negative. To explain the surface charge effects, a clear distinction must be made between the substrate binding site of the enzyme and its supersubstrate binding site (Brockerhoff, 1974). These two have so far not been distinguished in the literature because "substrate" has always been used ambiguously for the single substrate molecule as well as for the whole lipid phase. In the phospholipases, the two sites clearly are different. The neutral phosphatidylcholine molecule, for instance, must be accepted as such by the reactive site of the enzyme, regardless of the presence of the surfactants at the interface that influence the interfacial potential in this or that direction. However, such changes in the potential may have profound influence on the reaction rate. This must be an effect of the supersubstrate binding and the interfacial orientation of the enzyme. Since the substrate, located in an interfacial matrix, cannot move through the solvent to the enzyme, the enzyme must approach the supersubstrate, and it must also orient its reactive site toward the substrate molecules, leaving some space, all the same, for the access of the water that is indispensible for the reaction. This orientation is brought about by an asymmetric charge distribution on the enzyme, with the concentration of charges in a head that binds to an interface of the opposite charge. Figure VI-6 shows a model of a phospholipase requiring a negative potential, for example, pancreatic phospholipase 2. An examination of the structure of this enzyme (Fig. VI-3) suggests that the C-terminal end of the protein chain, with its abundance of lysine residues, might be a good candidate for the electrostatic head. We may surmise that this region forms an α-helix that may resist unfolding by interfacial forces (Lucy, 1968; Malcolm, 1968). The asymmetric charge distribution postulated in Figure VI-6 is also clearly indicated in Figure VI-3. In the phospholipase 2 of bee venom, which seems to have a more even charge distribution pattern (Shipolini *et al.,* 1971b), it is possibly the hydrophilic carbohydrate moiety that forms a tail that secures the proper interfacial orientation of the enzyme.

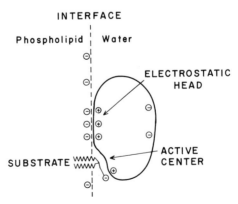

Fig. VI-6 Hypothetical orientation of porcine pancreatic phospholipase 2 at the supersubstrate–water interface.

Some puzzling observations can be explained within the frame of the orientation hypothesis. Phospholipase 1,2 ("B") needs a negative surface potential (Dawson, 1964), but spreading in a monolayer will also make the substrate susceptible to attack by the enzyme, and so may sonication (Kates *et al.,* 1965; Beare, 1967), although this is debated (Dawson and Hauser, 1967). Pancreatic phospholipase 2, which requires a negative surface charge as supplied by bile acids, can act without deoxycholate in ether (Hanahan, 1952) or on water-soluble substrates (Roholt and Schlamowitz, 1961; de Haas *et al.,* 1971) or on lecithin monolayers (Zografi *et al.,* 1971). These observations show, in the light of the orientation hypothesis, that the electrostatic orientation bonding is important at large liposomal interfaces, because with the wrong orientation the enzyme could never embrace a substrate molecule. In fine and concentrated dispersions, however, or in monomolecular substrate solutions, or under an uncharged layer of substrate molecules, the enzyme has a good chance of catching a substrate. This chance is still small compared with the certainty of fulfillment in the charge-oriented bed of the appropriate supersubstrate. This is attested by the low reaction rates found for water-soluble substrates.

H. Cofactors and Inhibitors

The role of calcium as an obligatory cofactor of phospholipases 2 has been confirmed in all recent reports on the enzymes of pancreas and in the majority of the reports on enzymes from snakes. Typically, it is found that the enzymes are completely inactive in the absence of calcium ion or in the presence of an excess of the chelating agent EDTA. Sporadic recent reports have thrown doubt on the necessity of calcium for some snake phospholipases, for instance, that of *Agkistrodon piscivorus* (Augustyn and Elliott, 1970), *Naja naja* and *Vipera russellii* (Salach *et al.,* 1971b), and *Haemachetus haemachetus* (Condrea *et al.,* 1970). In the last case, however, it is only the hydrolysis of red cell ghost phospholipids that apparently proceeds without calcium, whereas the hydrolysis of pure lecithin does require the metal. Salach *et al.* (1971b) report on the solubilization of enzymes from membranes by a phospholipase 2 as being independent of calcium. However, recent investigations on the action of phospholipases 2 on soluble substrates or monolayers have confirmed the earlier reports (Hughes, 1935; Roholt and Schlamowitz, 1961) that the requirement of snake venom enzymes for calcium is absolute (de Haas *et al.,* 1971; Zografi *et al.,* 1971; Wells, 1972). An explanation for the apparent exceptions may lie in the presence of small but sufficient amounts of calcium in the substrate or the enzyme preparation; perhaps the metal is held so tenaciously by some phospholipases 2 that it is not removed during their purification. Amylases can be cited as examples of enzymes from which even EDTA cannot remove the calcium.

Complementing the activation by calcium is the inhibition by other metal ions, e.g., Cu^{2+} (Long and Penny, 1957), Zn^{2+}, Cd^{2+}, and Ba^{2+} (Roholt and Schlamowitz, 1961; Tu *et al.,* 1970), Sr^{2+} (Uthe and Magee, 1971), all these inhibiting snake enzymes. The phospholipase of a scorpion is inhibited by Cu^{2+}, Fe^{2+}, Mn^{2+}, and Zn^{2+} (Kurup, 1966). Nickel has been claimed to activate the enzyme from *Crotalus atrox* (Brown and Bowles, 1965, 1966), and Mg^{2+} seems to be able to replace Ca^{2+} in some cases (Tu *et al.,* 1970; Uthe and Magee, 1971).

Wells (1972) has determined the binding constants K_{iCa} and K_{iBa} for Ca^{2+} and Ba^{2+} and the phospholipases 2 of *Crotalus adamanteus;* their values are 4×10^{-5} and 3×10^{-5} M.

The necessity of calcium classifies the phospholipases 2 as metalloenzymes. Their most general inhibitor is consequently EDTA. Paradoxically, this chelating agent is often added to an assay mixture, but in an amount insufficient to bind all calcium, in order to complex heavier metals which might otherwise inhibit the enzyme (Saito and Hanahan, 1962).

In the enzymic hydrolysis of lecithin in ether, sodium chloride is required for maximal activity (Saito and Hanahan, 1962). An extremely potent activation by sodium chloride has also been reported for aqueous micellar systems (de Haas *et al.,* 1971). It was suggested that the salt influences the interfacial structure of the substrate micelles; however, as in the case of pancreatic lipase, it is not unlikely that cations or anions have a polarizing or dehydrating influence on the enzyme–substrate complex itself.

The further list of inhibitors can be subdivided in two classes: those which prevent the enzyme–supersubstrate association or orientation; and those that react at the active site of the enzyme or compete with the substrate. Only the last class contains "inhibitors" in the proper sense.

Interfacial inhibition will obviously be exerted by any reagent that displaces substrate molecules from the surface of the supersubstrate or changes the interfacial potential in a direction opposite to that required by the particular enzyme. Consequently, an inhibitor for one enzyme may be an activator for another. This type of inhibition is displayed by detergents such as Triton and sodium dodecyl sulfate (Roholt and Schlamowitz, 1961; Dawson, 1963a; Dennis, 1973) or by long-chain quaternary ammonium salts (Rosenthal and Geyer, 1962; Shah and Schulman, 1967).

The inhibition of enzymes at the reactive center is often attempted with model compounds resembling the substrate, but in the case of the phospholipases, the effects of mock-substrate binding and rearrangements of the supersubstrate are not easy to distinguish. Rosenthal and Han (1970) tested a variety of choline and ethanolamine esters of long-chain phosphonates with *A. piscivorus* venom, but found that the inhibition depended on the nature of the reaction system, whether it was aqueous or activated by ether or deoxycholate. Only one substrate, which closely imitated the structure of lecithin, seemed to be a true competitive inhibitor in most systems. Hanahan (1971) suggested that phosphonolipids might also be true inhibitors. Kaneko and Hara (1967) reported competitive inhibition of lecithin hydrolysis by a trimethyleneglycol bisphosphatidic acid. Bonsen *et al.* (1972a,b) have synthesized a large number of potential analog inhibitors. They found that *sn*-1-phosphatidylcholines were inhibitors with K_i values identical to the K_m values of the corresponding *sn*-3 lipids. Lecithin analogs with an amide linkage at position 2 were very potent inhibitors; 2-sulfonyl ester analogs were not inhibitory; α-branched carboxyl ester analogs or compounds containing an additional CH_2 group between phosphate and carboxyl esters were competitive inhibitors. In such experiments, it is difficult to decide if the inhibition is due to enzyme–inhibitor binding or to changes in structure of the supersubstrate, e.g., substrate dilution. In view of the intractability of interfacial effects, further developments might be expected to come with the use of water-soluble substrate analogs rather than with inhibitors having the character of lipids.

Diisopropyl fluorophosphate (DFP), a reagent that acylates the "reactive" serine of many hydrolases, has no effect on phospholipases 2 (de Haas *et al.,* 1968a; Wells and Hanahan, 1969; Salach *et al.,* 1971b). Phospholipases 2, therefore, do not appear to be serine enzymes, although it is well to bear in mind that pancreatic lipase is also not inhibited by DFP and is yet probably a serine enzyme (Desnuelle, 1971).

Sodium fluoride has been reported to inhibit the phospholipase 2 of bee venom (Neumann and Habermann, 1954; Munjal and Elliott, 1971).

Saito *et al.* (1972) have studied the inhibitory effect of some antibiotics on snake venom enzyme.

If the pancreas protects itself from its phospholipase 2 by storing it as an inactive proenzyme, the venom glands of snakes seem to contain the active enzyme together with natural inhibitors. The inhibitor in the venom of *Naja naja* is a polypeptide with the molecular weight of 5000 and an isoelectric pH of 8.6 (Braganca *et al.,* 1970). It combines in a 1:1 ratio with the enzyme, probably by electrostatic linkage, since the isoelectric point of the enzyme is much lower (Table VI-3) and the complex can be split by electrophoresis. The inhibitor from *Bothrops neuwiedii* venom (Vidal and Stoppani, 1970, 1971a,b), also a polypeptide, binds to each of the two isoenzymes of this species, though with different affinities. Again the binding appears to be electrostatic, since it is strongly dependent on the pH. In contrast to the enzymes, the inhibitor contains free sulfhydryl and can be inactivated with mercury ions.

I. Reactive Sites and Mechanism of Catalysis

The reactivity of the phospholipases of the exocrine glands is dependent on the integrity of the secondary structure determined by the disulfide linkages, as might be expected. The enzyme from the scorpion *Heterometrus scaber* is unaffected by oxidizing agents such as H_2O_2 or $K_3Fe(CN)_6$ but inactivated by such reducing agents as cysteine and glutathione, which may be expected to split disulfide bonds (Kurup, 1965). Kornalik (1964) found that cysteine and thioglycolate destroy the phospholipase activity of several snake venoms. These observations indicate the absence of free sulfhydryl groups and the structural importance of disulfide bridges, and are in harmony with the molecular structure of the pancreatic phospholipase 2 (Fig. VI-3). N-Ethylmaleimide, also a potential sulfhydryl inhibitor, has been reported to inhibit the enzyme of bee venom (Munjal and Elliott, 1971), although this enzyme does not appear to possess any free sulfhydryl groups (Shipolini *et al.*, 1971a,b). The same reagent or the sulfhydryl-specific PCMB do not inhibit the enzymes of *Naja naja*, but partially inhibit the enzymes of *Vipera russellii*, and the reducing agent dithioerythritol inhibited one of the isoenzymes of *V. russellii* but had little effect on another one (Salach *et al.*, 1971b). The interpretation of results such as these, it should be mentioned, would require a detailed reinvestigation combined with kinetic analyses and the elimination of interfacial effects.

It has been found that photooxidation in the presence of a photosensitizing dye leads to rapid inactivation of the phospholipases 2 of snake venoms (*V. russellii, Crotalus atrox, Agkistrodon piscivorus*) (Kocholaty, 1966a,b; Kocholaty and Ashley, 1966; Tejasen and Ottolenghi, 1970; Salach *et al.*, 1971b); such an effect is usually considered as indicative of the destruction of essential histidine residues. Wells (1972) deducted from the pH-activity profile of the *Crotalus adamanteus* enzyme that a histidine might be involved in the catalytic mechanism, but the enzyme was not sensitive to photooxidation or to treatment with iodoacetate (Wells, 1973). Oxidation with N-bromosuccinimide rapidly destroyed two tryptophans with loss of activity; presumably, the tryptophans are located near the reactive site of the enzyme. Ethoxyformic anhydride inactivated by acylating one lysine residue per dimer. The remaining subunit could form another, active, dimer. Dissociation of the dimeric enzyme by 4 M urea made the lysine resistant; calcium also protected the lysine. It appears that the lysine residue of one subunit is exposed on dimerization but subsequently buried when calcium is bound to the enzyme.

The porcine pancreatic enzyme as well as its zymogen are completely inactivated by *p*-bromophenacyl bromide, with concomitant acylation of the histidine-53. Histidine-121 reacts also, but at a much slower rate. The deactivation can be inhibited by Ca^{2+} or Ba^{2+} (Volwerk *et al.* 1974).

Since DFP does not inhibit the enzymes, they are probably not serine enzymes. Further support for this conclusion comes from ^3H and ^{18}O exchange studies. Wells (1971b) showed that ^{18}O from water appears in the fatty acid that is liberated by the enzyme; this is evidence for an *O*-acyl cleavage, as it might be expected for enzymic hydrolysis.

However, there was no transfer of fatty acid to [^3H]methanol, a reaction that would be expected from a serine enzyme because the nucleophilic methoxy ion could react with the acyl enzyme intermediate. Transfer of acyl groups is, in fact, found in reactions of pancratic lipase. Reincorporation of free fatty acids, also found with lipase, was equally absent in the phospholipase 2 reaction (Wells, 1971b).

A reacylation of lysolecithin by *Crotalus* enzymes had been suggested in a series of investigations (Franck *et al.,* 1968, 1970; Franck, 1970), but it appears that an enzymatic breakdown of lecithin on the silicic acid used in these experiments can explain these unexpected findings (Goerke *et al.,* 1971; Franck, personal communication).

Two recent studies have been concerned with the sequence of steps in the enzymatic reaction. De Haas *et al.* (1971) studied the hydrolysis of diheptanoyllecithin, in micellar solutions, by pancreatic phospholipase 2. One of their objectives was to decide between two pathways already discussed by Roholt and Schlamowitz (1961): (I) the binding first of Ca^{2+}, then of the substrate, to the enzyme; (II) binding of the substrate first.

$$\text{(I)} \quad E \underset{}{\overset{Ca^{2+}}{\rightleftharpoons}} E \cdot Ca^{2+} \underset{}{\overset{S}{\rightleftharpoons}} E \cdot Ca^{2+} \cdot S$$

$$\text{(II)} \quad E \underset{}{\overset{S}{\rightleftharpoons}} E \cdot S \underset{}{\overset{Ca^{2+}}{\rightleftharpoons}} E \cdot Ca^{2+} \cdot S$$

De Haas *et al.* found an optimum near pH 6 for the hydrolysis of the micellar substrate. Ca^{2+} was a specific cofactor, Mg^{2+} was not bound by the enzyme, and Ba^{2+} and Sr^{2+} were competitive inhibitors. Kinetic analysis of experiments with changing Ca^{2+} and substrate concentration failed to support either pathway exclusively; de Haas *et al.* suggested a random mechanism of enzyme–calcium–substrate association:

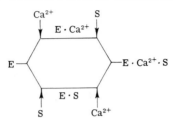

The random-binding model was supported by Ca^{2+}–enzyme equilibrium gel filtration, ultraviolet difference spectra, and fluorescence spectroscopy (Pieterson *et al.*, 1974a). Ca^{2+} seems to be bound by carboxyl groups and by an additional group, probably the histidine-53 imidazol ring. Binding of Ca^{2+} (and Ba^{2+} and Sr^{2+}) protected the enzyme against deactivation by *p*-bromophenacyl bromide and digestion by trypsin. Ca^{2+} was bound in the absence of substrate, and monomeric substrates were bound by enzyme and proenzyme in the absence of metal ions.

Experiments by Wells (1972) with monomolecular solutions of dibutyryllecithin and the *Crotalus adamanteus* enzymes led to different conclusions. The velocity patterns resulting from varying Ca^{2+} and substrate concentrations, and also inhibition studies with Ba^{2+}, were found to be consistent not with a random but with an ordered addition, first of the Ca^{2+} and then of the lecithin. Product inhibition studies with lysobutyryllecithin and dead-end inhibition with the butyric acid analog butyramide suggested an ordered release of the reaction products as well, first of butyric acid and then of the lyso compound.

In the nomenclature of Cleland (1970) the reaction is of the Ordered Bi Ter type:

In the diagram, A is calcium; B, lecithin; P, fatty acid; Q, lysolecithin. EAB–EAPQ is the active complex with its extreme equilibrium states, one containing the substrates, the other the products. Further studies by Wells (1973) have supported the ordered-addition model.

The differences between the results of de Haas *et al.* (1971) and Wells (1972) were discussed by Wells. A true difference in the enzymatic mechanisms of both enzymes was judged to be unlikely, but interfacial binding effects in the micellar system involving Ca^{2+} or the enzyme or both were believed to be possible causes of the discrepant results. However, the possibility of different mechanisms for the enzymes from venoms and pancreas cannot be discounted.

J. Intracellular Phospholipases 2

The existence of lecithin-hydrolyzing enzymes in animal tissues and in blood has been known for a long time (Thiele, 1913). In recent years, there has been an increase of activity in this field of research, partly owing to the work of Lands (1960), who found that lysophospholipids can be acylated by microsomal enzymes. Lands' findings implied a metabolic function for both 1-acyl and 2-acyl lysophospholipids. Analyses of tissue lysophosphatidylcholines (from bovine lung and plasma, human plasma, egg, and yeast) by Tattrie and Cyr (1963) suggested the presence of both types of isomers. Lloveras *et al.* (1963) found that an enzyme preparation from bovine spleen could hydrolyze lecithin at both positions. Robertson and Lands (1962), working with labeled phospholipids, found that these were broken down by various tissues, but the intermediate lyso compounds could not be isolated or characterized because the lysophospholipases of the tissues proved to be too active. By inhibiting these enzymes with deoxycholate, van den Bosch and van Deenen (1964, 1965), using rat liver homogenates, obtained lysolecithins from labeled lecithin, and by determining their radioactivity and also by degrading them to the isomeric monoglycerides, they could identify both the 1- and the 2-acyl isomer. Since then, research has progressed to determine the intracellular locations of the respective hydrolases.

The phospholipases of cells and body fluids are obviously much more interesting for the science of biology than the enzymes of snake venoms, or even those of the digestive glands; without doubt, books will be written one day that treat these intracellular enzymes alone. At present, however, as a branch of biochemistry the subject is yet in its infancy; we are, as it were, still at the stage of collecting and describing. However, the field begins to show outlines and structures, and some tentative generalizations can already be made—and, as elsewhere in this book, we shall not hesitate to make them.

In the membranes of mitochondria from animal tissues, alkaline phospholipases 2 occur that are reasonably heat-resistant, devoid of free sulfhydryl groups, and dependent on calcium. These properties of the enzymes are, of course, similar to those of the snake venoms and of the pancreas, and it can be noted with satisfaction that the enormous amount of work expended on these phospholipases has in no respect been wasted, and that our knowledge about them is probably, after all, very relevant to the study of the intracellular enzymes. The mitochondrial phospholipases obviously have an important role in the metabolism of the membrane, perhaps in a transport function.

The phospholipases 2 of "microsomes," i.e., the membranes of the endoplasmic reticulum, appear to be members of a different tribe; for instance, they may not require calcium. Together with the microsomal phospholipases 1, their role is that of catalysts in the Lands cycle of fatty acid exchange.

The lysosomal phospholipases 1 and 2 are, as might be expected, enzymes that are active at acidic pH. It is likely that structure and mode of action of the acidic phospholipases will be found to be quite different from that of the exocrine enzymes. The activity–pH profiles of some lysosomal preparations seen to announce the presence of a number of isoenzymes; doubtlessly, this speculation will soon be confirmed.

1. Phospholipases 2 of Mitochondria

Phospholipase activity localized in rat liver mitochondria was found by Rossi *et al.* (1965) and Scherphof and van Deenen (1965; Scherphof *et al.,* 1966). The latter authors determined the enzymatic optimum as around pH 8 and demonstrated that the enzyme was specific for the position 2 of phosphatidylethanolamine and phosphatidylcholine. The pH optimum and positional specificity have been confirmed in all subsequent investigations. Waite and van Deenen (1967) found that the enzyme was stable at temperatures up to 65°C and insensitive against PCMB, but completely inhibited by EDTA. This suggested a requirement for calcium, as it was also reported by Björnstad (1966). Nachbaur and Vignais (1968) prepared rat liver mitochondria free of lysosomes and also separated the inner and outer membranes of the organelles; they were then able to show that the phospholipase 2 activity at pH 8.1 was not due to lysosomal contamination, and that most of it was associated with the outer mitochondrial membrane. Vignais *et al.* (1968) also confirmed the requirement for calcium, and they showed that cardiolipin, the phospholipid typical for mitochondria, was hydrolyzed by the enzyme.

The mitochondrial phospholipase 2 of myocardial tissue of rat and dog was reported to have a pH optimum of 9.5 and a requirement for Ca^{2+} (Weglicki *et al.,* 1971). Intestinal musosa, which were known to contain phospholipase activity (Epstein and Shapiro, 1959; Gallai-Hatchard and Thompson, 1965), were found by Sarzala (1969) to contain a mitochondrial phospholipase 2, active at pH 7.5–8.0 and dependent on calcium. Bazán (1971) found similar activity in rat brain mitochondria.

A 160-fold purification of the enzyme from rat liver mitochondria was reported by Waite and Sisson (1971). Lipids had to be extracted from the organelles before the purification, and aqueous 0.5 M KCl was required to solubilize the enzyme. Even so, its high apparent molecular weight $>$ 2×10^5, suggested an aggregation of enzyme molecules. Maximal activity against phosphatidylethanolamine was found at pH 9.5, against phosphatidylserine at pH 7.4. The hydrolysis of the phospholipid liposomes responded to changes of the surface potential brought about by long-chain anions and cations, although the response was difficult to interpret. Hydrolysis rates decreased with the number of double bonds in the fatty acids of phosphatidylethanolamine; it is not clear whether or not the interfacial spacing and orientation of the substrate molecules is responsible for this effect.

2. Microsomes

Studies by Scherphof *et al.* (1966) and Waite and van Deenen (1967) showed that microsomal preparations from rat liver contained phospholipase 1 but were nearly devoid of phospholipase 2. However, this finding cannot be generalized for all animal tissues: the phospholipase 2 of rat intestinal mucosa seems to be partly located in the microsomal fraction (Subbaiah and Ganguly, 1970), and microsomes from rat myocardium also contain phospholipase 2 activity, which has a pH optimum of 7.5 and is only slightly stimulated by Ca^{2+}, and, more significantly, hardly inhibited by EDTA (Weglicki *et al.*, 1971). This enzyme, then, may be very different in structure and action from the phospholipases 2 of the exocrine glands or of mitochondria.

3. Lysosomes

Lysosomal enzymes are distinguished by being active at acidic pH. Phospholipase activity with an acidic optimum has been found in brain (Gatt *et al.*, 1966) and in rabbit alveolar macrophages (Elsbach, 1966). Activity was located in lysosomes from bovine adrenal medulla (Blaschko *et al.*, 1967), rat liver (Fowler, 1967; Mellors and Tappel, 1967; Stoffel and Greten, 1967; Fowler and de Duve, 1969), and macrophages (Franson and Waite, 1973; Franson *et al.*, 1973). Winkler *et al.* (1967), investigating the adrenal enzyme, reported both 1- and 2-activity, type 1 being active at pH 4.2, type 2 at 6.5. On the other hand, the lysosomes prepared from rat liver by Stoffel and Trabert (1969) did not show an optimum of phospholipase activity at the higher pH, but contained two acidic phospholipases, type 1 and 2. Waite *et al.* (1969) found, in their preparation of rat liver lysosomes, a 2-specific enzyme with a pH optimum around 6.5 and stimulated by Ca^{2+}, but a later study (Franson *et al.*, 1971) led to the conclusion that the activity had been due to an enzyme from a contaminating organelle, probably from the outer membranes of mitochondria which have sedimentation characteristics very similar to those of lysosomal membranes from Triton-treated rats (Vignais and Nachbaur, 1968). After osmotic rupture of the lysosomes, the acidic phospholipase 2 appeared almost completely in the soluble fraction (Waite *et al.,* 1969). The enzyme could be partially separated from the phospholipase 1 by gel filtration. Calcium did not stimulate; in fact, the inhibition by increasing concentrations of Ca^{2+} was severe, whereas the inhibition by EDTA was only slight.

Rahman *et al.* (1969) confirmed the presence of an acidic phospholipase in rat liver lysosomes, and they also described a phospholipase, presumably type 2, that was stimulated by Ca^{2+} and active at pH 7–8; this enzyme was bound to the membranes of lysosomes, and the activity was asserted not to stem from a contaminating organelle (Rahman and Verhagen, 1970). The existence of a membrane-bound lysosomal phospholipase 2 was confirmed be Colbeau *et al.* (personal communication) and Stoffel *et al.* (personal communication, 1972), but disputed by Franson *et al.* (1971, personal communication). An acid (pH 4–4.5) phospholipase 2 was concentrated 1000–1500-fold by ion-exchange and Sephadex chromatography (Franson *et al.,* 1972b); its apparent molecular weight was 32,000.

Lysosomes of rat myocardial tissue showed a Ca^{2+}-stimulated phospholipase 2 activity around pH 5 and an EDTA-stimulated activity at pH 4, but no activity in the neutral range (Franson *et al.,* 1972a).

4. Other Cell Fractions

A phospholipase 2 has been found in the 105,000 *g* supernate of rat lung (Ohta *et al.,* 1972). Its activity was optimal at pH 6.5 and unaffected by Ca^{2+} or EDTA. Victoria *et al.* (1971) found Ca^{2+}-dependent activity in plasma membranes from rat liver.

From the information so far available on the intracellular distribution of phospholipase 2, and anticipating the chapters on phospholipase 1 and on lysophospholipases, we can tentatively formulate the cellular activities of these enzymes as shown in Scheme VI-7. It remains to be seen how far Scheme VI-7 has to be corrected or elaborated.

5. Blood, Animal Tissues, and Microorganisms

Robertson and Lands (1962) could not demonstrate any phospholipase activity in human erythrocytes; however Paysant and Polonovski and their co-workers (Paysant and Polonovski, 1966; Paysant *et al.,* 1970; Bitran *et al.,* 1971) found a phospholipase in the erythrocytes as well as in the plasma of rat and man. The substrate used for much of the work reported in their studies was phosphatidylglycerol, a very minor component among animal phospholipids, and the question of the physiological substrate of

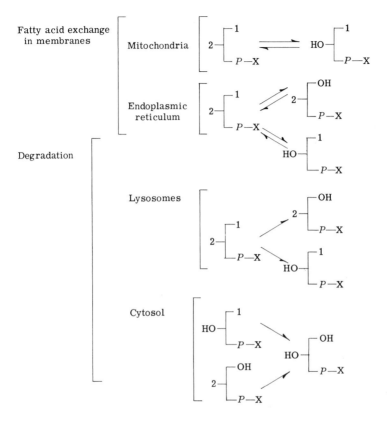

SCHEME VI-7

the enzyme must be raised. Possibly other phospholipids that are as strongly acidic as phosphatidylglycerol may be accepted, although cardiolipin reportedly is not.

The enzyme from rat plasma is a phospholipase 2; in red blood cells, both type 1 and 2 are found (Paysant and Polonovski, 1966; Paysant *et al.*, 1967). The rat plasma enzyme has been purified 1400-fold by precipitation with alcohol and fractionation on a Sephadex column (Paysant *et al.*, 1969a). To judge from its retention on the gel, the enzyme appears to have a low molecular weight. It requires Ca^{2+} and is stimulated by 0.25 *M* NaCl (Paysant *et al.*, 1969b). Serum albumin also promotes its action, and so does ethyl ether, both perhaps by removing inhibitory fatty acids; deoxycholate was found to be inhibitory. The pH optimum is 7.5. The enzyme is quite resistant to heat; at pH 7 it is stable up to 70–80°C, and at pH 4 even heating to 100°C for 3 minutes leaves it intact.

Most properties of the plasma phospholipase 2 are very similar to those of pancreatic phospholipase; one is almost inclined to speculate on their identity. The similarity has been underlined by the finding (Etienne *et al.,* 1969; Paysant *et al.,* 1969b) that the enzyme has a zymogen that can be activated by trypsin or by a factor, probably a proteinase, present in platelets (Duchesne *et al.,* 1972). It is in the form of this proenzyme that most of the phospholipase 2 occurs in the plasma. Etienne *et al.* (1967) and Osmond *et al.* (1973) have described the action of plasma phospholipase 2 on phosphatidylethanolamine.

Phospholipase 2 activity has been found in association with several commercial serum albumin preparations (Elsbach and Pettis, 1973), a finding to be remembered in experiments that call for albumin as a fatty acid acceptor.

The phosphatidylglycerol hydrolase of rat erythrocytes has been purified 230-fold by precipitation with ethanol from a lysate of red cells, followed by Sephadex chromatography (Delbauffe *et al.,* 1968). The enzyme is heat-resistant, requires Ca^{2+} and has a pH optimum of 7–8 (Polonovski *et al.,* 1969). It is located in the stroma of red cells, from which it is solubilized by 1 M NaCl. A similar enzyme is found in human erythrocyte lysate and stroma (Paysant *et al.,* 1970). This enzyme shows activity not only against phosphatidylglycerol but also against phosphatidylethanolamine and phosphatidylcholine. Its activity is weak compared to that of rat erythrocytes but can be increased by incubation with trypsin; it is not clear if this is a case of proenzyme activation or if the enzyme is otherwise unmasked.

The phospholipases 1 and 2 of bovine spleen were investigated by Lloveras and Douste-Blazy (1968a,b; Lloveras *et al.,* 1963). Activities 1 and 2 could be separated by chromatography on DEAE-cellulose. The phospholipase 2 was purified 166-fold. The activity was measured at pH 4.5; probably, then, the enzyme is of lysosomal origin.

Phospholipases of arterial tissue are of special interest in the research of atherosclerosis. Eisenberg *et al.* (1968) found a type 2 enzyme, with a pH optimum of 8.5, predominant in both rat and human aorta. The activity increases throughout the lifespan of the rat (Eisenberg *et al.,* 1971). In pig aorta, Patelski *et al.* (1971) found a heat-stable phospholipase (presumably 2), active at pH 8, inhibited by calcium and also, though not completely, by DFP.

A moderately thermostable phospholipase 2 from brain was described by Webster (1970) and Cooper and Webster (1970, 1972). It was partially purified by gel filtration, but could not be freed completely of position 1 activity; the authors suggested that both activities might belong to the same enzyme. The molecular weight of the preparation, as determined by Sephadex filtration, was 55,000.

Rabbit granulocytes contain a particle-bound phospholipase 2 that is optimally active at pH 7.5 and requires calcium (Elsbach *et al.,* 1972; Patriarca *et al.,* 1972b).

There is a large literature reporting phospholipase activity in micro-organisms; but the specificity of these enzymes has usually been left undetermined, either because the labeled substrates required in such studies were not available, or because the presence of powerful lysophospholipases prevented the accumulation of reaction products (Ferber *et al.,* 1970). It also appears that the phospholipases 1 rather than 2 may be the more typical microbial enzymes; however, evidence for the presence of a phospholipase 2 in *Escherichia coli* has been furnished by Proulx and van Deenen (1967) and Proulx and Fung (1969), and Bernard *et al.* (1972a,b,c) have found 2-specific activity in both the cytoplasm and the cell membrane of *E. coli* O118. Both enzymes, which may be identical, had an optimum of activity around pH 8, required Ca^{2+} and a (preferably neutral) detergent, were heat stable at neutral and basic pH and were stable in 40% ethanol.

K. Conclusion

This concluding discussion on phospholipase 2 is restricted to the enzyme family of exocrine glands, with the understanding that the mitochondrial enzymes may be close relatives, but that the microsomal and lysosomal enzymes are probably not.

Phospholipase 2 is a metalloenzyme with Ca^{2+} as a cofactor. In Chapter IV, Section II on pancreatic lipase, a presumed serine enzyme, we made much use of analogies with another serine enzyme, the peptide hydrolase chymotrypsin. For an enzyme analogous to phospholipase 2, we should look among the metal-requiring peptidases; and the best known of these is carboxypeptidase A.

Phospholipase 2 acts preferably on micellar substrates or at interfaces. The high reaction rates found under these conditions seem to be the result of supersubstrate binding and interfacial orientation of the enzyme, and they are thus, in the last analysis, substrate concentration effects; but the enzyme will also, at slow rates, hydrolyze substrates in monomolecular solution. Future studies in such systems may be especially rewarding.

The hydrolysis of the substrate proceeds by *O*-acyl cleavage. The enzyme probably requires a negative charge in its substrates that has to be five to six atoms away from the carboxyl carbon of the susceptible ester. The steric specificity can be explained by the postulate that only a hydrogen

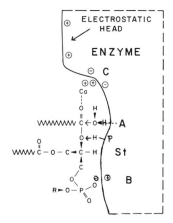

Fig. VI-7 Hypothetical model of the reactive site of porcine pancreatic phospholipase.

atom can find sufficient space on the side of the *sn*-glycerol-2 carbon that faces the surface of the enzyme.

Figure VI-7 is a model of the enzyme–substrate complex of pancreatic phospholipase 2 that incorporates these facts and ideas as well as some further speculations. The supersubstrate binding site is probably an electrostatic head; it serves to bind and orient the enzyme at the interface. However, hydrophobic supersubstrate binding may also occur. In the case of the pancreatic enzyme, this supersubstrate-binding site is created on tryptic activation of the zymogen. C is the electronegative binding site for Ca^{2+}; A, the nucleophilic group responsible for the attack on the ester; B, the positive site binding the phosphate; and St stands for the region of steric restriction on the surface of the enzyme that allows only a hydrogen atom to face it; P is a proton donor. The whole model, it should be pointed out, is stretched to accommodate it in a drawing of two dimensions; it should be visualized folded and compressed in three dimensions.

In our model, the group R on the phosphate, unnecessary for a substrate, is not involved in binding; neither is the C—1 glycerol ester group; the Ca^{2+} is not bound to the phosphate of the substrate, and the nucleophilic attack is mediated by water.

A comparison with the known amino acid grouping in the reactive center of carboxypeptidase A (Hartsuck and Lipscomb, 1971) will illuminate the background of our model and encourage some speculations on the nature of sites A, B, C, and P. First, it should be mentioned that the presence of several sites does not contradict our previous conclusion that two loci of fixation are sufficient to explain the steric specificity of the enzyme. Sites A, C, and P fixate the same side of the substrate, viewed from the anomeric

C—2 atom. Fixation that would influence the stereospecificity would have to be fixation to a third substituent of C—2. Likewise, of course, steric substrate fixation at two sites only does not preclude the involvement of many amino acid residues in the catalytic mechanism.

In carboxypeptidase, the metal, zinc, is believed to be bound ionically by glutamic acid and complexed by two histidines. Such residues, therefore, must be considered as candidates for site C in our model. The zinc does not bind to the free carboxyl of the peptidic substrate. This group, which corresponds to the phosphate in our model, seems to be salt-linked to arginine in the peptidase. Arginine, therefore, might constitute site B in phospholipase 2, although, in view of the stronger electronegativity of phosphate over carboxylate, a weaker basic residue, in particular lysine, might also serve. Group A, in carboxypeptidase, is thought to be a glutamic acid anion, and the nucleophilic attack may be relayed through a water molecule. In our model we have tried to illustrate why, for steric reasons, water can hydrolyze the ester but methanol cannot. Mediation of the attack by water would also explain the nonexistence of an acyl–enzyme intermediate. The site P, finally, believed to be tyrosine in carboxypeptidase, donates a proton to the alkoxy-O that appears on C—2 during the hydrolysis, and so prevents the reversal of the reaction.

II. PHOSPHOLIPASE 1

Phospholipase 1 is a latecomer among lipolytic enzymes; its existence was not recognized before 1960. This delay has in part been made up since then by a burst of publications that have described the occurrence of the enzyme in many organs, organelles, and microorganisms. However, no sources have been found that are as convenient for the preparation of phospholipase 1 as the snake venoms or the pancreas are for phospholipase 2; phospholipases 1 are not produced by exocrine glands. At present, therefore, there is little information available on their chemistry or mechanism of action. However, successful purifications of the enzymes from two microorganisms have recently been reported (Scandella and Kornberg, 1971; Raybin *et al.,* 1972), and we may expect that chemical and enzymological investigations on these or similar preparations will soon be under way.

A. Occurrence

The presence of phospholipid acyl hydrolases in tissues had been known for a long time (Thiele, 1913; Francioli, 1934), but the specificity of these enzymes could not be established until it became possible to distinguish the fatty acids of positions 1 and 2 in phosphoglycerides (Tattrie, 1959; de Haas and van Deenen, 1961a; Hanahan et al., 1960). In 1960, Hanahan reported that the two fatty acids of mammalian phospholipids turned over independently, and Lands (1960) found that both 1- and 2-acyl glycerylphosphorylcholine could be acylated by rat liver microsomes. The existence of a phospholipase 1 (as well as a 2) could be inferred from these observations. In 1963, evidence for such an enzyme in bovine spleen was presented by Lloveras, Douste-Blazy, and Valdiguié. This enzyme could later be separated from the accompanying phospholipase 2 (Lloveras and Douste-Blazy, 1968a,b).

Robertson and Lands (1962) had observed the total deacylation of phosphatidylcholine in several tissues of the rat, but could not detect the intermediate lysophospholipid that would have proved the activity of either phospholipase 1 or 2. Van den Bosch and van Deenen (1964), by inhibiting lysophospholipase activity with deoxycholate, succeeded in isolating a lysolecithin which was a mixture of both isomers. Fractionation of the subcellular components of rat liver tissue showed that at neutral pH the phospholipase 2 activity was strongest in mitochondria, whereas phospholipse 1 was associated mainly with microsomal particles (Scherphof et al., 1966; Waite and van Deenen, 1967). These observations have repeatedly been confirmed, although it seems that "microsomes," i.e., the endoplasmic reticulum, also contain considerable 2-activity (Weglicki et al., 1971). Microsomal phospholipase 1 has been demonstrated in rat brain (Webster and Cooper, 1968; Bazán, 1971) and heart (Weglicki et al., 1971), in cerebrospinal fluid (Illingworth and Glover, 1969), and in insects (Kumar et al., 1970; Rao and Subrahmanyam, 1970). Van Golde et al. (1971a,b) found the enzyme in the Golgi complex from bovine and rat liver.

A second cellular locus of phospholipases 1 are the lysosomes. The lysosomal enzymes are active under acidic conditions at pH 4–5. The phospholipase 1 from spleen (Lloveras et al., 1963) must probably be counted among them; likewise, an acidic enzyme from brain (Gatt et al., 1966; Gatt, 1968; Webster, 1970; Woelk et al., 1972) and perhaps an enzyme from guinea pig pancreas reportedly active at pH 6 (White et al., 1971). The isolation of lysosomes (usually as tritosomes, i.e., lysosomes loaded with the surfactant Triton WR-1339) has brought unequivocal proof that phospholipase 1 (together with 2) is among the lysosomal enzymes. Lysosomes of bovine adrenal medulla show an optimum of activity 1 at pH 4.2 (Blaschko et al., 1967; Winkler et al., 1967; Smith and Winkler, 1968). Rat liver lysosomes (Stoffel and Trabert, 1969; Waite et al., 1969; Franson et al., 1971), heart lysosomes (Franson et al., 1972a), and lysosomes of alveolar macrophages (Franson and Waite, 1973) also contain the enzyme. An acidic phospholipase 1 (optimum pH 4.8) is also found in human blood platelets (Smith and Silver, 1973).

The occurrence of phospholipase 1 in the plasma membrane of rat liver cells is somewhat less well documented. Newkirk and Waite (1971) found a calcium-dependent type 1 enzyme as the principal phospholipase, but Victoria *et al.* (1971) described type 2 as the major activity, although they believed an enzyme 1 also to be present. Contamination of the membrane preparations with lysosomes or microsomes may account for the discrepancy.

The presence of phospholipase 1 activity in post-heparin human blood plasma was reported by Vogel *et al.* (1965) and Vogel and Bierman (1967). Infante *et al.* (1968) have reported type 1 activity in the post-heparin plasma of the rat. There were indications that the enzyme might be identical with lipoprotein lipase (Vogel and Bierman, 1965; Doizaki and Zieve, 1968), a suspicion that was strengthened when it was shown by de Haas *et al.* (1965) that the triglyceride lipase of pancreas could also hydrolyze phospholipids in position 1. However, Greten (1972) has reported that in patients with hyperlipoproteinemia type 1 a post-heparin phospholipase activity of mixed 1 and 2 character remained normal, although lipase activity was deficient. Since the disease is caused by a deficiency in lipoprotein lipase of extrahepatic origin (LaRosa *et al.,* 1970a,b), the possibility remains that the phospholipase 1 is identical with the lipoprotein lipase originating in liver plasma membranes (Assmann *et al.,* 1973a). The phospholipase does, in fact, reside in plasma membranes (Waite, 1973), and it is reduced by 80% in hepatectomized rats (Zieve and Zieve, 1972). Zieve and Zieve (1972) however, found that the purified enzyme hydrolyzed mono- and diglycerides but not triglycerides.

Phospholipase 1 may be the predominant lipolytic enzyme of bacteria. It was first identified in *Escherichia coli* (Fung and Proulx, 1969; Proulx and Fung, 1969); optima of activity 1 were found at pH 5 and 8.4; at alkaline pH, smaller levels of activity 2 were also detected. A "detergent-resistant," methanol-activated phospholipase "A" (Okuyama and Nojima, 1969; Doi *et al.*, 1972) is perhaps identical with the pH 8.4 phospholipase 1. Patriarca *et al.* (1972a) found type 1 activity with two pH maxima in *E. coli* spheroblasts, and Scandella and Kornberg (1971) have succeeded in purifying the enzyme from *E. coli* cell membranes. Bell *et al.* (1971) and Albright *et al.* (1973) localized the enzyme in the wall of *E. coli.* Kent and Lennarz (1972) detected type 1 activity in an osmotically fragile mutant of *Bacillus subtilis,* while the wild-type bacteria contained a protein that specifically inhibited the enzyme. A soluble phospholipase 1 has been isolated from the spores and sporangia of *Bacillus megaterium* (Raybin *et al.,* 1972).

B. Detection and Assay

Phospholipase activity is easily detected on incubation of phosphoglycerides by the appearance of free fatty acids or lysophosphoglycerides, most conveniently on thin-layer chromatography of the lipid extract. In order to assess the type of activity, 1 or 2, the fatty acid distribution on the phospholipid that serves as the substrate must be known. The fatty acids must be different, or the acid in one position must be labeled, or each fatty acid must be labeled with a different tracer, for instance, ^{14}C and ^{3}H.

Numerous studies have been published, and some are still being published, that describe the activity of a phospholipase "A" in tissues and organisms without establishing its specificity for position 1 or 2 of phosphoglycerides. Such studies have been of value before the different activities could be distinguished, and they may still be of value in certain biological or physiological investigations. The biochemist, however, can no longer be content with long and detailed reports on phospholipases which leave the question unanswered. In view of the ubiquitous occurrence of phospholipases, the mere discovery of a phospholipase "A" is usually not more interesting than the discovery of "proteolytic activity" in a digestive gland. It has been possible now for over a decade to distinguish the two activities, and a biochemist should make every effort to avail himself of this possibility. In many instances, it will not even be necessary to synthesize radioactive substrates for this purpose.

We hope that our proposal to replace the trivial names phospholipase A_1 and A_2 with phospholipase 1 and 2 will eventually have the effect of forcing authors to state clearly in the titles and summaries of their papers whether they have determined these specificities or not. The necessity of spelling out phospholipase "A" as phospholipase 1 or 2 might then stimulate the small additional effort needed to clarify the matter, unless there are good reasons why this should not be possible. It must be granted that such reasons may exist, for instance, in systems where lysophospholipase activity cannot be suppressed (Ferber *et al.*, 1970).

In their first identification of phospholipase 1, Lloveras *et al.* (1963) used egg lecithin, which is known to have saturated fatty acids in position 1 and unsaturated acids in position 2. Such a natural substrate can still be used. The possible objection that a phospholipase might exist that is specific for certain types of fatty acids, and might therefore mimic positional specificity where none exists, can be countered with the reply that (with the single exception of the *Geotrichum* lipase) no lipolytic enzyme with "fatty acid specificity" has yet been found. An important consideration, however, is the low sensitivity of assays with unlabeled substrates. The phospholipase activities 1 and 2 may be many thousandfold lower in tissue homogenates or microorganisms than the activities in the excretions of glands.

Phosphoglycerides with radioactive fatty acids have been prepared by biosynthesis with rat liver microsomes (Robertson and Lands, 1962; van den Bosch *et al.,* 1965; Waite and van Deenen, 1967; Bazán, 1971), in the intact rat (Gatt, 1968), or with microorganisms, and also by chemical synthesis (Stoffel and Greten, 1967; Stoffel and Trabert, 1969). In most of the labeled substrates so far prepared, the fatty acids in position 1 have been chemically different from those in 2, just as in egg lecithin, which leaves any doubts concerning fatty acid specificity unresolved. Substrates having the same fatty acid in both positions, but differently labeled, would be more commendable. A [1-^{14}C]oleoyl [2-^{3}H]oleoyl phosphoglyceride, for instance, would not be very difficult to prepare by biosynthetic as well as chemical procedures, and it would have ideally average physicochemical properties. A chemically synthesised [1-^{3}H]linoleyl [2-^{14}C]linoeoyl glyceryl-phosphorylcholine has been used by Greten (1972), [1-^{3}H]stearoyl [2-^{14}C]stearoyl glycerylphosphorylcholine by Winkler *et al.* (1967).

It is important to realize that the liberation of a fatty acid from position 1 or 2 of a phosphoglyceride does not prove the existence of the corresponding phospholipase. For instance, fatty acid 1 may well be hydrolyzed by a lysophospholipase from the lysophosphoglyceride that is the product of a phospholipase 2. Lysophospholipases do, in fact, almost always accompany the phospholipases. Especially with singly labeled substrates, the danger of misinterpreting the experimental results is great. In many cases, lysophospholipases can be inhibited by the addition of deoxycholate (Magee *et al.,* 1962); however, this method is not always effective (Ferber *et al.,* 1970). In any case, it is necessary to identify the resulting lyso-phosphoglyceride in order to prove the presence of a particular phospholipase. Paradoxically, then, a substrate for the assay of phospholipase 1 should carry a labeled fatty acid in position 2.

After the identification of the lysophosphoglyceride, it is still necessary to establish that the enzyme is indeed a phospholipase. This statement will appear less nonsensical if it is remembered that pancreatic lipase, although undoubtedly designed to hydrolyze triglycerides, will also hydrolze phospholipids in position 1 (de Haas *et al.,* 1965). The identification of a phospholipase 1 is therefore usually bolstered by a demonstration that the preparation does not hydrolyze triglycerides. For this purpose, a triglyceride should be chosen with physical properties similar to those of the phospholipid. Labeled tripalmitin, which is solid, is less suitable than triolein, which is liquid. To insure that phospholipid and triglyceride are equally available to the enzyme, Waite *et al.* (1969) have employed these substrates in mixed emulsions.

The most convincing identification of the enzyme employs a synthetic *sn*-1-phosphoglyceride in comparison with a natural *sn*-3 phospholipid (Scandella and Kornberg, 1971). Since these compounds differ only in the configuration around C-2, the resistance of the first substrate proves the action of a stereospecific phospholipase 1. The *sn*-1 phospholipid can be prepared by exhaustive enzymatic hydrolysis of a racemic phospholipid with the phospholipase 2 from snake venom.

Once the identity of the enzyme is established, less rigorous methods can be employed to assay its activity or follow the course of a purification procedure; e.g., the liberation of a radioactive acid from position 1 may be followed (Gatt and Barenholz, 1969a). Scandella and Kornberg (1971) have employed the partition of phosphatidylglycerol and its lyso derivative between an aqueous and an organic phase for a convenient and fast assay.

C. Isolation

Phospholipases 1 of animal tissues have so far not been isolated in pure form, although some partially purified preparations have been obtained. Lloveras and Douste-Blazy (1968a), starting with acetone powder of spleen and employing ammonium sulfate precipitation and chromatography on DEAE-cellulose, could increase the specific activity of the enzyme 126-fold. Gatt (1968) extracted the enzyme from rat brain particles (the mito-chondrial–lysosomal fraction) with sodium cholate under sonic disintegration. Dialysis of the supernate against buffer of pH 4.2 yielded a soluble enzyme 54-fold enriched. An extract from calf brain, after adjustment to pH 3.8, ammonium sulfate precipitation, and Sephadex chromatography, gave a 46-fold purified preparation. Woelk *et al.* (1972) purified the enzyme from human brain 80-fold, starting from acetone–butanol powder, by extraction at pH 7.4 and consecutive precipitations with acid (pH 3.2), ammonium sulfate, and taurocholate, followed by Sephadex filtrations.

Attempts to isolate phospholipase 1 from bacteria have been more successful. Table VI-6 summarizes the purification of the membrane-bound calcium-dependent enzyme of *E. coli* (Scandella and Kornberg, 1971). Cell paste was homogenized and centrifuged; the precipitate, suspended in water, was left to autolyze for 1 hour; it then yielded, on centrifugation, the particulate fraction II. The enzyme was solubilized with 1% sodium dodecyl sulfate (which dissolves the entire cell envelope); butanol precipitated half the protein. Addition of sodium acetate, at pH 5.4, removed more protein, and acetone precipitated the enzyme from fraction IV. Extraction and acetone fractionation gave further purification. Before the final step, preparative electrophoresis (VIII), some lipid material associated with the protein had to be removed by extraction with butanol (VII). Fig-

TABLE VI-6

Purification of the Phospholipase 1 of *Escherichia coli* **Membrane**[a]

	Fraction	Total (gm)	Specific activity × 10^{-3}	Yield
I.	Cell homogenate	50	0.17	(100)
II.	Particulate	10	0.67	80
III.	SDS–butanol supernate	4.0	2.3	110
IV.	Acetate supernate	0.4	23	110
V.	SDS–butanol extract	0.08	92	88
VI.	Acetone fraction	0.013	400	63
VII.	Lipid extraction	0.010	500	57
VIII.	Electrophoresis	0.002	2000	45

[a] Scandella and Kornberg (1971).

ure VI-8 shows the progressive purification of the enzyme as it appears on disk gel electrophoresis at three different stages. The authors estimate that the final enzyme preparation, which had a specific activity of 2.0 (μmoles/minute/mg), was over 80% pure. Table VI-6 gives the weights starting from 1 pound of cell paste, which yielded 2 mg of fraction VIII, corresponding to a 45% yield. A large-scale run with 100 pounds of cell paste gave a similar yield. It is thus possible to prepare this enzyme in quantities sufficient for chemical investigations.

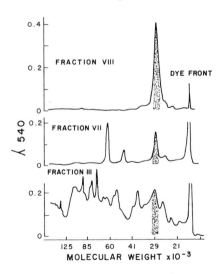

Fig. VI-8 Sodium dodecyl sulfate–acrylamide gel electrophoresis of the phospholipase 1 of *E. coli* membrane at three stages of purification. From Scandella and Kornberg (1971).

A calcium-independent, water-soluble phospholipase 1 was isolated from spores of *Baccillus megaterium* by Raybin *et al.* (1972). The spores were germinated by freezing and thawing and extracted at pH 10.9. Consecutive ammonium sulfate fractionation, DEAE-cellulose chromatography, preparative electrophoresis, and hydroxyapatite chromatography led to a preparation with a specific activity of 1560, corresponding to a 170-fold purification. The yield was 5.5% and the enzyme was estimated to be 75% pure.

D. Properties

During the short history of phospholipase 1, researchers have mainly occupied themselves with demonstrating the identity of the enzyme and determining its distribution in subcellular fractions, and few other results have so far come forth; with the recent discovery of the enzyme in microorganisms, this state of affairs is certain to change soon. It can already be seen that the phospholipases 1 are a very heterogeneous family of enzymes, perhaps comparable in diversity to proteinases. There are species acting at neutral or basic pH and species acting under acidic condition; some require calcium, others do not. We have, then, at least four groups of different character and probably quite dissimilar structure and mechanism. Perhaps it will soon be possible to discuss the enzymes within the frames of such groups; at present, the outlines of the framework are not sharp enough, and we must group the enzymes according to their natural sources.

1. Phospholipase 1 of Microsomes and Plasma Membrane

The phospholipase 1 of rat liver microsomes has its optimum of activity around pH 8 (Waite and van Deenen, 1967). It is not stimulated by Ca^{2+} nor inhibited by EDTA; in fact, Ca^{2+} at 5×10^{-3} inhibits the activity by 50%. The sulfhydryl reagent *p*-chloromercuribenzoate (PCMB) does not inhibit. Sonication of the microsomes partly destroys the enzymatic activity, heating above 65°C destroys it completely. Phosphatidylethanolamine and phosphatidic acid are better substrates than phosphatidylcholine. Waite and van Deenen (1967) found triglyceride lipase activity in liver microsomes almost equalling the activity of the phospholipase 1, but the lipase was partially inhibited by PCMB; this indicated the presence of two distinct enzymes.

The microsomal enzymes from rat and dog myocardial tissue seem to be very similar to the enzymes from liver (Weglicki *et al.,* 1971). A phospholipase 1 of mosquito larvae was active at pH 8.6 and in the presence of bile salts (Rao and Subrahmanyam, 1970). A preparation from housefly larvae, with an optimum near pH 8.0 was inhibited by 1 mM $HgCl_2$ and 2.5 mM lauryl sulfate or 5 mM deoxycholate (Kumar *et al., 1970*).

The phospholipase 1 found by Newkirk and Waite (1971) in plasma membranes of rat liver was optimally active against phosphatidylethanolamine around pH 9. It was stimulated by Ca^{2+} and inhibited by EDTA.

2. Lysosomes

The lysosomal phospholipases 1 have optima between pH 4 and 5. The enzyme from rat liver does not seem to require Ca^{2+}, nor is it inhibited by it (Stoffel and Trabert, 1969; Franson *et al., 1971*), but for myocardial lysosomes it has been reported that activity at pH 5 is stimulated by Ca^{2+}, whereas the activity at pH 4 is not (Franson *et al., 1972a*).

The (presumably lysosomal) phospholipase 1 of rat brain (Gatt, 1968) had a pH optimum of 4.0, did not require metal ions and was not inhibited by EDTA. The neutral detergent Triton X-100 increased the rates of reaction. The enzyme had no lysolecithinase activity; 1-palmitoylglycerylphosphorylcholine was hardly attacked. Fatty acids acted as strong competitive inhibitors, with the apparent K_i of palmitic acid being 7×10^{-5} in comparison to an apparent K_m (phosphatidylcholine) of 8×10^{-4} M. Phosphatidylcholine was a better substrate than phosphatidylethanolamine, but the reaction rates became equal in a system containing optimal amounts of Triton and deoxycholate. The V_{max} was 0.05 μmoles/minute/mg of the preparation.

Cooper and Webster (1970) investigated the phospholipases 1 and 2 of human brain. Neither required Ca^{2+}, and both were inhibited by the serine-blocking reagent DFP. Most of the type 1 activity was destroyed by heating to 70°C; a small remainder cochromatographed on Sephadex with the heat-stable phospholipase 2 with an estimated molecular weight of 55,000. The authors speculated that activity 2 and part of activity 1 might pertain to one enzyme. A similar suggestion was made by Greten (1972) for the phospholipase of post-heparin human blood. The phospholipase preparation of Woelk *et al.* (1972) from brain had a specific activity of 0.04 at pH 4.2; the molecular weight, determined by gel filtration, was 75,000.

3. Bacteria

The best-characterized phospholipases 1 are those that have been isolated from bacteria. In *E. coli*, Proulx and Fung (1969) found activity optima at pH 5 and 8.4, both requiring Ca^{2+}. Magnesium stimulated to a lesser degree. The "detergent-resistant" enzyme of Doi *et al.* (1972), which may be identical with the pH 8.4 phospholipase 1, was stimulated by methanol and produced methyl esters of the substrate's fatty acids. The purified enzyme of the membrane of *E. coli* (Scandella and Kornberg, 1971) has an optimum around 8.4. It hydrolyzes the 1-ester bond of phosphatidylcholine, phosphatidylethanolamine, and phosphatidylglycerol (the major phospholipid of *E. coli*) at comparable rates. Its molecular weight, determined by electrophoresis in sodium dodecyl sulfate–acrylamide gel, is 29,000. The enzyme is remarkably stable in a 3% sodium dodecyl sulfate solution, which normally denatures proteins; the authors suggest that a hydrophobic exterior of the enzyme molecule is responsible for this stability. In the absence of sodium dodecyl sulfate and lipid, the enzyme tends to aggregate. In a crude extract, the enzyme is stable to heating at 98°C; if purified, it denatures. The obligatory calcium cannot be replaced by Mg^{2+} or other divalent metals; Hg^{2+} at 1 mM causes no appreciable inhibition. The affinity of the enzyme for phosphatidylglycerol seems to be very strong; the apparent K_m is 3.4×10^{-7}.

The *E. coli* enzyme was identified as phospholipase 1 by its action on synthetic substrates. More than 90% of the breakdown products of phosphatidylcholine and phosphatidylethanolamine were the 2-acyl lyso compounds. Triolein was not hydrolyzed. The enzyme is stereospecific: *sn*-1-phosphatidylcholine was not noticeably hydrolyzed. There was no fatty acid specificity: phosphatidylglycerol with cyclopropane fatty acids or branched acids was also attacked. The requirement for a free acidic function on the phosphate does not seem to be absolute: 1-palmitoyl-2-oleoyl-*sn*-glycerol-3-dibenzyl phosphate was hydrolyzed, though only at 2% of the rate attained with natural phosphoglycerides. The enzyme hydrolyzes 1-acyl glyceryl phosphorylcholine at twice the rate of 1,2-diacyl glycerylphosphorylcholine. It might, therefore, be described as a lysophospholipase; but the 1-acyl lysophospholipid, which is not the product of its action, is probably also not among its natural substrates.

The second well defined microbial phospholipase 1, isolated from the spores of *Bacillus megaterium* by Raybin *et al.* (1972), is a soluble enzyme with a molecular weight around 26,000 which does not require Ca^{2+} or other metal ions. Its pH optimum was either 5 or 6.5, depending on the buffer. The specific activity, 1560, is almost a thousand times higher than that of the *E. coli* enzyme. Its apparent K_m for phosphatidylglycerol was 6×10^{-5} M. Fatty acids inhibited the enzyme. The activity against tributyrin was only 0.2% of that against phosphatidylcholine. In contrast to the *E. coli* enzyme, the enzyme hardly attacked 1-acyl lysophospholipid.

The enzyme from *B. megaterium* is very responsive to the micellar potential of the substrate. Phosphatidylglycerol, which is strongly acidic, is a very good substrate, but phosphatidylethanolamine and phosphatidylcholine are attacked only in the presence of anionic detergents. It may be surmised that the enzyme has a cationic electrostatic head, which governs its orientation at the supersubstrate–water interface (Brockerhoff, 1973).

The role that phospholipases play in microorganisms is not clear; for a discussion on the subject, the papers of Scandella and Kornberg (1971), Raybin *et al.* (1972), and Ferber *et al.* (1970) may be consulted.

E. Conclusion

The experimental identification of a phospholipase 1 is not altogether easy, and before it is attempted the definition for such an enzyme should be made clear. A phospholipase 1 hydrolyzes the 1-acyl bond of a 1,2-diacyl-*sn*-glycerol phosphate and its esters; in addition, such phosphoglycerides should be its natural substrates. The term "natural" is intentionally vague because it is not possible to give a more quantitative and still general criterium. In particular, relative reaction rates are not useful for the purpose of nomenclature. Pancreatic lipase, for instance, can hydrolyze phospholipids at specific rates that may be higher than those of many phospholipases, but Darwinian fate clearly did not design this enzyme for phospholipids. On the other hand, the phospholipase 1 of *E. coli* is reported to be more active against 1-acyl glycerylphosphorylethanolamine than against the diacyl compound (Scandella and Kornberg, 1971), but the lysophospholipid is not likely to be its natural substrate. The "true identity" of such enzymes must be decided by common sense after the weighing of all the evidence.

A definition of phospholipases 1 should perhaps include the stereospecific requirement for substrates with the *sn*-glycerol 3-phosphate skeleton; the enzyme from *E. coli* has indeed been shown to be stereospecific. It is quite possible, however, that nonstereospecific enzymes will be found that still deserve the designation phospholipase 1.

Using our definition, the identity of the microbial and lysosomal type 1 enzymes can be considered as established, although the lysosomal enzymes are in dire need of further purification and characterization; for the microsomal enzymes this need is even greater.

The phospholipases 1 of mammalian intracellular organelles fall into two groups; the microsomal type, active around pH 8, and the lysosomal type, active between pH 4 and 5, both metal-independent. About the microsomal enzymes little is known except their existence. They do not seem to be sulfhydryl enzymes; perhaps they are serine enzymes with an esterolytic mechanism resembling that of chymotrysin; but this is a mere guess. The lysosomal enzyme of the brain is reported to be competitively inhibited by fatty acids (Gatt, 1968) and irreversibly by DFP (Cooper and Webster, 1970). These observations suggest the presence of a reactive serine and the formation of an acyl enzyme intermediate. Since fatty acids are in their undissociated form at the pH optimum of the lysosomal enzymes, they could well be accepted as competitive substrates by the enzyme, as they are by pancreatic lipase at low pH and in the absence of calcium (Borgström, 1954). The mechanism of the activation of the serine (if it exists) must be different from that of chymotrypsin, because histidine—which is believed to partake in a charge relay system in the proteinase (Blow *et al.,* 1969)—is completely protonated at the pH, 4–5, of the phospholipase.

The pH optimum around 8 and the calcium-dependence of the phospholipase 1 of *E. coli* (Scandella and Kornberg, 1971) suggest that its mechanism may be similar to that of the phospholipase 2 of exocrine glands. The suggestion may not be too helpful because that mechanism is not yet elucidated, but we can permit ourselves to glance at the hypothetical model in Fig. VI-7. Perhaps the only major difference might be that the cationic site B is one carbon diameter farther away from the other binding sites; this would bring the C—1 ester into the position occupied by the C—2 ester in the figure. The stereospecificity of the *E. coli* enzymes could be explained in analogy to the specificity of the phospholipase 2 (Fig. VI-4) by a necessity of having the C—2 ester group point away from the enzyme.

Is there an enzyme that can attack either one or the other of the carboxyl esters of a phospholipid, a phospholipase 1-or-2? If we stipulate that a phospholipase, by definition, must bind the phosphate group of the substrate, then we have already fixed the distance of the C—1 and C—2 ester group from this binding site, and it follows that in a type 1-or-2 enzyme the actually attacking enzymatic center must be midway between the two esters and able to react in either one of two directions. This is an unbelieveable picture; but admitting its possibility, we would expect the enzyme to attack the lysophospholipid also, in other words, to be a phospholipase 1,2 or "B." We shall return to this question in the chapter on lysophospholipases.

For the mechanism of the calcium-dependent, pH 5 optimal enzyme of *E. coli* (Proulx and Fung, 1969) and the calcium-independent, pH 6 optimal enzyme from *B. megaterium* (Raybin *et al.,* 1972), not even speculations can be offered at this time.

III. LYSOPHOSPHOLIPASE

Lysophospholipases remove the single fatty acid from lysophosphoglycerides. Since these lipids have lytic properties, i.e., disrupt biological membranes, the lysophospholipases have the obvious function of protecting organisms against an accumulation of these compounds. Consequently, lyso-phospholipases are universally distributed in cells of animals, plants, and microorganisms, and their activity regularly exceeds that of the phospho-lipases 1 and 2 that supply the toxic substrates. The high activity has often prevented the characterization of the cellular phospholipases because no accumulation of lysophospholipids could be observed. This experimental problem is often solved by inhibiting the lysophospholipase with deoxycho-late (Magee *et al.*, 1962).

Since both 1-acyl and 2-acyl lysophospholipids occur in cells, two kinds of lysophospholipase, 1 and 2, might be expected. At present, however, there is no evidence for the existence of two distinct types, and we shall see that it is doubtful that separate lysophospholipases 1 and 2 will be found. Indeed, the existence of "lysophospholipases" as such may be doubted—or rather, the answer to the question whether lysophospholipases exist depends on how such an enzyme is defined. Further discussion of the question is reserved for the conclusion of this chapter.

A. Occurrence

By and large, lysophospholipases are found wherever they are looked for. In the older literature, the complete deacylation of a phospholipid is often ascribed to "phospholipases B" (phospholipase 1 + 2) activity, but since the lysophospholipid is an obligatory intermediate in the total deacy-lation we can interpret such reports as proving lysophospholipase activity. In some cases the phospholipase and lysophospholipase activity have not yet been disentangled. This is true, for instance, for the enzymes reported in rice bran and *Aspergillus niger* (Contardi and Ercoli, 1933), castor beans (Contardi and Latzer, 1928), barley (Acker and Mueller, 1965), and *Penicillium notatum* (Fairbairn, 1948). The enzyme from *Penicillium* has repeatedly been investigated (e.g., Bangham and Dawson, 1960; Kates *et al.*, 1965; Saito and Sato, 1968a), but the different activities have not yet been separated. A purified phospholipase from *Sclerotium rolfsii* is re-ported to display both activities (Tseng and Bateman, 1969), as is an en-zyme from mouse intestine (Ottolenghi, 1973).

Lysophospholipase activity has been found in many microorganisms: *Serratia plymuthicum* (Hayaishi and Kornberg, 1954), Vibrio El Tor (Chatterjee and Mitra, 1962), *Sclerotinia sclerotiorum* (Oi and Satomura, 1963a,b), *E. coli* (Proulx and van Deenen, 1967; Okuyama and Nojima, 1969), *Saccharomyces cerevisiae* (van den Bosch *et al.*, 1967), *Mycobac-terium phlei* (Ono and Nojima, 1969), *Tetrahymena pyriformis* (Thomp-son, 1969), *Dictyoscelium discoideum* (Ferber *et al.*, 1970), *Myco-plasma laidlawii* (Van Golde *et al.*, 1971a). In most of these studies, a lysophosphoglyceride was used as a substrate; in some cases, the identity of the enzyme as differing from that of a phospholipase was established by additional criteria, for instance, by a different response to inhibitors (e.g., Ferber *et al., 1970*).

In all cases where the enzymes of animal tissues have been investigated it has been found that phospholipases 1 and 2 and lysophospholipase are distinct entities. Lysophospholipase was first characterized in beef pancreas (Shapiro, 1953), where it appears as an intracellular enzyme that has no digestive functions; its activity is very much lower than that of the digestive phospholipase 2. Similar enzymatic activity has been found in rat intestinal mucosa (Epstein and Shapiro, 1959) and in liver (Dawson, 1956). Marples and Thompson (1960) detected relatively high levels of the enzyme in intestine, lung, spleen, liver, and pancreas of the rat, low levels in muscle, kidney, testis, brain, and blood. Lysophospholipases were found in fish muscle (Yurkowski and Brockerhoff, 1965; Cohen *et al.*, 1967), in erythrocytes (Mulder *et al.*, 1965), in the housefly (Khan and Hodgson, 1967), in a mosquito (Rao and Subrahmanyan, 1969b; Kumar *et al.*, 1970), and, interestingly, also in the venoms of snakes and invertebrates (Doery and Pearson, 1964; Mohamed *et al.*, 1969; Shiloah *et al.*, 1973). This might be a matter of practical concern for many lipid biochemists, because the snake venoms are routinely used as a source of phospholipase 2 for the analysis of the fatty acid distribution in phospholipids; obviously, the action of a lysophospholipase would invalidate the analysis. However, in most venoms (with the exception of the venoms of some Australian snakes) the lysophospholipase is active at a higher pH, above 8, than the phospholipase 2; furthermore, the enzymatic activity is suppressed by ether as well as by deoxycholate (Doery and Pearson, 1964), and one of these agents is usually employed in the degradation of phospholipids by the snake venoms. Shiloah *et al.* (1973) have presented evidence that the phospholipase 2 of *Vipera palestinae* has also lysophospholipase activity, but at a higher pH (10–10.5) and a much lower rate (0.1%).

The subcellular distribution of lysophospholipase has frequently been studied, usually in conjunction with the distribution of other lipolytic enzymes. Plama membranes of rat liver showed no detectable activity (Victoria *et al.*, 1971), and rat liver mitochondria also seem to contain little or no lysophospholipase (Shibko and Tappel, 1964; Björnstad, 1966; Waite and van Deenen, 1967; Hoertnagl *et al.*, 1969; Weglicki *et al.*, 1971; however, Stoffel and Greten, 1967). Activity is reportedly associated with the mitochondrial fraction of a homogenate of housefly larvae (Khan and Hodgson, 1967), but Kumar *et al.* (1970) found the enzyme in a microsomal fraction of the larvae. In mammalian tissues, the microsomal fractions have been reported to contain the bulk of the lysophospholipase in rat liver (Shibko and Tappel, 1964; Stoffel and Greten, 1967), brain (Leibovitz and Gatt, 1968), heart (Weglicki *et al.,* 1971), and adrenal medulla (Hoertnagl *et al.,* 1969). On the other hand, Lands (1960), Waite and van Deenen (1967), and van den Bosch *et al.* (1968) found most of the lysophospholipase activity of rat liver to be associated with the soluble fraction. In rat intestinal mucosa, Subbajah and Ganguly (1970) found the enzyme distributed over the microsomal and soluble fractions.

Lysosomes of rat liver have been reported to contain an active lysophospholipase (Shibko and Tappel, 1964; Mellors and Tappel, 1967; Fowler and de Duve, 1969). In another study, rat liver lysosomes showed only low activity (Stoffel and Greten, 1967) and the enzyme seemed to be absent from the lysosomes of bovine adrenal medulla (Blaschko *et al.,* 1967; Hoertnagl *et al.,* 1969).

In microorganisms, lysophospholipase may be associated with cell membranes, e.g., in *Mycoplasma laidlawii* (Van Golde *et al.,* 1971a) or *Mycobacterium phlei* (Ono and Nojima, 1969); with the mitochondria in yeast (Vignais *et al.,* 1970); or with particulate cell fractions, as in *Serratia plymuthicum* (Hayaishi and Kornberg, 1954) or *Dictyostelium discoideum* (Ferber *et al.,* 1970); but it may also be released from cell cultures as a soluble enzyme, as in *Sclerotium rolfsii* (Tseng and Bateman, 1969) and Vibrio El Tor (Chatterjee and Mitra, 1962).

B. Detection and Assay

The assay methods for lysophospholipases are similar to those for the phospholipases, especially to the methods described for phospholipase 1, with the difference that a lysophospholipid is used as a substrate. 1-Acyl *sn*-glycerol-3-phosphorylcholine and 1-acyl *sn*-glycerol-3-phosphorylethanolamine are most frequently used; these compounds can be prepared conveniently from the diacyl phosphoglycerides with the help of the phospholipase 2 of snake venom. The 2-acyl isomers are accessible from the phosphoglycerides through degradation with pancreatic lipase (Slotboom *et al.,* 1970b).

The activities of the intracellular lysophospholipases are very much smaller than those of the lipolytic enzymes of digestive fluids and venoms, and their determination by continuous automatic titration is therefore not practicable. Incubation times of 60 minutes or longer may be required for the quantitation of the enzyme. The methods used can be divided into those that determine the formation of free fatty acid and those that measure the liberated glycerophosphate ester. Fatty acids can be titrated directly after the addition of ethanol to the incubation mixture or after extraction. In recent years, substrates with radioactive fatty acids have most often been employed; after the extraction of the lipids, the fatty acid is isolated by thin-layer chromatography and counted. More directly, the incuation mixture can be extracted with a solvent mixture of hexane–isopropanol–sulfuric acid (Dole and Meinertz, 1960; Ibrahim, 1967), which leaves the lysophospholipid in the polar phase (Hoertnagl *et al.,* 1969; van den Bosch *et al.,* 1968).

The second group of assay methods makes use of the solubility in water of the nonlipid reaction products, *sn*-glycerol-3-phosphorylcholine or *sn*-glycerol-3-phosphorylethanolamine. Lysophospholipid and fatty acid can be precipitated, together with albumin, by trichloroacetic acid, and the soluble phosphate ester can be determined by sensitive assays for choline or phosphorus (Hayaishi, 1955; Dawson, 1956). Alternatively, and more effectively, *sn*-glycerol-3-phosphorylcholine or *sn*-glycerol-phosphoryletha-moleamine can be isolated by distribution between chloroform and methanol–water (Rao and Subrahmanyam, 1969b).

C. Isolation

The first attempt to isolate a lysophospholipase was made by Shapiro (1953), who obtained a 40-fold purified enzyme from beef pancreas by extraction, pH adjustment, and ammonium sulfate fractionation. A lysophospholipase from rat brain was purified 20-fold by Leibovitz and Gatt (1968). The enzyme was solubilized by sonication of a brain homogenate, phospholipase activity was precipitated with protamine sulfate, and an ammonium sulfate sediment of the supernate was chromatographed on Sephadex. The final specific activity was 0.016.

A new effort has recently been made (van der Bosch *et al.,* 1973) to isolate the lysophospholipase of beef pancreas, and a nearly homogeneous preparation has been obtained. The procedure is outlined in Table VI-7.

A lysolecithinase with phospholipase 1,2 activity that is elaborated by mouse intestine on infection with the tapeworm *Hymenolepis nana* has been purified 14-fold by ammonium sulfate precipitation and fractionation on calcium phosphate gel (Ottolenghi, 1973); the final specific activity against lysolecithin was 79.

Attempts to purify the enzyme from *Sclerotina sclerotiorum* (Oi and Satomura, 1963a) and Vibrio El Tor (Chatterjee and Das, 1965) with the help of chromatography on DEAE-cellulose resulted in large losses of enzyme but little increase in specific activity, although the final preparation from Vibrio El Tor appeared to be considerably purified. An extracellular protein believed to be a phospholipase 1,2 ("B") has been purified 68-fold from *Sclerotium rolfsii* by ammonium sulfate fractionation and

TABLE VI-7

Purification of Lysophospholipase from Beef Pancreas[a]

Purifica- tion step	Fraction	Specific activity	Recovery (%)	Purifica- tion (fold)
1	Cell-free homogenate	0.014	—	—
		0.037	—	—
2	pH 4.0 supernatant	0.230	100	6
3	0–50% $(NH4)_2SO_4$; 20% n-Butanol	0.400	60	11
4	SE-Sephadex-C50	2.43	55	65
5	DEAE-cellulose	3.92	33	106
6	Second SE-Sephadex-C50	5.58	30	151
7	DEAE-Sephadex-A50	6.1	29	164

[a] Van den Bosch *et al.* (1973).

Sephadex chromatography (Tseng and Bateman, 1969). It behaved as a single component when subjected to starch gel electrophoresis.

The lysolecithinase from *Penicillium notatum* has been purified by Kawasaki and Saito (1973). An extract of the mycelia was adjusted to pH 4; from the supernatant, the activity was precipitated with 75% saturated $(NH_4)_2SO_4$. Repeated gel filtration on Sephadex G-200, ion exchange on DEAE-Sephadex, electrophoresis on starch block and, finally, isoelectric focusing yielded a preparation of the specific activity 3610 in 8.6% yield. During the whole course of the purification, the lysophospholipase activity was accompanied by a phospholipase 1,2 ("B") activity at the constant rate of 9%.

D. Molecular Weight, Stability, pH Optima

Molecular weights have been reported for the enzyme from Vibrio El Tor, 58,000 (Chatterjee and Das, 1965); rat brain, 15,000–20,000 (Leibovitz and Gatt, 1968); and beef pancreas, 65,000 (van den Bosch *et al.*, 1973); the amino acid composition has been determined for the last enzyme. The lysophospholipase from *Pencillium notatum* had an apparent molecular weight of 116,000 and contained 30% carbohydrate (Kawasaki and Saito, 1973).

Lysophospholipases are generally less stable than phospholipases, especially in comparison to the calcium-dependent phospholipases 2. The intestinal lysophospholipase activity is lost on heating to 60°C or by treatment with organic solvents (Epstein and Shapiro, 1959; Subbajah and Ganguly,

1970). The enzymes of liver (Dawson, 1956; Waite and van Deenen, 1967) and *Penicillium* (Noguchi, 1944) are also destroyed at 60°C. However, the lysophospholipases of some microorganisms, such as Vibrio El Tor (Chatterjee and Mitra, 1962) and *Serratia plymuthicum* (Hayaishi and Kornberg, 1954), will survive short-term heating to 100°C.

The lysophospholipases of animal tissues generally have activity optima around the point of neutrality. A pH of 6–7 has been reported for the bovine pancreatic enzyme (Shapiro, 1953; van den Bosch *et al.*, 1973), for liver (Dawson, 1956; van den Bosch *et al.*, 1968), rat intestine (Subbajah and Ganguly, 1970), and bovine adrenals (Hoertnagl *et al.*, 1969); pH 8 for cod muscle (Yurkowski and Brockerhoff, 1965), rat brain (Leibovitz and Gatt, 1968), mosquito larvae (Rao and Subrahmanyam, 1969b), and housefly (Kumar *et al.*, 1970). All these enzymes are presumably of microsomal or supernatant origin. The lysosomal lysophospholipases (Shibko and Tappel, 1964; Mellors and Tappel, 1967; Fowler and de Duve, 1969) are active between pH 4 and 5, and this is probably their region of maximal activity.

Lysophospholipases from microbial sources have widely differing pH optima: *Pennicillium notatum,* pH 4 (Fairbairn, 1948); *Sclerotinia sclerotiorum,* around pH 5 (Oi and Satomura, 1963a,b); *Serratia plymuthicum,* pH 6 (Hayaishi and Kornberg, 1954); Vibrio El Tor, pH 8 (Chatterjee and Das, 1965); *Mycoplasma laidlawii,* a broad optimum with a plateau over pH 7.5 to 9 (Van Golde *et al.*, 1971a).

The lysophospholipases of the venoms of two Australian snakes of the genus *Pseudechis* were active around pH 7, but in *Agkistrodon* and *Naja* and in bee venom there was no activity below pH 8.5, and the optima were around pH 10 (Doery and Pearson, 1964). The optima of *Naja haje* and a scorpion venom were between 8 and 11 (Mohamed *et al.*, 1969). Such a high experimental pH gives cause for wonder under what physiological conditions these lysophospholipases could possibly function. Furthermore, as for their role in the venoms, they would seem to annul the action of the phospholipase 2, because they would hydrolyze the lysophospholipid that this enzyme produces and thus abolish the lytic effect. As a speculation, we suggest that these "lysophospholipases" are identical with the phospholipase 2 of the venoms, and act on lysophosphoglycerides in which the fatty acid ester, because of the higher pH, has migrated from position 1 to position 2. This speculation is supported by the heat-stability of the enzymes, which contrasts with the lability of other animal lysophospholipases, and by their absolute requirement for Ca^{2+}, which is not shared by any other lysophospholipase. A purified phospholipase 2 of *Vipera palestinae* has, in fact, been shown to have lysophopholipase activity (Shiloah *et al.*, 1973).

E. Activation and Inhibition

Lysophospholipases do not require any cofactors; in particular, they are not activated by Ca^{2+} nor inhibited by EDTA. The only exceptions are the lysophospholipases of venoms discussed in the previous paragraph; and for the enzyme from Vibrio El Tor, some depression of the activity by EDTA has been reported (Chatterjee and Mitra, 1962).

Concerning their response to various enzyme inhibitors, the lysophospholipases from various sources show some differences. Heavy metals, especially Hg^{2+}, inhibit the activity of all preparations; disulfide reducing agents such as cysteine and glutathione do not affect them. The lysophospholipase of *Penicillium* is strongly inhibited by 10^{-3} M KCN and slightly by fatty acids (Fairbairn, 1948). The Vibrio El Tor enzyme is inhibited by 50% by 10^{-3} M PCMB (Chatterjee and Mitra, 1962). This sulfhydryl reagent caused also partial inhibition of the membrane-bound enzyme of *Mycoplasma laidlawii* (Van Golde *et al.*, 1971a), but did not affect the lysophospholipase of *Dictyostelium discoideum* (Ferber *et al.*, 1970). The enzyme of rat intestinal mucosa was completely blocked by PCMB (Subbajah and Ganguly, 1970). Rat liver enzyme was not affected by diisopropyl fluorophosphate (DFP) (Dawson, 1956) and only slightly by PCMB (van den Bosch *et al.*, 1968), but appreciably by diphenyl chloroarsine and bromoacetophenone (Dawson, 1956). This enzyme displays a distinct substrate inhibition effect, but is not inhibited by fatty acid (van den Bosch *et al.*, 1968). Rat brain lysophospholipase was inhibited 50% by 10^{-5} M PCMB or 10^{-4} fatty acid (Leibovitz and Gatt, 1968). The enzyme from cod muscle (Yurkowski and Brockerhoff, 1965) showed 50% inhibition with 2×10^{-5} PCMB and 70% inhibition with 10^{-4} M oleic acid. It was not affected by iodoacetate, NaF or KCN, but slightly inhibited by phenylmethylsulfonyl fluoride (a reagent that may attack a reactive serine); and the histidine reagent L-1-tosylamido-2-phenylchloromethylketone, at 3×10^{-5} M, inhibited by 40%. The enzyme of mosquito larvae (Rao and Subrahmanyam, 1969b) was unaffected by KCN, NaF, PCMB, iodoacetate, or oleic acid. The best characterized lysophospholipase, that of bovine pancreas, is completely inhibited by 1 mM DFP and is therefore probably a serine enzyme (Table VI-8).

A universal property of lysophospholipases is their inhibition by detergents. Epstein and Shapiro (1959) found the intestinal lysophospholipase activity inhibited by Tween 80, saponin, and cholic acid. Dawson (1959b) reported inhibition by hexadecyltrimethylammonium bromide, and Magee *et al.* (1962) reported that the activity in many mammalian tissues could be suppressed with deoxycholate. This reagent has often been used since then. Hexadecyltrimethylammonium bromide as well as deoxycholate were

TABLE VI-8
Effect of Various Agents on Lysophospholipase Activity[a]

Addition	Concentration	Inhibition (%)
Dithionitrobenzoic acid	0.4 mM	0
N-Ethylmaleimide	10 mM	0
Iodoacetamide	40 mM	0
p-Chloromercuribenzoate	20 μM	0
Diisopropyl fluorophosphate	1 mM	100
Sodium deoxycholate	12 μM	4
	120 μM	66
	1.2 mM	99
Palmitate	40 μM	0

[a] Van den Bosch *et al.* (1973).

found to inhibit the lysophospholipase of the brain (Leibovitz and Gatt, 1968).

The inhibition of enzymatic activity by cationic as well as anionic and neutral detergents suggests that it is not the enzyme that is affected by the inhibitor, but the substrate. It is known that the micelles of bile salts can incorporate single-chain lipids, such as fatty acids and monoglycerides (Borgström, 1967). Lysophospholipids may, we suggest, be imbedded in detergent micelles, and the ester group may thus become inaccessible to the enzyme. This mechanism can also explain the inhibitory effect that phosphatidylcholine and phosphatidylethanolamine have on lysophospholipase (Rao and Subrahmanyan, 1969b); ether, which also inhibits (Dawson, 1956; Doery and Pearson, 1964), may protect lysophospholipids in a similar manner. It follows, then, that differential inhibition of phospholipase and lysophospholipase activity by a detergent does not prove the existence of two different enzymes. The same enzyme might attack a diacyl or lysophospholipid that is not buried in a detergent micelle, but not a lysophospholipid that is. The inhibition by detergents of the heat-stable calcium-dependent venom lysophospholipases (Doery and Pearson, 1964) does not, therefore, exclude the possibility that they are identical with phospholipase 2.

F. Substrate Specificity

The first systematic attempts to define the minimal substrate requirements of lysophospholipases (van den Bosch *et al.*, 1968; de Jong *et al.*, 1973) have led to the conclusion that these enzymes have the character of "nonspecific" carboxyl esterases, which will hydrolyze a large spectrum of carboxyl esters with no apparent common features.

The enzyme of the supernate of rat liver (van den Bosch *et al.*, 1968) hydrolyzes a variety of 1-acyl *sn*-glycerol-3-phosphorylcholines; 2-acyl analogues are split at half the velocity. A 1-acyl-propanediol-3-phosphoryl-choline (in which the secondary hydroxyl of glycerylphosphorylcholine is missing) and an acyl ethyleneglycolphosphorylcholine are also attacked. Dipalmitoyl glycerylphosphorylcholine is completely resistant, but 1-acyl-2-linoleoyl glycerylphosphorylcholine is attacked, although at a rate less than 10% of that of the monoacyl compounds. 1-Benzyl-2-stearoyl-*sn*-glycerol-3-phosphorylcholine and triolein are almost resistant. The enzyme seems to make some distinction between optical isomers: the 1- and 2-stearoyl esters of the (natural) *sn*-glycerol-3-phosphorylcholine derivatives are hydrolyzed twice as fast as the corresponding *sn*-glycerol-1-phosphorylcholine derivatives, and 1-stearoyl-*sn*-glycerol-2-phosphorylcholine is hydrolyzed faster than 3-stearoyl-*sn*-glycerol-2-phosphorylcholine.

The lack of all-or-none specificity is also displayed by the lysophospho-lipase of beef pancreas (de Jong *et al.*, 1973) (Table VI-9). This enzyme had been purified to homogeneity on disk gel electrophoresis. It attacks not only lysophosphoglycerides but also short-chain lecithins, triglycerides, and aromatic esters; these substrates are, in fact, hydrolyzed faster than palmitoyllysolecithin, although none of them is a physiological compound. Surprisingly, octanoyllysolecithin is not attacked at all. The reactivity of the enzyme toward long-chain monoglycerides has not yet been tested.

The enzyme from rat brain (Leibovitz and Gatt, 1968) did not attack triolein, long-chain diacyl phosphoglyceride, or *p*-nitrophenyl laurate. Kinetic studies have indicated that the enzyme attacks monomolecularly dissolved lysophospholipids but not micellar substrates (Gatt *et al.*, 1972; Leibovitz-Ben Gershon *et al.*, 1972).

G. Conclusion

A discussion on lysophospholipases must be preceded by a discussion whether any enzymes exist that deserve the name of lysophospholipase. The answer depends on whether substrate specificity or biological function are considered for the purpose of definition.

If we define a lysophospholipase as a lysophosphoglyceride hydrolase that, analogous to phospholipases, requires the *sn*-glycerol-3-phosphate structure (or at least a degenerate form of it) in its substrates, then we find that only the exocrine enzymes which are known as phospholipases 2 fulfill this requirement, or the phospholipase 1 of *E. coli* which also hydrolyzes lysophosphatidylethanolamine (Scandella and Kornberg, 1971).

TABLE VI-9

**Specific Activity of Pancreatic Lysophospholipase
toward Various Substrates**[a]

Substrate	Concentration (mM)	Specific activity (units/mg)
1-Palmitoyllysolecithin	0.4	6.2
Triacetin	500[b]	8.2
Tributyrin	30[b]	49
p-Nitrophenyl acetate	2.4	20
p-Nitrophenyl caprylate	0.16	28
Rat-liver lecithin	1.0	0
Dihexanoyllecithin	2.5	66
Dioctanoyllecithin	1.0	193
1-Octanoyllysolecithin	10	0
Acetylcholine	7	0
Glycerylphorylcholine	4	0
Cholesterol palmitate	1.5[c]	0

[a] De Jong *et al.* (1973).
[b] Dispersed by vigorous stirring.
[c] Dispersed by sonication.

The "lysophospholipases" of mammalian tissues are much too unspecific; they require neither the glycerol structure nor even a phosphate group. They hydrolyze any carboxyl ester that is not too hydrophobic or hydrophilic; it is possible that they may be identical with the "monoglyceride lipases."

An appropriate name for these enzymes might be "amphiphilic ester hydrolases" rather than "lysophospholipases." If we consider, however, that the only amphiphilic carboxyl esters that are liable to occur in any quantity in nonadipose cells are, in fact, the lysophosphoglycerides, and we wish to name the enzymes after their principal physiological function, then we can justify naming them lysophospholipases despite their nonspecificity.

Most lysophospholipases are inhibited only partially or not at all by PMCB; only for one enzyme, that of rat intestinal mucosa, has 100% inhibition been reported (Subbajah and Ganguly, 1970). The enzymes are therefore probably not sulfhydryl enzymes. The natural assumption that these enzymes are of the serine–histidine type is supported by the finding that the lysophospholipase from pancreas is inhibited by DFP (van den Bosch *et al.,* 1973). Partial inhibition by the potential serine-blocking agent diphenylchloroarsine and PMSF and by the histidine agent bromoacetophenone and PTCK has also been reported (Dawson, 1956; Yurkowski and Brockerhoff, 1965). On the other hand, DFP does not seem to inhibit the enzyme from rat liver (Dawson, 1956). It should be remembered, however, that DFP also does not inhibit pancreatic lipase, which is probably a serine enzyme nevertheless. Since the immunity of pancreatic lipase toward DFP is probably due to steric hindrance on the "acid" side of the inhibitor (Brockerhoff, 1973), it might be rewarding to investigate the reactivity of lysophospholipases against esters of branched or aromatic acids.

Lysophospholipases share the character of nonspecific esterases with lipases. The principal difference is that lipases prefer hydrophobic esters, whereas lysophospholipases prefer their substrates soluble, hydrated, or in loose micellar structures. It seems that lysophospholipases are also less discriminating against steric hindrance in the alcohol moiety of the substrate and less dependent on electrophilic substrate activation.

The "nonspecific" lysophospholipases of animal tissues cannot be expected to occur in distinct types 1 and 2. It remains to be seen if a position-specific lysophospholipase will be found; such an enzyme would probably bind to the phosphate group of the substrate. The "lysophospholipase" of the Australian snakes (Doery and Pearson, 1964) might, on further investigation, turn out to be a lysophospholipase 1. For a phospholipase 1,2 we can predict that it should be an unspecific carboxyl esterase that does not bind the phosphate group of the phosphoglyceride. It should have the properties of a lysophospholipase except for a greater ability to attack dehydrated esters in compact micellar structures, for instance, diglycerides or long-chain nitrophenyl esters. In general, the spectrum of substrates should extend farther into the hydrophobic region than is the case with "lysophospholipases." The "phospholipases B" of *Penicillium* (Saito and Sato, 1968a), *Sclerotium rolfsii* (Tseng and Bateman, 1969), and mouse intestine (Ottolenghi, 1973), if they are not mixtures, may be such enzymes.

VII

Phospholipases: Phosphohydrolases

I. PHOSPHATIDATE PHOSPHOHYDROLASE

Phosphatidate phosphohydrolase (EC 3.1.3.4), which catalyses the hydrolysis of phosphatidic acid to diglyceride and inorganic phosphate, is an enzyme of great interest to biochemists, because both phosphatidic acid and diglyceride are key intermediates in lipid metabolism (Kennedy, 1961). The sn-1,2-diglyceride that results from the dephosphorylation of phosphatidic acid is the precursor of the phospholipids phosphatidylcholine and phosphatidylethanolamine as well as of triglycerides. However, despite its importance, phosphatidate phosphohydrolase is one of the least well understood lipolytic enzymes. This is due in part to the difficulties in purifying the particle-bound enzyme, in part to the circumstance that biochemists investigating the enzyme have usually been more concerned with its regulatory role in metabolism than with its molecular properties.

A. Occurrence and Distribution

Indications that the enzyme occurs in plants (Kates, 1955) have not been further pursued. The first definitive identification was that of the phosphatidate phosphohydrolase in chicken liver (Weiss *et al.,* 1956; Smith *et al.,* 1957). Since then, the activity has been studied in mammalian tissues—erythrocyte membranes (Hokin and Hokin, 1961; Hokin *et al.,* 1963); brain (Agranoff, 1962; Strickland *et al.,* 1963; McCaman *et al.,*

1965); liver (Coleman and Hübscher, 1962; Wilgram and Kennedy, 1963); intestine (Johnston *et al.,* 1967).

In pig brain, the highest activity was found in the microsomal fraction (Agranoff, 1962). In rat liver, activity was concentrated in an "intermediate" fraction, probably lysosomes (Wilgram and Kennedy, 1963); Sedgwick and Hübscher (1965) found it in mitochondria, lysosomes, and microsomes; and Stoffel and Trabert (1969) found that the membranes as well as the soluble fraction of the lysosomes contained the enzyme. In intestinal mucosa, the microsomes, mitochondria, and the supernate contained activity (Johnston and Bearden, 1965; Johnston *et al.,* 1967). In pig kidney, most of the activity is in the microsomal fraction (Coleman and Hübscher, 1962); in small-intestinal epithelium, it was found in microsomes and supernate (Bickerstaffe and Annison, 1969). These reports taken together indicate that the intracellular distribution of the enzyme in animals extends over mitochondria, microsomes, lysosomes, and the cytosol.

B. Assay and Purification; Properties

For the detection and assay of phosphatidate phosphohydrolase, the release of water-soluble inorganic phosphate from phosphatidic acid is measured (Hajra and Agranoff, 1969). The substrate is best prepared from a natural phosphatidylcholine (e.g., from egg) with the help of phospholipase 4; synthetic phosphatidic acid may not react, perhaps because it is poorly dispersible (Agranoff, 1962). After the reaction, the product, diglyceride, and residual phosphatidic acid are precipitated with trichloroacetic acid, preferably after addition of albumin or charcoal (Coleman and Hübscher, 1962), or else removed by solvent extraction. For the identification of the enzyme, the diglyceride formed should be isolated by thin-layer chromatography. Dodecyl phosphate in an emulsion stabilized with albumin has been used as a substrate for the enzyme (Hajra and Agranoff, 1969). This compound is more stable than phosphatidic acid and easier to obtain, but it cannot be taken for granted that the same enzyme hydrolyzes both substrates (Sedgwick and Hübscher, 1967).

Phosphatidate phosphohydrolase of pig brain can be concentrated fourfold by ammonium sulfate precipitation from an acetone powder extract (Agranoff, 1962; Hajra and Agranoff, 1969). Sedgwick and Hübscher (1967) purified the "soluble" enzyme of rat liver mitochondria sixteenfold. After solubilizing the activity by freezing and thawing of the mitochondrial fraction, the proteins precipitated with 30–65% saturated ammonium sulfate were subjected to chromatography on Sephadex and DEAE-cellulose. From Sephadex, a fast and a slow fraction with activity emerged, the fast one perhaps an enzyme–lipid conglomerate; the slow peak corresponded to a molecular weight of 100,000–200,000. This material was again split into two active peaks by DEAE-cellulose. The activities against phosphatidic acid and hexadecyl phosphate were partially separated by Sephadex chromatography. The specific activity against phosphatidic acid was 0.05.

Phosphatidic acid is hydrolyzed by microsomal, lysosomal and mitochondrial preparations between pH 5 and 9, with optima around pH 6 and 7 (Sedgwick and Hübscher, 1965; Hajra and Agranoff, 1969).

Divalent cations, especially Mg^{2+}, inhibit the enzyme, but some preparations are stimulated by Mg^{2+} (Hokin *et al.,* 1963; Sedgwick and Hübscher, 1967). EDTA does not inhibit; fluoride strongly inhibits, at 10 mM, the hydrolysis of phosphatidic acid, but not of hexadecyl phosphate (Sedgwick and Hübscher, 1967). Detergents are general inhibitors. The enzyme is only partially inhibited by PCMB (Coleman and Hübscher, 1962). Treatment of a preparation from pig brain with organic solvents caused a reduction in activity that could be reversed by the addition of lipids (Coleman and Hübscher, 1963).

II. PHOSPHOLIPASE 3

Phospholipases 3 are phosphodiesterases; they hydrolyze the ester bond between the diglyceride and the substituted phosphoric acid of natural phosphoglycerides. The name "lecithase D" was proposed for the group by Contardi and Ercoli (1933). When the first such enzyme was discovered, it was named lecithinase C by MacFarlane and Knight (1941), who were unaware of Contardi and Ercoli's paper. The term phospholipase C is now generally accepted, but confusion reigned for a long time; D was still used in 1960 (Kates, 1960) and 1962 (Rosenthal and Geyer, 1962). We suggest the name phospholipase 3 because it is the *sn*-glycerol 3-phosphate bond that is split in natural phospholipids.

The individual enzymes of the group have quite stringent substrate requirements, especially in relation to the hydrophilic alcohol that forms the nonreacting second ester of the phosphate. The enzymes known so far fall into two subgroups: (1) enzymes hydrolyzing phosphatidylcholine (and also phosphatidylethanolamine, or phosphatidylserine), phosphatidylcholine cholinephosphohydrolases, EC 3.1.4.3; we shall refer to them as phospholipases 3(PC) or 3(PE); (2) the enzymes hydrolyzing PI or PI mono- or diphosphate, phosphatidylinositol inositolphosphohydrolases, or phospholipases 3(PI). These enzymes may actually be cyclizing phosphotransferases. The sphingomyelin cholinephosphohydrolases, or sphingomyelinases, are sometimes also considered as "phospholipase C," and bacterial phospholipases 3 of the PC type have been reported to have sphingomyelinase activity. However, it has been found that the sphingomyelinases of animal tissues do not hydrolyze phosphatidylcholine, and distinct bacterial enzymes attacking preferentially phosphatidylcholine or sphingomyelin have been separated (Pastan *et al.,* 1968). The most highly purified bacterial phospholipase 3(PC) is devoid of sphingomyelinase activity (Zwaal *et al.,* 1971). Even if, in enzymes from other sources, the 3(PC) and sphingomyelinase activities may be found to overlap, it seem advisable for now to retain a separate group name for the sphingomyelinases and treat these enzymes separately.

The phosphatidic acid phosphohydrolases also split a phosphate ester in position 3 of a phosphoglyceride and might be classified as phospholipases 3, but since their specificity is quite different from that of the diesterases they have also been treated in a separate chapter.

A. Occurrence

Phospholipases 3(PC) are typically found as extracellular bacterial enzymes. First discovered by MacFarlane and Knight (1941) in *Clostridium perfringens* (*C. welchii*), they were then detected in filtrates of other *Clostridium* species (MacFarlane, 1948; Lewis and MacFarlane, 1953), in *Bacillus cereus, B. mucoides* and *B. anthracis* (Chu, 1949), in *Acinetobacter calcoaceticus* (Lehmann, 1971, 1972), in *Pseudomonas aeruginosa* (Kurioka and Liu, 1967), and *P. fluorescens* (Doi and Nojima, 1971). A preparation from *Staphilococcus aureus* attacks sphingomyelin but also lysophosphatidylethanolamine (Doery *et al.,* 1965). The enzyme of *C. perfringens* is identical with the α-toxin of the organism (MacFarlane and Knight, 1941); in *B. cereus,* however, phospholipase 3, hemolysin, and lethal toxin are different proteins (Molnar, 1962; Johnson and Bonventre, 1967). A soluble phospholipase 3 has been found in a marine alga, *Monochrysis lutheri* (Antia and Bilinski, 1967; Bilinski *et al.,* 1968), and there is evidence that the enzyme occurs in plants (Kates, 1955).

Phospholipase 3 of the phosphatidylcholine-hydrolyzing type has not yet been identified in animals. The PI type, on the other hand, occurs in mammals as well as in bacteria. Sloane-Stanley (1953) and Rodnight (1956) found that guinea pig brain homogenates liberated organic phosphate from inositol phosphoglycerides, and Kemp *et al.* (1959) identified inositol phosphate and diglyceride as the products of the hydrolysis of phosphatidylinositol by an enzyme from rat liver. The enzyme from brain was further investigated by Thompson (1967) and Friedel *et al.* (1967). Guinea pig intestinal mucosa (Atherton and Hawthorne, 1968) and bovine pancreas (Dawson, 1959a) contain a similar enzyme. In ox brain, Thompson and Dawson (1964a,b) detected an enzyme splitting phosphatidylinositol and a separate enzyme hydrolyzing only phosphatidylinositol mono-

and diphosphate. Rat kidney cortex also contains an enzyme specific for di- and triphosphoinositide (Tou *et al.,* 1973). Most of the triphospho-inositide-splitting enzyme of rat brain appears before and at the beginnings of myelination (Keough and Thompson, 1970).

From *B. cereus,* Slein and Logan (1965) separated in part the activities of phospholipase 3(PC), phospholipase 3(PI), and sphingomyelinase. Doery *et al.* (1965) reported on a phospholipase 3 in *S. aureus* that preferentially hydrolyzed phosphatidylinositol. Braganca and Khandeparkar (1966) have reported phospholipase 3 activity, toward phosphatidyl-ethanolamine and phosphatidylserine, in cobra venom.

B. Detection and Assay

Phospholipase 3 was first detected with the lecithovitellin test, i.e., the increase in the turbidity of an egg yolk emulsion that results from the appearance of diglycerides (Van Heyningen, 1941). MacFarlane and Knight (1941) established the character of the enzyme from *C. perfringens* by isolating both breakdown products of phosphatidylcholine, diglyceride and phosphorylcholine. Phospholipase 3(PI) was characterized by Kemp *et al.* (1959) by the isolation of diglyceride and inositol phosphate. In order to identify the enzyme in a new source, a complete analysis for both components should be performed, since the accumulation of just one of the hydrolysis products might result from the combined action of other esterases.

Once the identity of a phospholipase 3(PC) ic established, the egg yolk turbidimetry assay can still be used with advantage to quantitate the enzyme since it is the easiest method to perform (Ottolenghi, 1969). The possibility of an enzymatic lag period must be considered (Ottolenghi, 1963). A number of other indirect methods have been used. In an assay exploiting the antithromboplastic effect, the destruction of thromboplastin by the enzyme is measured in a blood-clotting test (Gollub *et al.,* 1953; Otnaess *et al.,* 1972). The lethal and hemolytic properties of the enzymes have also been exploited (Van Heyningen, 1941; Ikezawa *et al.,* 1964). Such methods must be used with discrimination; for instance, while the phospholipase 3 of *C. perfringens* is hemolytic (MacFarlane and Knight, 1941), the *B. cereus* enzyme is not (Zwaal *et al.,* 1971).

Assaying the enzyme by indirect methods may in some cases be justified for reasons of convenience. These methods, however, have the drawback, even if the identity of the enzyme has been established, that the activity is measured in arbitrary units, such as the increase in optical density or in clotting time, which can only with difficulty or not at all be translated into standard activity units. In some recent reports, the purification of phospholipases 3 cannot be judged because of this deplorable circumstance. Since the measurement of the specific activity in standard units (micromoles per milligram per minute) requires no more than a simple titration or a determination of phosphorus, there exists hardly an excuse for omitting it, and editors and reviewers should insist that standard units be given in any report on the purification of the enzyme.

The most straightforward methods measure either the liberated acidic group, most conveniently by continuous titration (Rosenthal and Geyer, 1962; Zwaal *et al.,* 1971), or the "water-soluble" phosphorus, e.g., phosphorylcholine or inositol phosphate, which can be obtained by precipitating the unhydrolyzed lipid (together with added albumin) with trichloroacetic acid or $HClO_4$ (Kemp *et al.,* 1959; Ottolenghi, 1965), by removing the lipids by ether extraction (MacFarlane and Knight, 1941; Haverkate and van Deenen, 1964) or with chloroform (Atherton *et al.,* 1966), or by mixing the incubation mixture with chloroform–methanol and determining the phosphorus in the aqueous phase (Pastan *et al.,* 1968; Kleiman and Lands, 1969). Solvent extraction rather than acid precipitation must be employed if cyclic 1,2-inositol phosphate is to be identified and quantitated, because this compound is hydrolyzed by trichloroacetic acid (Dawson *et al.,* 1971). For identification, the hydrolysis products can be analyzed by paper or thin-layer chromatography. The sensitivity of the solvent-distribution methods can be increased by the use of labeled substrates. Phosphatidylcholine labeled with ^3H in the choline (Diner, 1970) and phosphatidylethanolamine labeled with [^{32}P]- or [^{14}C]linoleic acid (Doi and Nojima, 1971) have been employed.

The phospholipases 3 tolerate ether, even if they are not stimulated by it, and ether has been included in assay systems to facilitate the dispersion of the substrate (Kleiman and Lands, 1969; Zwaal *et al.,* 1971). As in the case of the phospholipases 2, the enzymatic hydrolysis proceeds also in a homogeneous system of moist ether (Hanahan and Vercamer, 1954).

Assays have been described in which the liberated phosphorylcholine is further degraded by a phosphatase and the inorganic phosphorus is measured (Kurioka and Liu, 1967; Ohsaka and Sugahara, 1968). *p*-Nitrophenylphosphorylcholine has been proposed as a general substrate for phospholipases 3 (Kurioka, 1968), but its suitability cannot be judged as long as the substrate requirements of the enzymes are not known.

C. Isolation

The bacterial phospholipases 3 can be precipitated from the supernate of culture centrifugates by 75–85% saturation with ammonium sulfate. A large apparent purification is often achieved in this stage, but it consists mainly of the removal of the soluble ingredients of the culture broth, and reports of final enzyme purifications of 5000 or 10,000 have to be interpreted with this in mind. Further enrichment of the protein precipitate has been achieved with ethanol fractionation (Ottolenghi, 1969). The latest methods have employed ion exchange chromatography and gel filtration.

The phospholipase 3 of *Clostridium perfringens* has been concentrated in a complicated series of DEAE-cellulose and Sephadex chromatographic preparations to a degree of purity of perhaps 60% by Shemanova *et al.* (1965). Subsequent work (Shemanova *et al.*, 1968) led to a purer preparation, although with a yield of only 0.5%. Ispolatovskaya (1971) found that butanol and Zn^{2+} stabilized the enzyme and could obtain preparations of 80–90% purity. A concentration of 244-fold and a yield of 23% were obtained by Macchia and Pastan (1967); both values relate to the original activity of a 50–85% ammonium sulfate protein fraction from the culture filtrate. A single DEAE-cellulose and Sephadex separation each were employed in the procedure, and separation from the sphingomyelinase activity was achieved. The degree of purity of the final preparation was not determined. The specific activity was 366. Activity was slowly lost even at $-20°C$.

Diner (1970), starting from a commercial preparation of *C. perfringens* protein, obtained, after a series of chromatographic separations, an enzyme judged to be 60–70% pure, but in a yield of less than 1% and with a specific activity of 72. It is clear that the enzyme is rather unstable and loses activity either because of chromatographic removal of stabilizers or through denaturation. Rapid surface denaturation of the enzyme had already been described by MacFarlane and Knight (1941) and Smith and Gardner (1950).

The phospholipases of *B. cereus* (strain 6464) were partially separated by DEAE-cellulose chromatography into three fractions with activities of phospholipase 3(PC,PE), (PI), and sphingomyelinase (Slein and Logan, 1965). It should be mentioned here that not all strains may contain all enzymes; for *C. perfringens* it has been reported (Pastan *et al.*, 1968) that some preparations from the same strain may contain sphingomyelinase while others do not. Kleiman and Lands (1969) purified the phospholipase 3 (probably the PC,PE type alone) by chromatography on polyethyleneiminecellulose columns and also by protamine sulfate precipitation and DEAE-cellulose chromatography to a specific activity of 9 toward phosphatidylethanolamine and 15–20 toward monomethyl phosphatidylethanolamine and phosphatidylcholine. A complete purification of the enzyme was reported by Zwaal *et al.* (1971). The protein obtained from a *B. cereus* supernate with ammonium sulfate was dialyzed and then precipitated with 40% ethanol. The following steps were performed in 50% glycerol-in-water to avoid inactivation during the purification. Half of the contaminating protein was removed by precipitation with protamine sulfate, and subsequent chromatography on Sephadex, DEAE-Sephadex, and carboxymethyl-Sephadex yielded a preparation that was homogeneous as judged by disk electrophoresis, with a recovery of 23% and 450-fold purification over the growth medium. Table VII-1 summarizes the procedure.

The enzyme had a specific activity against purified egg lecithin of 1010, more than 50 times higher than the preparation of Kleiman and Lands (1969); the ammonium sulfate precipitate of Zwaal *et al.* (1971) was already much more active than the final enzyme of the latter authors. The conditions of assay were almost identical in both studies. It is clear that in the earlier study the enzyme, though much enriched, must have been largely inactivated or denatured. The importance of expressing the activity in standard units is underlined by these reports; because of the lability of the enzymes, the mere physical homogeneity of the final preparation does not guarantee that an active enzyme has been isolated.

A simpler procedure for purification involving ammonium sulfate precipitation, DEAE-Sephadex batch absorption of accompanying proteins and Sephadex filtration has been described by Otnaess *et al.* (1972). A 9000-fold purification (over the culture medium) and a 50% yield were achieved. It was reported that the inactivation of the enzyme could be prevented by performing the gel filtration in 0.1 mM $ZnCl_2$. The final preparation was homogeneous in ultracentrifugation, gel electrophoresis at pH 2.5 and 9.3, and sodium dodecyl sulfate electrophoresis. The activity could be increased by 50% by freezing and thawing the preparation, and the enzyme was stable for months even at room temperature. Specific activities in standard units were unfortunately not given.

A phospholipase 3(PE) from *Pseudomonas fluorescens,* an enzyme more active against phosphatidylethanolamine than against phosphatidylcholine, has been purified 2500-fold over the culture medium by ammonium sulfate fractionation and chromatography on Sephadex and DEAE-Sephadex (Doi and Nojima, 1971). The yield was 17% and the specific activity 36.5, but the final preparation was still not homogeneous on disk gel electrophoresis.

The phospholipases 3(PI) that hydrolyze the inositolphosphoglycerides have not yet been purified to any degree. In both guinea pig brain (Friedel *et al.,* 1967) and intestinal mucosa (Atherton and Hawthorne, 1968) such enzymes are located in the 105,000 *g* supernatant fraction of tissue homogenates and can be precipitated with ammonium sulfate. Membrane-bound enzymes in addition to soluble enzymes are found in ox brain (Keough and Thompson, 1972) and rat brain (Lapetina and Michell, 1973).

TABLE VII-1

Purification of Phospholipase 3 (PC) from B. cereus[a]

Step	Total activity (10³ units)	Protein (mg)	Specific activity	Recovery (%)	Purification
B. cereus supernate	130	57800	2–2.5	100	1
Ammonium sulfate precipitate (77% saturation)	83.2	1140	73	64	32
Dialysis	75.4	1140	66	58	29
Ethanol precipitate (40%; −27°C)	58.5	462	127	45	56
Protamine sulfate supernate (pH 7)	58.5	246	238	45	106
Sephadex G-100	46.8	62	754	36	336
DEAE-sephadex A-50	37.7	39.7	950	29	422
Carboxymethyl-Sephadex C-50	29.9	29.6	1010	23	450

[a] Zwaal et al. (1971).

D. Molecular Properties; pH Optima

The molecular weight of the phospholipase 3 or *C. perfringens* has been reported as 51,200 (Shemanova *et al.*, 1968), 90,000 (Casu *et al.*, 1971), or 100,000 (Ikezawa *et al.*, 1964); the first value was obtained from the probably purest preparation. For the molecular weight of the phospholipase 3(PC) of *B. cereus* there is better agreement: 20,000 (Ottolenghi, 1969), 21–25,000 (Zwaal *et al.*, 1971), 23,000 (Otnaess *et al.*, 1972), 24,000 (Lysenko, 1972).

The isoelectric point of the *B. cereus* enzyme has been reported as 8.0–8.1 (Ottolenghi, 1969), between 7.5 and 8.2 (Zwaal *et al.*, 1971), and 6.5 (Otnaess *et al.*, 1972). An amino acid analysis of the *C. perfringens* 3(PC) has been presented by Shemanova *et al.* (1968). The enzyme contains no methionine; 6 out of 309 residues are cysteine, probably linked in disulfide bridges, since the enzyme is resistant to oxidizing agents such as hydrogen peroxide (MacFarlane and Knight, 1941).

All reports agree on the instability of the bacterial enzymes. MacFarlane and Knight (1941) and Smith and Gardner (1950) noted that the enzyme was destroyed by aeration of *C. perfringens* cultures. The surface denaturation has been studied quantitatively (Miller and Ruysschaert, 1971). Great losses of enzymatic activity have been reported in most purification procedures; the deactivation proceeds more rapidly with increasing purification (Kleiman and Lands, 1969), and final preparations deteriorate even when kept at —20°C (Macchia and Pastan, 1967).

Protection has been achieved by the addition of bovine albumin (Doi and Nojima, 1971), butanol (Ispolatovskaya, 1971), or glycerol (Doi and Nojima, 1971; Zwaal *et al.*, 1971). Otnaess *et al.* (1972) report that the *B. cereus* enzyme is completely protected by 0.1 mM $ZnCl_2$, so that the metal serves not only as an obligatory cofactor (Ottolenghi, 1965) but also as a stabilizer. Zinc also protects the *Clostridium* enzyme (Ispolatovskaya, 1971).

The bacterial phospholipases 3(PC) have activity optima around neutrality: *C. perfringens,* pH 7.0–7.6 (MacFarlane and Knight, 1941); *B. cereus,* 7 (Chu, 1949). For the second enzyme, Otnaess *et al.* (1972) found two pH optima, 6.6 and 8.0. *B. anthracis* was active at pH 7.5 (Costlow, 1958); *P. fluorescens,* around 7.5 (Doi and Nojima, 1971). The enzyme of the marine alga *M. lutheri* is active at pH 5 (Bilinski *et al.*, 1968).

The pH optima of most phospholipases 3(PI) are in the acidic range: pH 5.4 for the rat liver (PI) enzyme (Kemp *et al.*, 1959); pH 5.9 for guinea pig intestinal mucosa (Atherton *et al.*, 1966; Atherton and Hawthorne, 1968); pH 5.4 for brain (Thompson, 1967); pH 4.5–5.0 for a staphylococcal enzyme hydrolyzing phosphatidylinositol (Doery *et al.,* 1965). The activity of the brain enzyme hydrolyzing di- and triphosphoinositide (PIP, PIP_2) was highly dependent on the nature of the buffer (Thompson and Dawson, 1964a,b).

E. Activation and Inhibition

In many studies on the phospholipases 3 of bacteria, Ca^{2+} has been reported as being required for maximal activity. Moskowitz *et al.* (1956) found that the *C. perfringens* enzyme was inhibited by EDTA and reactivated not by Ca^{2+} but by Zn^{2+}, Co^{2+}, and Mn^{2+}. Ottolenghi (1965) reported that the *B. cereus* enzyme was slowly inactivated by EDTA and more rapidly by *o*-phenanthroline, and that the inhibition could be reversed by Zn^{2+}, and to a lesser degree by Mn^{2+}, but not by other cations, including Ca^{2+}. The complete reactivation of the *B. cereus* enzyme by 0.1 mM $ZnCl_2$ was confirmed by Otnaess *et al.* (1972); Mg^{2+} and Mn^{2+} (10 mM) restored 30% and 40% of the activity. Dyatlovitskaya *et al.* (1967) and Ispolatovskaya (1970) confirmed that the *C. perfringens* enzyme was inactivated by EDTA and reactivated by Zn^{2+} but not by Ca^{2+}. Nevertheless, Ca^{2+} was also required in the lecithovitellin reaction; egg yolk phospholipids were not split in the absence of Ca^{2+}. It is possible that Zn^{2+} is an obligatory cofactor of the enzyme, whereas Ca^{2+} may be required for a supersubstrate–enzyme binding or orientation. Ispolatovskaya (1970, 1971) reported that inhibition by EDTA occurred only when the reagent was added before the substrate. This finding may explain why Johnson and Bonventre (1967), investigating the *B. cereus* enzyme, could not observe inhibition when EDTA was added to the substrate and the enzyme was then introduced. It seems that the binding of the enzyme to the substrate, or to the micellar supersubstrate, makes the Zn^{2+} cation (and perhaps also the Ca^{2+}) inaccessible to the inhibitor.

Dependence on Zn^{2+} has so far been reported only for the *C. perfringens* and *B. cereus* phospholipases 3. The enzyme from *P. fluorescens* was activated by many divalent cations, but especially by 10 mM $CaCl_2$; that of *M. lutheri* was unaffected by EDTA and α,α'-dipyridyl, but *o*-phenanthroline caused a concentration-dependent inhibition which was not reversed by divalent cations (Bilinski *et al.,* 1968). Zn^{2+}, Cu^{2+}, and Hg^{2+} were found inhibitory.

The phospholipases 3(PI) of intestinal mucosa and brain have an absolute requirement for divalent cations; Ca^{2+} is the best activator (Atherton and Hawthorne, 1968; Thompson, 1967). Activity of the brain enzyme lost on dialysis could be restored by an undefined diffusable factor (Thompson, 1967). For the enzyme hydrolyzing triphosphoinostide,

Thompson and Dawson (1964b) found that it could be stimulated by Ca^{2+} or Mg^{2+}, but equally well by cationic amphiphilic compounds such as long-chain amines or hexadecyltrimethylammonium bromide.

Bangham and Dawson (1962; Dawson, 1968) found that the *C. perfringens* phospholipase 3 was activated by long-chain amines as well as by Ca^{2+}, Mg^{2+}, or $(UO_2)^{2+}$. The activation was a function of the electrophoretic mobility of the supersubstrate particles, and it was concluded that phospholipases 3 need a positive ζ potential and that it is the function of the cations to supply the necessary charge.

The enzyme of *C. perfringens* is inhibited by low concentrations of sodium dodecyl sulfate (MacFarlane and Knight, 1941), probably by denaturation, in accord with its general instability. The enzymes of *P. fluorescens* (Doi and Nojima, 1971) and *M. lutheri* (Bilinski *et al.,* 1968) are equally sensitive to sodium dodecyl sulfate.

The phospholipase 3 of *B. cereus* is stable against 5,5'-dithiobis(2-nitrobenzoic acid), PCMB, *N*-ethylmaleimide, or iodoacetamide (Otnaess *et al.,* 1972). Dithiothreitol and PCMB caused partial inhibition of the algal enzyme (Bilinski *et al.,* 1968). The *B. cereus* enzyme was inhibited only partially by dithiothreitol (Kleiman and Lands, 1969). The triphosphoinositide-hydrolyzing enzyme was sensitive to cystein, PCMB, phenylmercuric acetate, and mercuric chloride (Thompson and Dawson, 1964b). The (PI) type enzyme of guinea pig brain and intestinal mucosa were blocked by PCMB (Thompson, 1967; Atherton and Hawthorne, 1968).

Several inhibitory substrate analogs have been prepared by Rosenthal and co-workers. A long-chain ammonium salt was a powerful inhibitor (Rosenthal and Geyer, 1962), perhaps because of interfacial effects. Phosphonate (Rosenthal and Pousada, 1968) and phosphinate (Rosenthal *et al.,* 1969) analogs of phosphatidylcholine behaved as simple competitive inhibitors in a possibly more specific manner.

F. Substrate Specificity

No systematic studies on the substrate requirements of phospholipases 3 have yet been produced, and much of the sporadic information found in the literature must be discounted because it has been obtained with unpurified enzymes, and it now appears that bacteria such as *C. perfringens* (Pastan *et al.,* 1968), *B. cereus* (Slein and Logan, 1965) or *S. aureus* (Doery *et al.,* 1965), as well as mammalian tissues (Thompson and Dawson, 1964b), all contain mixtures of enzymes with similar but different specificities. The phospholipase 3(PC) of *B. cereus* has, at last, been prepared in a pure state (Zwaal *et al.,* 1971; Otnaess *et al.,* 1972), and we may hope for studies on its specificity in the near future.

It is known that the *B. cereus* enzyme hydrolyzes phosphatidylcholine and also phosphatidylethanolamine but leaves sphingomyelin unattacked (Zwaal *et al.*, 1971). Phosphatidylglycerol and diphosphatidylglycerol are split (Haverkate and van Deenen, 1964; de Haas *et al.*, 1966), and phosphatidylserine is partially attacked (Zwaal *et al.* 1971), but not lysophosphoglycerides (Kleiman and Lands, 1969). Ethanolamine plasmalogen is hydrolyzed (Ansell and Spanner, 1965).

The phospholipase 3 of *C. perfringens* has little or no activity against phosphatidylethanolamine or phosphatidylserine (van Deenen *et al.*, 1961; Pastan *et al.*, 1968), but a preparation that had been separated from the sphingomyelinase of the organism still showed some activity against sphingomyelin (Pastan *et al.*, 1968). Diacetyl- and dibutyryllecithin were not hydrolyzed, perhaps because these phosphoglycerides are completely water soluble (van Deenen *et al.*, 1961). Phosphonolecithins, in which deoxycholine is bound to the phosphorus in a P—C linkage, are hydrolyzed, while the corresponding ethylamine derivative is not (Baer and Stanacev, 1966). Dyatlovitskaya *et al.* (1967) reported that under certain conditions some phosphatidylethanolamine can be hydrolyzed. In general, however, the *C. perfringens* enzyme seems to differ from the *B. cereus* enzyme in not tolerating a net negative charge in the substrate.

The enzyme of *P. fluorescens* preferred phosphatidylethanolamine to phosphatidylcholine (Doi and Nojima, 1971), but hydrolyzed phosphatidylglycerol and lysophosphatidylethanolamine very slowly and diphosphatidylglycerol not at all. The *M. lutheri* enzyme degraded lysophosphatidylcholine at 10–12% of the rate of phosphatidylcholine. The bacterial phospholipases 3 have been used to achieve the controlled breakdown of biological membranes (de Gier *et al.*, 1961).

The phospholipases 3(PI) have not yet been purified but the preparations obtained did not hydrolyze phosphatidylcholine, phosphatidylethanolamine, phosphatidylserine, or lysophospholipids (Kemp *et al.*, 1959; Atherton and Hawthorne, 1968). The phospholipase 3(PI) of guinea pig intestinal mucosa did not attack phosphatidylglycerol in spite of the structural similarity of this lipid to the proper substrate (Atherton and Hawthorne, 1968). We may conclude that the inositol ring is necessary for substrate–enzyme binding.

G. Phospholipase 3(PI) as Cyclizing Phosphotransferase

Jungalwala *et al.* (1971) found a soluble, acidic (pH 5.4), calcium-dependent enzyme in pig thyroid that split phosphatidylinositol into diglyceride and water-soluble products of which the major one (65%) was not inositol phosphate but a related compound, which was then characterized as myoinositol 1,2-cyclic phosphate (Dawson *et al.,* 1971). A membrane-bound enzyme from rat cerebral cortex catalyzed the formation from phosphatidylinositol of 55% cyclic phosphate and 45% 1-phosphate (Lapetina and Michell, 1973). This enzyme was active at pH 7, also activated by Ca^{2+}, and completely inhibited by EDTA. Dawson *et al.* (1971) suggest that the enzymes should be called phosphatidylinositol-2-inositol phosphotransferase (cyclizing) rather than phosphatidylinositol inositol-phosphohydrolase. It is not clear, at present, why such a substantial portion of inositol 1-phosphate is formed in the reaction; this compound does not seem to be an artifact (Lapetina and Michell, 1973). The enzyme appears to be associated with a specific subunit of microtubular protein in rat brain (Quinn, 1973). Its function may be the production of cyclic inositol phosphate, for which a role as a neurotransmitter has been postulated (Michell and Lapetina, 1972).

F. Conclusion

Phospholipases 3 hydrolyze amphiphilic phosphodiesters on the side of the hydrophobic alcohol; however, within this group of possible substrates the structural requirements are quite narrow. The hydrophilic alcohol may be choline, ethanolamine, glycerol, inositol, or inositol mono- or diphosphate, but the individual enzymes are usually specific for only one or a few of these structures. The hydrophobic alcohol is a diglyceride. The ceramide group of sphingomyelins, physicochemically very similar, requires a distinct group of enzymes, although some overlap may possibly occur. Lysophosphoglycerides are poor substrates. It remains to be seen, in future studies with pure enzymes, how much the diglyceride structure can be modified.

Some, at least, of the bacterial enzymes are metalloenzymes requiring zinc. In this respect they resemble some well known monophosphatases, the alkaline phosphatases (Fernley, 1971; Reid and Wilson, 1971). The zinc enzymes also have some vaguely defined requirement for additional divalent cations, in particular Ca^{2+}, a situation also reminiscent of the mammalian alkaline phosphatases, which are stimulated by magnesium (Fernley, 1971). The Ca^{2+} may be needed for supplying an adequate charge to the micellar supersubstrate–water interface (Bangham and Dawson, 1962).

Phospholipases 3, if the enzyme of *C. perfringens* is a typical example, seem to require a positive interfacial ζ potential (Dawson, 1968). In terms of the enzyme-orientation hypothesis (Brockerhoff, 1974), we can expect that they have a negatively charged electrostatic head. Contrary to the phospholipases that hydrolyze carboxyl esters, the phospholipases 3 are easily inactivated by denaturation. An amino acid analysis for the *C. per-*

fringens enzyme indicated that the enzyme has only few disulfide bridges and owes its tertiary structure mainly to ionic ammonium carboxylate links. This peculiarity, it has been suggested (Shemanova *et al.,* 1968; Ispolatovskaya, 1971), is related to the extracellular nature of the enzyme, which has to pass through the cell wall of the bacterium. The stabilization of the enzymes by zinc might be due to a strengthening of the ionic binding, the stabilization by glycerol and butanol due to a reduction of the dissociating effect of solvent polarity.

The phospholipases 3(PI) are calcium-dependent. The enzymes splitting 1-phosphatidylinositol act as cyclizing phosphotransferases and form diglyceride and cyclic 1,2-inositol phosphate. It is not known if the enzymes acting on di- and triphosphoinositides act in a similar manner.

III. SPHINGOMYELINASE

Sphingomyelinase (sphingomyelin cholinephosphoryhydrolase) splits the bond between the *N*-acylsphingosine (ceramide) group and the phosphoric acid of sphingomyelin. Because of the close structural similarity between sphingomyelin and phosphatidylcholine, it might be expected that the same enzymes can hydrolyze both substrates, and the "phospholipase C" of *Clostridium perfringens* has long been believed to be such an enzyme. However, it is now known that the bacterium contains two distinct enzymes, a phospholipase 3(PC) and a sphingomyelinase (Pastan *et al.,* 1968). Other bacteria also contain phospholipases 3 that will not hydrolyze sphingomyelin (Slein and Logan, 1965; Zwaal *et al.,* 1971), and the mammalian sphingomyelinases will not attack phosphatidylcholine. At present, therefore, the sphingomyelinases must be considered as a group distinct from the phospholipases 3. If future research should establish a closer relationship between the two groups, the sphingomyelinases might be renamed phospholipases 3(Sph).

A. Occurrence

Sphingomyelinase is found in bacteria such as *C. perfringens* (Pastan *et al.,* 1968; Saito and Mukoyama, 1968), *B. cereus* (Slein and Logan, 1965), *S. aureus* (Doery *et al.,* 1963), and also in many mammalian tissues: rat liver (Heller and Shapiro, 1966; Kanfer *et al.,* 1966; Kanfer and Brady, 1969), rat brain (Barenholz *et al.,* 1966; Gatt and Barenholz, 1969b; Klein and Mandel, 1972), human spleen (Schneider and Kennedy, 1967), rat pancreas (Nilsson, 1969), and arterial tissue (Blatt *et al.,* 1971). In liver and kidney, most of the enzyme is located in the lysosomes

(Weinreb *et al.*, 1968). The enzymes of brain and spleen, which have pH optima around 5, are probably also lysosomal, but rat intestinal mucosa contain a sphingomyelinase active at pH 8.8 (Nilsson, 1969).

In some varieties of Niemann–Pick disease, a human lipidosis characterized by the accumulation of sphingomyelin in body tissues, especially in the spleen, the levels of the enzyme are markedly reduced from their normal level (Brady *et al.*, 1966; Schneider and Kennedy, 1967; Brady, 1969, 1972).

B. Assay

The methods are identical to those used for phospholipase 3 except that, of course, sphingomyelin is used for a substrate (Gatt and Barenholz, 1969b; Kanfer and Brady, 1969; Sloan, 1972).

C. Purification

The complete isolation of a sphingomyelinase has not yet been achieved, but preparations of considerable degree of concentration have been obtained. The sphingomyelinase of *C. perfringens* has been prepared from the 50–80% ammonium sulfate precipitate of the culture supernatant (Pastan *et al.*, 1968). Chromatography on Sephadex and DEAE-cellulose led to a preparation with a specific activity of 3.1, a 190-fold purification over the crude protein.

Sphingomyelinase from rat liver has been purified 60-fold from a particulate fraction of the homogenate (Kanfer and Brady, 1969). The enzyme was solubilized by treatment with Cutscum, dialyzed at pH 4.5 and precipitated with ammonium sulfate at 0.4–0.55 saturation. Freezing and thawing removed some extraneous protein and resulted in a preparation with the specific activity of 0.2.

The lysosomal sphingomyelinase from human liver has been enriched 120-fold over the 10,000 *g* particulate fraction (Sloan, 1972). The enzyme was solubilized with 0.5% Triton X-100, precipitated with ammonium sulfate, delipidized by treatment with butanol, and finally purified by affinity chromatography on a column of agarose to which sphingosine phosphorylcholine had been coupled. The enzyme was eluted with 0.1% Triton X-100 at pH 6.1. The specific activity of the final preparation was 0.011.

Rat brain sphingomyelinase has been concentrated 18.5-fold from a homogenate (Barenholz *et al.*, 1966; Gatt and Barenholz, 1969b). The enzyme was solubilized with sodium cholate and precipitated at pH 5.5. The enzyme from human spleen has been purified 76-fold with a recovery of 7% by ethanol precipitation and ion exchange chromatography (Schneider and Kennedy, 1967).

D. Properties

The sphingomyelinase from *C. perfringens* (Pastan *et al.,* 1968) showed optimal activity between pH 7.8 and 8.8. It was completely inhibited by EDTA and partially by 2-mercaptoethanol. Iodoacetate, *N*-ethylmaleimide, and DFP did not inhibit, nor did the reaction products, ceramide and phosphorylcholine. The enzyme was activated by Mg^{2+}, but completely inhibited by Ca^{2+}, which stimulated the phospholipase 3 of the organism. Lecithin was hydrolyzed at 10% of the rate toward sphingomyelin; this activity was also suppressed by Ca^{2+}, and was therefore probably due to the sphingomyelinase itself rather than to contaminating phospholipase 3. Phosphatidylethanolamine and phosphatidylinositol were not attacked. The enzyme did not catalyze the exchange of phosphorylcholine.

The *C. perfringens* enzyme is slightly stimulated by hexadecyltrimethylammonium ion and by ether (Saito and Mukoyama, 1968). It hydrolyzes phosphonolipid containing a deoxycholine–phosphorus C—P bond but does not attack sphingomyelins that contain α-hydroxy fatty acids (Hori *et al.,* 1969).

The sphingomyelinases of rat brain (Barenholz *et al.,* 1966), rat liver (Heller and Shapiro, 1966; Kanfer and Brady, 1969), human spleen, and human liver (Sloan, 1972) have optimal activity around pH 5.0. They are highly specific for sphingomyelin and will not attack phosphatidylcholine. The enzymes hydrolyze synthetic sphingomyelins having the (unnatural) L-erythro or threo structures at much reduces rates. The enzyme from brain hydrolyzed sonicated liposomes of sphingomyelin, but further activation was achieved by the inclusion of detergents, especially neutral ones such as Triton X-100, but also cationic or anionic detergents (Gatt *et al.,* 1973). The spleen enzyme did not hydrolyze *O*-acetylsphingomyelin and sphingosylphosphorylcholine. It was not affected by Mg^{2+}, Mn^{2+}, and Ca^{2+}, and not inhibited by EDTA nor by PCMB and *N*-ethylmaleimide.

IV. PHOSPHOLIPASE 4; CERAMIDASE

Phospholipase 4 hydrolyzes the ester bond between the phosphate group and the hydrophilic alcohol of phosphoglycerides, and also the corresponding bond of sphingomyelin. Contardi and Ercoli (1933) proposed that the enzyme should be named "lecithase C," and after the discovery of such enzymatic activity (Hanahan and Chaikoff, 1947a,b, 1948), the name "phospholipase C" was used until the mid 1960s (e.g., in the review by Kates, 1960), but at the same time the enzyme was described by other authors as "phospholipase D" because the term "phospholipase C" had been assigned (MacFarlane and Knight, 1941) to the bacterial enzyme that we have named phospholipase 3. At present, "phospholipase D" is generally used for the enzyme described in this chapter. Since we have proposed the terms phospholipase 1, 2, and 3 for the hydrolases acting at these positions of *sn*-glycerol in phosphoglycerides, we propose "phospholipase 4" for the enzyme hydrolyzing the remaining ester bond.

A. Occurrence

The enzyme was first detected by Hanahan and Chaikoff (1947a, 1948) in carrot root and spinach leaves and has since been found in many plants such as beets (Kates, 1954), brussels sprout and Savoy cabbage (Davidson and Long, 1958) and many others (Vaskvosky *et al.,* 1972). A phospholipase 4 has also been found in a unicellular red alga (Antia *et al.,* 1970) and in corynebacteria (Fossum and Hoyem, 1963; Souček and Souckova, 1966). An enzyme found in *Corynebacterium pseudotuberculosis* (Souček *et al.,* 1971) hydrolyzes the choline ester bond of sphingomyelin and lysophosphatidylcholine but not that of diacyl phosphoglycerides. A phospholipase 4 from *Haemophilus parainfluenzae* hydrolyzes preferentially diphosphatidylglycerol (Ono and White, 1970a,b; Astrachan, 1973).

The enzymes of carrot root, spinach, and cabbage leaves have been found in the plastid fractions (chloroplasts, chromoplasts) (Kates, 1953), and attempts to solubilize them from the particles were not successful (Kates, 1960). Carrot and cabbage have also yielded soluble phospholipases 4 (Davidson and Long, 1958; Einset and Clark, 1958; Kates and Sastry, 1969), as have other plants (barley, Acker and Bücking, 1956; Hevea latex, Smith, 1954; cotton-seed, Tookey and Balls, 1956; peanut, Heller *et al.,* 1968). In *Scorzonera* (Black salsify) rhizomes, activity is found in the mitochondria (Douce, 1966). The algal enzyme is particle-bound (Antia *et al.,* 1970), *Corynebacterium pseudotuberculosis* produced an extracellular enzyme (Souček *et al.,* 1971), the enzyme of *H. parainfluenzae* is bound to membranes (Ono and White, 1970b). The distribution of phospholipase 4 in the parts of developing and mature plants has been studied by Quarles and Dawson (1969a).

B. Detection and Assay

The action of phospholipase 4 on phosphatidylcholine leads to the appearance of free, water-soluble choline which can be quantitatively precipitated and determined as its reineckate, choline$^+$[Cr(NH$_3$)$_2$(SCN)$_4$]$^-$ (Kates, 1960), or its enneaiodide (Shapiro, 1953; Kates and Sastry, 1969). The remaining phosphoglycerides, phosphatidylcholine and phosphatidic acid, must be removed by precipitation with albumin and HClO$_4$ or by solvent distribution in a chloroform–methanol–water system. If a choline–labeled phosphatidylcholine has been employed, the radioactivity in the aqueous phase can be measured (Heller *et al.,* 1968). Alternatively, the other product of the enzymic reaction, phosphatidic acid, can be assayed after isolation as calcium salt (Kates and Sastry, 1969), by thin-layer chromatography (Yang, 1969), or paper chromatography (Clermont, 1972); quantitation is achieved by phosphorus analysis. For establishing the identity of a phospholipase 4, the isolation of phosphatidic acid is obligatory because free choline might also arise as the end product of a different enzymatic breakdown sequence, e.g., through the action of phospholipases 1,2 and a phosphodiesterase.

The phosphatidyl transferase activity of phospholipase 4 can be determined by adding 4% ethanolamine to the incubation mixture and isolating the phosphatidylethamolamine formed from the phosphatidylcholine by thin-layer chromatography (Yang, 1969) or by paper-chromatographic analysis of the deacylation products of the phosphoglycerides (Dawson, 1967).

The reaction of phosphoglycerides with phospholipase 4 is usually started by the addition of ether to the otherwise complete assay mixture (Kates and Sastry, 1969; Yang, 1969) since most preparations, though not all, are almost inactive in the absence of an organic solvent or a detergent.

C. Purification

A pure phospholipase 4 has not yet been isolated. A recent preparation of high specific activity contained 80%, or more, extraneous protein (Tzur and Shapiro, 1972), and it is unlikely that earlier preparations have been of higher purity. However, considerable degrees of concentration of the soluble enzymes have been achieved.

The soluble enzyme from Savoy cabbage leave extract has been concentrated ten times by heat coagulation and acetone precipitation (Davidson and Long, 1958). Further purification by phosphate gel elution and chromatography on DEAE-cellulose led to a 110 times enriched preparation with specific activity 11.0 in a yield of 21% (Yang *et al.,* 1967). Fractionation of the acetone precipitate by density gradient electrophoresis gave a purified enzyme (Dawson, 1967), which had to be protected from denaturation by the addition of albumin.

The soluble phospholipase 4 of dry peanuts has been concentrated 1000-fold to specific activity 234 (Heller *et al.,* 1968; Tzur and Shapiro, 1972). An extract of the seeds was centrifuged and the enzyme was precipitated from the post-microsomal supernate with 20% ammonium sulfate. This step was followed by DEAE-cellulose, Sephadex G-200, and again by DEAE-cellulose chromatography; the final recovery was 36%. Albumin inactivated rather than protected the enzyme. In spite of the high specific activity, it was found on gel electrophoresis that 80% of the protein appeared in bands devoid of activity. The extracellular phospholipase 4 of *Corynebacterium pseudotuberculosis* has been purified 176-fold, to specific activity 2.6, by precipitation with methanol and chromatography on carboxymethylcellulose (Souček *et al.,* 1971).

D. Optimal pH, Stability, Molecular Weight

The plastid-bound enzymes are maximally active at pH 4.7 (spinach, sugar beet) or 5.6–5.8 (cabbage, carrot) (Kates, 1953). A maximum of around pH 5.6 has been found for the soluble enzymes of all higher plants; however, the addition of surfactants such as sodium dodecyl sulfate can shift the apparent pH optimum from 4.9 up to 6.6, due to interface effects which are not yet well understood (Quarles and Dawson, 1969b). For the phospholipase 4 of *Haemophilus parainfluenzae* the optimum is at pH 7.5–8.0 (Ono and White, 1970b), for *Corynebacterium pseudotuberculosis* pH 8 (Souček *et al.,* 1971), for the alga *Porphyridium cruentum* pH 7.0 (Antia *et al.,* 1970).

The enzymes are not particularly thermostable and do not survive short-term heating at 70°C (Kates, 1955; Einset and Clark, 1958; Tookey and Balls, 1956). They are also easily deactivated by surface denaturation (Quarles and Dawson, 1969b).

The molecular weight of the enzyme of *Corynebacterium pseudotuberculosis* is around 90,000 (Souček *et al.,* 1971; calculated from gel filtration data given by the authors); that of peanut phospholipase 4 seems to be much higher (Tzur and Shapiro, 1972).

E. Activation and Inhibition

Phospholipases 4 are strongly inhibited by EDTA and have an absolute requirement for divalent cations, usually Ca^{2+}, which can not be satisfied by surfactants (Davidson and Long, 1958; Dawson and Hemington, 1967). Monolayer experiments have confirmed that Ca^{2+} is essential to enzymatic activity and does not merely play a role as supersubstrate activator, as it may do for phospholipases 3 (Quarles and Dawson, 1969c). In the phospholipase of a red algae, Sr^{2+} can replace Ca^{2+} (Antia *et al.*, 1970); Ba^{2+} as well as Sr^{2+} can partially activate the cabbage enzyme (Davidson and Long, 1958). The diphosphatidylglycerol-specific enzyme of *Haemophilus parainfluenzae* is activated by Mg^{2+} rather than by Ca^{2+} (Ono and White, 1970b).

The phospholipases 4 of plants are stimulated by ether and by linear ketones and esters (Kates, 1953, 1957). In the case of the plastid-bound enzymes, the organic solvent promotes the coalescence of plastids and substrate particles (Kates and Gorham, 1957); in general, ether appears to function through a spacing effect in the substrate micelle. Sonication also activates the supersubstrate (Dawson and Hemington, 1967). Cationic surfactants are inhibitory, anionic surfactants stimulatory (Kates, 1957). For this reason crude phospholipid mixtures may be attacked under conditions where purified phosphatidylcholine is resistant (Einset and Clark, 1958; Weiss *et al.*, 1959). Dawson and his co-workers (Dawson and Hemington, 1967; Quarles and Dawson, 1969b,c) found that phosphoinositides and hexadecyl phosphate, but especially phosphatidic acid, were potent activators; however, the degree of activation did not bear a direct relation to the potential that the activators conferred to the substrate micelles.

In view of the inhibition of phospholipases 4 by cationic and the activation by anionic surfactants, it is clear that the electrostatic properties of the supersubstrate are important for the enzymatic activity. Several considerations may explain why a straightforward account of these charge effects cannot yet be given. The anionic surfactants undoubtedly confer a negative charge to the micellar surfaces, but since they also attract the calcium ions that are obligatory for activity, a charge reversal (electrical double layer) may result and the supersubstrate particles may acquire a cationic electrophoretic mobility (Quarles and Dawson, 1969b). There is, therefore, a polar electric field surrounding the aggregated substrate whose potential and depth depend on the nature and concentrations of the surfactants as well as the cations. These factors are likely to influence not only the adsorption of the enzyme to the interface but also its orientation (Dawson, 1968; Brockerhoff, 1973); at present, these effects cannot be quantitatively evaluated.

Chen and Barton (1971) have suggested that both an appropriate surface charge and a gel to liquid–crystalline phase transition of the substrate are required for the action of phospholipase 4.

The algal phospholipase 4 is not activated by ether or by anionic surfactants (Antia *et al.*, 1970). The enzyme of peanut is stimulated by the neutral detergent Triton X-100 and by the anionic sodium dodecyl sulfate but also by hexadecyltrimethylammonium ion at low concentrations (Heller and Arad, 1970).

The phospholipase 4 of cabbage is nearly completely inhibited by 10^{-4} M PCMB, though not by N-ethylmaleimide, and it is protected by reduced glutathione; DFP does not affect the enzyme. On the basis of these results (and the transferase activity) Yang *et al.* (1967) have concluded that phospholipase 4 is an SH enzyme that forms a phosphatidyl-S-enzyme intermediate.

F. Substrate Specificity

The phospholipases of plants hydrolyze most phosphoglycerides: phosphatidylcholine, phosphatidylethanolamine, phosphatidylserine with decreasing facility in this order (Davidson and Long, 1958; Kates, 1956); phosphatidylglycerol (Haverkate and van Deenen, 1964); lysophosphatidylcholine (Long *et al.*, 1962); diphosphatidylglycerol (Stanacev and Stuhne-Sekalec, 1970; Heller *et al.*, 1968). Sphingomyelin can also be attacked, though less easily (Negishi *et al.*, 1971; Tzur and Shapiro, 1972). Phosphoglycerides of *sn*-1 and even *sn*-2 structure are also hydrolyzed, but very slowly (Davidson and Long, 1958). Surprisingly, 1-alkenyl ether phosphoglycerides (plasmalogens) are essentially resistant against the phospholipase 4 from cabbage (Lands and Hart, 1965).

The enzyme of *Corynebacterium pseudotuberculosis* is reported to degrade sphingomyelin and lysophosphatidylcholine but not diacyl phosphoglycerides (Souček *et al.*, 1971); perhaps it requires a hydroxyl function in the hydrophobic alcohol. The *Corynebacterium pyogenes* enzyme, however, does attack phosphatidylcholine (Fossum and Hoyem, 1963). A phospholipase 4 from *Haemophilus parainfluenzae* seems to be highly specific for diphosphatidylglycerol (Ono and White, 1970a,b); it is the phosphate bond at the *sn*-3 position of the central glycerol that is hydrolyzed (Astrachan, 1973).

G. Phosphatidyl Transferase Activity

Phospholipase 4 can transfer the phosphatidyl group of phospholipids to alcohols (Bartels and van Deenen, 1966; Dawson, 1967; Douce *et al.*, 1966; Yang *et al.*, 1967). This property stays with the enzyme during purification (Yang *et al.*, 1967; Tzur and Shapiro, 1972). Phosphatidyl transfer to alcohols rather than to water begins to be the preferred reaction at alcohol-in-water concentrations of a few percent. Only primary alcohols (or amino alcohols) serve as acceptors (Dawson, 1967); glycerol is acylated in *sn*-1 as well as *sn*-3 position (Yang *et al.*, 1967). Phosphatidylglycerol can be converted to diphosphatidylglycerol by intermolecular phosphatidyl transfer (Stanacey *et al.*, 1973). The transferase activity is inhibited by PCMB; this is interpreted as evidence that the phosphatidyl group acylates a sulfhydryl group of the enzyme (Yang *et al.*, 1967).

H. Ceramidase

The last step in the hydrolytic degradation of sphingolipids is the hydrolysis of *N*-acylsphingosine, or ceramide, by *N*-acylsphingosine acylhydrolase, or ceramidase. Such an enzyme from rat brains has been studied by Gatt (1963, 1966; Yavin and Gatt, 1969). The activity was measured by the distribution of the reaction products between aqueous and organic solvent—with *N*-oleoylsphingosine, labeled in either the fatty acid or the sphingosine—as a substrate. Alternatively, since the reaction is reversible, the formation of ceramide from oleic acid and sphingosine can be measured.

The enzyme has been purified 200-fold from the 25,000 g sediment of rat brain homogenate. The activity was solubilized with 0.5% sodium cholate, and the protein precipitated from the dialyzed supernate with 30–60% saturated ammonium sulfate was subjected to Sephadex chromatography. The most active fraction emerging from the column had a specific activity of 0.012.

The purified enzyme preparation required cholate; Triton X-100 further stimulated the activity. The optimal pH for both hydrolysis and synthesis was 4.8, but there was still about 50% activity at pH 8.0. The hydrolytic reaction was inhibited by palmitic acid and by sphingosine at pH 5, but not at pH 8. The best substrate for the enzyme was *N*-oleoylsphingosine; *N*-palmitoyl- and *N*-stearoylsphingosine and *N*-palmitoyldihydrosphingosine were less active; and *N*-acetyl and *N*-tetracosanoylsphingosine were not attacked.

Judging from its low pH optimum the enzyme is probably lysosomal; however, Fowler and de Duve (1969) could not detect ceramidase activity in rat liver lysosomes.

VIII

Synopsis

Lipolytic enzymes can be divided into two groups, the first containing enzymes of broad substrate specificity and not requiring cofactors, the second containing metalloenzymes with very narrow substrate requirements. The protagonist of the first group is pancreatic lipase; of the second group, phospholipase 2.

In the following, we shall try to give a broad view of the two groups of enzymes and then conclude with a discussion of the special nature of lipolytic enzymes, namely, their adaptation to activity at the lipid–water interface. The picture we present is synthetic and to a large degree constructed from circumstantial evidence, and while we think that it is coherent and convincing, it will certainly not be found acceptable in its entirety by every specialist in the field. Future research will show if it needs only minor revisions or if it has to be fundamentally changed.

I. THE NONSPECIFIC LIPOLYTIC ENZYMES

These hydrolases can be ordered on a ladder, or in a hierarchy, which starts with "esterases," i.e., enzymes attacking water-soluble substrates, and culminates in "lipases," i.e., enzymes specific for completely insoluble substrates. At the bottom rung, we find the lysophospholipase of the brain, which hydrolyzes only monomolecularly dissolved lysophospholipid (Gatt *et al.,* 1972); we may suspect that such an enzyme will hydrolyze any water-soluble carboxyl ester. At the top, we find pancreatic lipase, which acts at the sheer surface of an oil droplet on a completely anhydrous substrate. At intermediate levels appear hydrolases that seem to require substrates in hydrated, or "loose," interfaces, as in the surface of micelles (cholesterol esterase, monoglyceride hydrolases, some phospholipases, perhaps also "galactolipases"), or in a surface modified, perhaps "loosened," by phospholipids and proteins (lipoprotein lipase). The relative position of these enzymes in the hierarchy cannot yet be ascertained.

Present evidence indicates that the catalytic mechanism of most nonspecific lipolytic hydrolases involves the acylation of a reactive serine, but it must be kept in mind that the evidence is not yet conclusive for any of the enzymes.

The hierarchy of lipolytic enzymes suggests a biological evolution from esterases to lipases. The highly potent lipases of microorganisms and higher animals are esterases that became specialized for the rapid digestion of insoluble esters when triglycerides started to become important as energy depots. How this specialization was effected will be discussed in the last paragraphs.

Some of the unspecific lipolytic enzymes have been reported to require cofactors, but under certain conditions only. For instance, pancreatic lipase in the presence of bile salts is stimulated by co-lipase; cholesterol esterase needs bile salts; and lipoprotein lipase needs serum protein. However, these effects are most likely not due to cofactor–enzyme interactions, but to physicochemical supersubstrate activation or to the removal of surface inhibition.

Apart from their different reactivity toward substrates found in solutions or in loose or dense interfaces, the nonspecific lipolytic enzymes differ in their steric substrate requirements: some will hydrolyze primary and secondary, some only primary esters. Such discrimination is due, as far as known, not to substrate–enzyme binding effects, but to simple steric hindrance that prevents the approach of secondary ester groups to the reactive site. Steric restriction seems to parallel the ladder of the lipolytic hierarchy: lysophospholipases and cholesterol esterases can hydrolyze secondary esters, but pancreatic lipase and many microbial lipases are strictly position specific, i.e., they hydrolyze only the primary ester groups of triglycerides. This parallel is probably not accidental. The necessity, for the lipases, to approach the dense surface of an oil very closely with their reactive sites may necessitate a more compact architecture around this site, and this may entail additional steric restrictions.

The peculiar substrate specificity of the *Geotrichum* lipase, which prefers *cis* Δ9 unsaturated fatty acid esters, remains still unexplained.

There is no evidence at present that any lipolytic enzyme requires lipophilic binding in its enzyme–substrate complex, i.e., hydrophobic binding of the aliphatic chains of the substrate to a receptor site that is part of the reactive site. In fact, there is strong circumstantial evidence indicating that in the most lipophilic of hydrolases, pancreatic lipase, no lipophilic enzyme–substrate binding takes place. We shall see below how this paradox may be resolved.

II. METAL-DEPENDENT LIPOLYTIC ENZYMES

Among the carboxyl ester hydrolases of this group, the calcium dependent phospholipases 2 of venoms and pancreas are the best known, but phospholipases 1 that are metalloenzymes also appear to exist. The metal is possibly complexed by histidine residues, and the catalytic mechanism may be similar to that of carboxypeptidase; but this is only a speculation.

The phospholipases 2 have very narrow substrate requirements; they will hydrolyze the ester in position 2 in *sn*-3 phospholipids only, but not the antipodes. Since the only phospholipids in nature are those of *sn*-3 structure, why is there a need for stereospecificity on the part of the enzyme? This stereospecificity is perhaps a by-product of the geometry of the enzyme–substrate complex. If only the 2-ester, but not the 1-ester, is to be hydrolyzed, the substrate must be anchored on the enzyme at an appropriate distance from the point of attack, probably by binding of the phosphate group, and a steric restriction is thus introduced which is absent in lipases and other unspecific lipolytic enzymes.

The lipolytic phosphohydrolases are also most likely metalloenzymes; e.g., bacterial phospholipase 3 requires zinc. Nothing is known about the catalytic mechanism of these enzymes.

Most phospholipases require loosely structured substrate–water interfaces; they cannot hydrolyze *in vitro* phospholipid-in-water dispersions unless these have been treated with ultrasonification or by the addition of surfactants. Furthermore, most phospholipases respond to the surface potential of the lipid layers or micelles.

III. NATURE OF LIPOLYTIC ENZYMES

Lipolytic enzymes hydrolyze water-insoluble substrates, and they must perform at the lipid–water interface. Before the formation of an enzyme–substrate complex can take place, two requirements have to be fulfilled: the enzyme must move from the aqueous phase to the interface, and it must position itself with its reactive site facing the lipid, since the substrate molecules cannot move into the aqueous phase. How these requirements may be met can be deduced from the properties of pancreatic lipase and phospholipases.

Pancreatic lipase has a high affinity for oil–water interfaces; from *in vitro* experiments (Brockerhoff, 1971a) it can be estimated that the enzyme can successfully compete for the interface with the 1000-fold amount of bovine albumin. Since the enzyme is not exceptionally rich in nonpolar amino acid residues, it must be assumed that only part of its surface is hydrophobic enough to form a region of lipophilic affinity. This region, the hydrophobic head, attaches itself to the surface of the oil droplet (the supersubstrate); the remainder of the enzyme extends into the aqueous phase. The reactive site of the lipase is near the hydrophobic head, but not identical with it. In this manner, the head provides not only the enzyme–supersubstrate attachment, but also that interfacial orientation of the enzyme which brings the reactive site into the reach of the substrate molecules. This orientation is probably further stabilized by a hydrophilic tail on the enzyme, which may consist of polar amino acid and carbohydrate residues.

The necessity for digestive lipases to maintain separate heads, tails, and reactive sites in the same protein molecule probably accounts for their size, which is large compared to that of proteinases or nucleases; a more elaborate infrastructure is needed to keep the three sites apart.

The distinction between the supersubstrate binding site and the reactive site of lipolytic enzymes becomes even clearer for the phospholipases. Most of them are sensitive to the surface potential of the supersubstrate, and this effect is independent of the substrate requirements. The phospholipase 2 of porcine pancreas, for example, attacks the electrically neutral phosphatidylcholine only if bile salts or other amphiphilic anions are included in the substrate micelles. Since this activation can take place at a pH higher than the isoelectric point of the enzyme, i.e., under conditions where the enzyme molecule as a whole is itself negatively charged, it is clearly an attachment to only a part of the enzyme's surface, a positively charged, and perhaps also hydrophobic, site that binds the enzyme to the supersubstrate (the micelle) and also establishes its correct orientation. The supersubstrate binding site of the pancreatic enzyme seems to be formed when the prophospholipase is converted to the active enzyme.

The specialization of hydrolases into lipolytic enzymes can thus be regarded as their adaptation to activity at lipid–water interfaces by evolution of supersubstrate binding sites and orientative architectural features.

Bibliography

Abita, J. P., Lazdunski, M., Bonsen, P. M., Pieterson, W. A., and de Haas, G. H. (1972). *Eur. J. Biochem.* **30**, 37.

Acker, L., and Bücking, H. (1956). *Z. Lebensm. -Unters. -Forsch.* **104**, 423.

Ackman, R. G. (1972). *Progr. Chem. Fats Other Lipids* **12**, 165.

Adamson, A. W. (1967). "Physical Chemistry of Surfaces," 2nd ed. Wiley, New York.

Agranoff, B. W. (1962). *J. Lipid Res.* **3**, 190.

Albright, F. R., White, D. A., and Lennarz, W. J. (1973). *J. Biol. Chem.* **248**, 3968.

Alford, J. A., and Pierce, D. A. (1961). *J. Food Sci.* **26**, 518.

Alford, J. ., and Pierce, D. A. (1963). *J. Bacteriol.* **86**, 24.

Alford, J. A., and Smith, J. L. (1965). *J. Amer. Oil Chem. Soc.* **42**, 1038.

Alford, J. A., Pierce, D. A., and Suggs, F. G. (1964). *J. Lipid Res.* **5**, 390.

Anfinsen, C. B., and Quigley, T. W., Jr. (1953). *Circulation* **8**, 435.

Anonymous. (a). Lipase My Tech. Serv. Bull. No. 1. Meito Sangyo Co., Tokyo.

Anonymous. (b). Lipase My Tech. Serv. Bull. No. NED-ENZ-2. Monsanto Co., St. Louis, Missouri.

Anonymous. (1972) *In* "Worthington Enzyme Manual," p. 75. Worthington Biochem. Corp., Freehold, New Jersey.

Ansell, G. B., and Spanner, S. (1965). *Biochem. J.* **97**, 375.

Antia, N. J., and Bilinski, E. (1967). *J. Fish. Res. Bd. Can.* **24**, 201.

Antia, N. J., Bilinski, E., and Lau, Y. C. (1970). *Can. J. Biochem.* **48**, 643.

Arnesjö, B., and Grubb, A. (1971). *Acta Chem. Scand.* **25**, 577.

Arnesjö, B., Barrowman, J., and Borgström, B. (1967). *Acta Chem. Scand.* **21**, 2897.

Askew, E. W., Emery, R. S., and Thomas, J. W. (1970). *J. Dairy Sci.* **53**, 1415.

Assmann, G., Krauss, R. M., Fredrickson, D. S., and Levy, R. I., (1973a). *J. Biol. Chem.* **248**, 1992.

Assmann, G., Krauss, R. M., Fredrickson, D. S., and Levy, R. I. (1973b). *J. Biol. Chem.* **248**, 7184.

Association of Official Agricultural Chemists. (1965). "Official Methods of Analysis," 10th ed., p. 423. AOAC, Washington, D.C.

Astrachan, L. (1973). *Biochim. Biophys. Acta* **296**, 79.

Atherton, R. S., and Hawthorne, J. N. (1968). *Eur. J. Biochem.* **4**, 68.

Atherton, R. S., Kemp, P., and Hawthorne, J. N. (1966). *Biochim. Biophys. Acta* **125**, 409.

Attwood, D., and Saunders, L. (1965). *Biochim. Biophys. Acta* **98**, 344.

Augusteyn, R. C., de Jersey, J., Webb, E. C., and Zerner, B. (1969). *Biochim. Biophys. Acta* **171**, 128.

Augustyn, J. M., and Elliott, W. B. (1969). *Anal. Biochem.* **31**, 246.

Augustyn, J. M., and Elliott, W. B. (1970). *Biochim. Biophys. Acta* **206**, 98.

Baer, E., and Stanacev, N. Z. (1966). *Can. J. Biochem.* **44**, 893.

Bagdade, J. D., Hazzard, W. R. and Carlin, J. (1970). *Metab. Clin. Exp.* **19**, 1020.

Baker, S. P. (1957). *Circulation* **15**, 889.

Balls, A. K., and Matlack, M. B. (1938). *J. Biol. Chem.* **123**, 679.

Bangham, A. D., and Dawson, R. M. C., (1960). *Biochem. J.* **75**, 133.

Bangham, A. D., and Dawson, R. M. C. (1962). *Biochim. Biophys. Acta* **59**, 103.

Barenholz, Y., Roitman, A., and Gatt, S., (1966). *J. Biol. Chem.* **241**, 3731.

Barford, R. A., Luddy, F. E., and Magidman, P. (1966). *Lipids* **1**, 287.

Barron, E. J., and Hanahan, D. J. (1958). *J. Biol. Chem.* **231**, 493.

Barron, E. S. G., and Singer, T. P. (1943). *Science* **97**, 356.

Barrowman, J. A. (1969). *Biochim. Biophys. Acta* **184**, 653.

Barrowman, J. A., and Borgström, B. (1968). *Gastroenterology* **55**, 601.

Barrowman, J. A., and Darnton, S. J. (1970). *Gastroenterology* **59**, 13.

Barry, J. M., Bartley, W., Linzell, J. W., and Robinson, D. S. (1963). *Biochem. J.* **89**, 6.

Bartels, C. T., and van Deenen, L. L. M. (1966). *Biochim. Biophys. Acta* **125**, 395.

Barth, F. (1934). *Biochem. Z.* **270**, 63.

Baskys, B., Klein, E., and Lever, W. F. (1963). *Arch. Biochem. Biophys.* **102**, 201.

Bazán, N. G., Jr. (1971). *Acta Physiol. Lat. Amer.* **21**, 101.

Beare, J. L. (1967). *Can. J. Biochem.* **45**, 101.

Behrman, H. R., and Armstrong, D. T. (1969). *Endocrinology* **85**, 474.

Belfrage, P., and Vaughan, M. (1969). *J. Lipid Res.* **10**, 341.

Bell, R. M., Mavis, P. R., Osborn, M. J., and Vagelos, P. R. (1971). *Biochim. Biophys. Acta* **249**, 628.

Belleville, J., and Clément, J. (1966). *Arch. Sci. Physiol.* **20**, 249.

Belleville, J., and Clément, J. (1968a). *Bull. Soc. Chim. Biol.* **50**, 1419.

Belleville, J., and Clément, J. (1968b). *C. R. Acad. Sci., Ser. D* **266**, 959.

Bender, M. L., and Kézdy, F. J. (1965). *Annu. Rev. Biochem.* **34**, 49.

Benzonana, G. (1968). *Biochim. Biophys. Acta* **151**, 137.

Benzonana, G. (1969). *Biochim. Biophys. Acta* **176**, 836.

Benzonana, G. (1974). *Lipids* (in press).

Benzonana, G., and Desnuelle, P. (1965). *Biochim. Biophys. Acta* **105**, 121.

Benzonana, G., and Desnuelle, P. (1968). *Biochim. Biophys. Acta* **164**, 47.

Benzonana, G., and Esposito, S. (1971). *Biochim. Biophys. Acta* **231**, 15.

Benzonana, G., Entressangles, B., Marchis-Mouren, G., Pasero, L., Sarda, L., and Desnuelle, P., (1964). *Metab. Physiol. Significance Lipids, Proc. Advan. Study Course, 1963* p. 141.

Berg, R. N., Green, S. S., Goldberg, H. S., and Blenden, D. C. (1969). *Appl. Microbiol.* **17**, 467.

Bergström, S., Borgström, B., Tryding, N., and Westöö, G. (1954). *Biochem J.* **58**, 604.

Bernard, C. (1856). "Mémoire sur le pancréas et sur le rôle du suc pancréatique dans les phénomènes digestifs, particulièrement dans la digestion des matières grasses neutres." Baillière *et* Fils, Paris.

Bernard, M. C., and Denis, F. (1969). *C. R. Soc. Biol.* **163**, 1926.
Bernard, M. C., Brisou, J., Denis, F., and Rosenberg, A. J. (1972a). *Biochimie* **54**, 261.
Bernard, M. C., Brisou, J., Denis, F., and Rosenberg, A. J. (1972b). *Biochimie* **54**, 297.
Bernard, M. C., Brisou, J., and Denis, F. (1972c). *C. R. Acad. Sci., Ser. D* **274**, 1964.
Berndt, W., Masche, E., and Mueller-Wieland, K. (1968). *Z. Gastroenterol.* **6**, 28.
Berner, D. L., and Hammond, E. G. (1970a), *Lipids* **5**, 558.
Berner, D. L., and Hammond, E. G. (1970b). *Lipids* **5**, 572.
Biale, Y., and Shafrir, E. (1969). *Clin. Chim. Acta* **23**, 413.
Bickerstaffe, R., and Annison, E. F. (1969). *Biochem. J.* **111**, 419.
Bilinski, E., Antia, N. J., and Lau, Y. C. (1968). *Biochim. Biophys. Acta* **159**, 496.
Bird, P. R., de Haas, G. H., Heemskerk, C. H. T., and van Deenen, L. L. M. (1965). *Biochim. Biophys. Acta* **98**, 566.
Bitran, M., Wald, R., Paysant, M., and Polonovski, J. (1971). *C. R. Soc. Biol.* **165**, 1847.
Bjorklund, A. (1970). "Some Lipolytic Psychrophilic Pseudomonas Bacteria and their Hydrolysis of Edible Fats." State Inst. Tech. Res., Finland, Julkaisu 156 Publ., Helsinki.
Björnstad, P. (1966). *J. Lipid Res.* **7**, 612.
Blanchette-Mackie, E., and Scow, R. D. (1971). *J. Cell Biol.* **51**, 1.
Blaschko, H., Smith, A. D., Winkler, H., van den Bosch, H., and van Deenen, L. L. M. (1967). *Biochem. J.* **103**, 30C.
Blatt, J. M., Eisenberg, S., Stein, O., and Stein, Y. (1971). *Biochim. Biophys. Acta* **231**, 327.
Blobel, H., Shah, D. B., and Wilson, J. B. (1961). *J. Immunol.* **87**, 285.
Bloomstrand, R., Tryding, N., and Westöö, G. (1956). *Acta Physiol. Scand.* **37**, 91.
Blow, D. M. (1971). *In* "The Enzymes" (P. D. Boyer, ed.), 3rd ed., Vol. 3, p. 185. Academic Press, New York.
Blow, D. M., Birktoft, J. J., and Hartley, B. S. (1969). *Nature (London)* **221**, 337.
Blum, A. L., and Linscheer, W. G. (1970). *Proc. Soc. Exp. Biol. Med.* **135**, 565.
Boberg, J. (1970). *Lipids* **5**, 452.
Bollade, D., Paris, R., and Moulins, M. (1970). *J. Insect Physiol.* **16**, 45.
Bonsen, P. P. M., Burbach-Westerhuis, G. J., de Haas, G. H., and van Deenen, L. L. M. (1972a). *Chem. Phys. Lipids* **8**, 199.
Bonsen, P. P. M., de Haas, G. H., Pieterson, W. A., and van Deenen, L. L. M. (1972b). *Biochim Biophys. Acta* **270**, 364.
Borgström, B. (1953). *Acta Chem. Scand.* **7**, 557.
Borgström, B. (1954). *Biochim. Biophys. Acta* **13**, 491.
Borgström, B. (1957). *Scand. J. Clin. Lab. Invest.* **9**, 226.
Borgström, B. (1964a). *Biochim. Biophys. Acta* **84**, 228.
Borgström, B. (1964b). *J. Lipid Res.* **5**, 522.
Borgström, B. (1967). *J. Lipid Res.* **8**, 598.
Borgström, B., and Erlanson, C. (1971). *Biochim. Biophys. Acta* **242**, 509.
Borgström, B., and Ory, R. L. (1970). *Biochim. Biophys. Acta* **212**, 521.

Bottino, N. R., Vandenburg, G. A., and Reiser, R. (1967). *Lipids* **2,** 489.

Boyer, J. (1967). *Biochim. Biophys. Acta* **137,** 59.

Boyer, J., LePetit, J., and Giudicelli, H. (1970). *Biochim. Biophys. Acta* **210,** 411.

Bradshaw, W. S., and Rutter, W. J. (1972). *Biochemistry* **12,** 1517.

Brady, R. O. (1969). *Neurosci. Res.* **2,** 301.

Brady, R. O. (1972). *In* "Current Topics in Biochemistry" (C. B. Anfinsen, R. F. Goldberger, and A. V. Schlechter, eds.), p. 1. Academic Press, New York.

Brady, R. O., Kanfer, J. N., Mock, M. B., and Fredrickson, D. S. (1966). *Proc. Nat. Acad. Sci., U.S.* **55,** 366.

Braganca, B. M., and Khandeparkar, V. G. (1966). *Life Sci.* **5,** 1911.

Braganca B. M., Sambray, Y. M., and Ghadially, R. (1969). *Toxicon* **7,** 151.

Braganca, B. M., Sambray, Y. M., and Sambray, R. Y. (1970). *Eur. J. Biochem.* **13,** 410.

Bragdon, J. H., Havel, R. J., and Boyle, E. (1956). *J. Lab. Clin. Med.* **48,** 36.

Braun, T., and Hechter, O. (1970). *Proc. Nat. Acad. Sci. U.S.* **66,** 995.

Breckenridge, W. C., and Kuksis, A. (1968). *J. Lipid Res.* **9,** 388.

Brockerhoff, H. (1965a). *Arch. Biochem. Biophys.* **110,** 586.

Brockerhoff, H. (1965b). *J. Lipid Res.* **6,** 10.

Brockerhoff, H. (1966a). *J. Fish. Res. Bd. Can.* **23,** 1835.

Brockerhoff, H. (1966b). *Lipids* **1,** 162.

Brockerhoff, H. (1967). *J. Lipid Res.* **8,** 167.

Brockerhoff, H. (1968). *Biochim. Biophys. Acta* **159,** 296.

Brockerhoff, H. (1969a). *Arch. Biochem. Biophys.* **134,** 366.

Brockerhoff, H. (1969b). *Biochim. Biophys. Acta* **191,** 181.

Brockerhoff, H. (1970). *Biochim. Biophys. Acta* **212,** 92.

Brockerhoff, H. (1971a). *J. Biol. Chem.* **246,** 5828.

Brockerhoff, H. (1971b). *Lipids* **6,** 942.

Brockerhoff, H. (1973). *Chem. Phys. Lipids* **10,** 215.

Brockerhoff, H. (1974). *Bioorg. Chem.* **3,** 176.

Brockerhoff, H., and Hoyle, R. J. (1965). *Biochim. Biophys. Acta* **98,** 435.

Brockerhoff, H., Stewart, J. E., and Tacreiter, W. (1967). *Can. J. Biochem.* **45,** 421.

Brockerhoff, H., Hoyle, R. J., and Hwang, P. C. (1970). *Anal. Biochem.* **37,** 26.

Brockman, H. L., Do, U. H., Law, J. H., and Kézdy, F. J. (1973). *Fed. Proc., Fed. Amer. Soc. Exp. Biol.* **32,** 561 (abstr.).

Brot, R., Lossow, W. J., and Chaikoff, I. L. (1963). *Proc. Soc. Exp. Biol. Med.* **114,** 786.

Brown, H. D., ed. (1971). "Chemistry of the Cell Interface," Part A. Academic Press, New York.

Brown, J. H., and Bowles, M. E. (1965). *U. S. Army Med. Res. Lab., Rep.* **630,** 1–11.

Brown, J. H., and Bowles, M. E. (1966). *Toxicon* **3,** 205.

Brown, J. L., and Johnston, J. M. (1964). *Biochim. Biophys. Acta* **84,** 448.

Bull, H. B., (1947). *Advan. Protein Chem.* **3,** 95.

Burke, J. A., and Schubert, W. K. (1972). *Science* **176,** 309.

Byron, J. E., Wood, W. A., and Treadwell, C. R. (1953). *J. Biol. Chem.* **205,** 483.

Caillat, J. M., and Drapron, R. (1970). *Bull. Soc. Chim. Biol.* **52,** 59.

Campbell, L. B., Watrons, G. H., and Keeney, P. G. (1968). *J. Dairy Sci.* **51**, 910.

Carpenter, D. L., and Jensen, R. G. (1971). *Conn. Agr. Exp. Sta., Bull.* **414.**

Carter, J. R., Jr. (1967). *Biochim. Biophys. Acta* **137**, 147.

Casu, A., Pala, V., Monacelli, R., and Nanni, H. (1971). *Ital. J. Biochem.* **20**, 166.

Chakrubarty, M. M., Bhattacharyya, D., and Kundu, M. K. (1969). *J. Amer. Oil Chem. Soc.* **46**, 473.

Chandan, R. C., and Shahani, K. M. (1963a). *J. Dairy Sci.* **46**, 275.

Chandan, R. C., and Shahani, K. M. (1963b). *J. Dairy Sci.* **46**, 597.

Chandan, R. C., and Shahani, K. M. (1964). *J. Dairy Sci.* **47**, 471.

Chandan, R. C., and Shahani, K. M. (1965). *J. Dairy Sci.* **48**, 1413.

Chandan, R. C., Shahani, K. M., Hill, R. M., and Scholz, J. J. (1963). *Enzymologia* **46**, 82.

Chandan, R. C., Parry, R. M., Jr., and Shahani, K. M. (1968). *J. Dairy Sci.* **51**, 606.

Chatterjee, G. C., and Das, S. K. (1965). *Enzymologia* **28**, 346.

Chatterjee, G. C., and Mitra, S. (1962). *Biochem. J.* **83**, 384.

Chen, J.-S., and Barton, P. G. (1971). *Can. J. Biochem.* **49**, 1362.

Chen, L., and Morin, R. (1971). *Biochim. Biophys. Acta* **231**, 194.

Cherry, I. S., and Crandall, L. A., Jr. (1932). *Amer. J. Physiol.* **100**, 266.

Chino, H., and Gilbert, L. I. (1965). *Anal. Biochem.* **10**, 395.

Chorvath, B., and Bakoss, P. (1972). *J. Hyg., Epidemiol., Microbiol., Immunol.* **16**, 352.

Chorvath, B., and Benzonana, G. (1971a). *Biochim. Biophys. Acta* **231**, 277.

Chorvath, B., and Benzonana, G. (1971b). *J. Hyg., Epidemiol., Microbiol., Immunol.* **15**, 123.

Chorvath, B., and Fried, M. (1970a). *Folia Microbiol. (Prague)* **15**, 303.

Chorvath, B., and Fried, M. (1970b). *J. Bacteriol.* **102**, 879.

Chu, H. P. (1949). *J. Gen. Microbiol.* **3**, 255.

Chung, J., and Scanu, A. M. (1973). *Fed. Proc., Fed. Amer. Soc. Exp. Biol.* **32**, 363 (abstr.).

Chung, J., Scanu, A. M., and Reman, J. (1973). *Biochim. Biophys. Acta* **296**, 116.

Clarenburg, R., Steinberg, A. B., Asling, J. H., and Chaikoff, I. L. (1966). *Biochemistry* **5**, 2433.

Claycomb, W. C. (1972). *Ala. J. Med. Sci.* **9**, 180.

Claycomb, W. C., and Kilsheimer, G. S. (1971). *J. Biol. Chem.* **246**, 7139.

Cleland, W. W. (1970). *In* "The Enzymes" (P. D. Boyer, ed.), 3rd ed., Vol. 2, p. 1. Academic Press, New York.

Chatterjee, B. G. C., and Das, S. K. (1965). *Enzymologia* **28**, 346.

Clément, G., Clément, J., and Bézard, J. (1962a). *Biochem. Biophys. Res. Commun.* **8**, 238.

Clément, G., Clément-Champougny, J., Bézard, J., Di Costanzo, G., and Paris, R. (1962b). *Arch. Sci. Physiol.* **16**, 237.

Clermont, H. (1972). *Physiol. Veg.* **10**, 153.

Cohen, H., Hamosh, M., Atia, R., and Shapiro, B. (1967). *J. Food Sci.* **32**, 179.

Cohen, M., Morgan, R. G. H., and Hofmann, A. F. (1971). *Gastroenterology* **60**, 1.

Colacicco, G. (1969). *J. Colloid Interface Sci.* **29**, 345.

Colacicco, G., and Rapport, M. M. (1966). *J. Lipid Res.* **7**, 258.

Coleman, M. H. (1963). *Advan. Lipid Res.* **1**, 1.

Coleman, R., and Hübscher, G. (1962). *Biochim. Biophys. Acta* **56**, 479.
Coleman, R., and Hübscher, G. (1963). *Biochim. Biophys. Acta* **73**, 257.
Condrea, E., and de Vries, A. (1965). *Toxicon* **2**, 261.
Condrea, E., de Vries, A., and Mager, J. (1962). *Biochim. Biophys. Acta* **58**, 389.
Condrea, E., Barzilay, M., and Mager, J. (1970). *Biochim. Biophys. Acta* **210**, 65.
Constantin, M. J., Pasero, L., and Desnuelle, P. (1960). *Biochim. Biophys. Acta* **43**, 103.
Contardi, A., and Ercoli, A. (1933). *Biochem. Z.* **261**, 275.
Contardi, A., and Latzer, T. (1928). *Biochem. Z.* **197**, 222.
Cooper, M. F., and Webster, G. R. (1970). *J. Neurochem.* **17**, 1543.
Cooper, M. F., and Webster, G. R. (1972). *J. Neurochem.* **19**, 333.
Corbin, J. D., Soderling, T. R., and Park, C. R. (1973). *J. Biol. Chem.* **248**, 1813.
Costlow, R. D. (1958). *J. Bacteriol.* **76**, 317.
Coutts, J. R. T., and Stansfield, D. A. (1967). *Biochem. J.* **103**, 55P.
Coutts, J. R. T., and Stansfield, D. A. (1968). *J. Lipid Res.* **9**, 647.
Crabtree, B., and Newsholme, E. A. (1972). *Biochem. J.* **130**, 697.
Crum, L. R., and Calvert, D. N. (1971). *Biochim. Biophys. Acta* **225**, 161.
Crum, L. R., and Lech, J. J. (1969). *Biochim. Biophys. Acta* **178**, 508.
Crum, L. R., Harbecke, R. G., Lech, J. J., and Calvert, D. N. (1970). *Biochim. Biophys. Acta* **198**, 229.
Currie, B. T., Oakley, D. E., and Broomfield, C. A. (1968). *Nature (London)* **220**, 371.
Dailey, R. E., Swell, L., and Treadwell, C. R. (1962). *Arch. Biochem. Biophys.* **99**, 334.
Dailey, R. E., Swell, L., and Treadwell, C. R. (1963). *Arch. Biochem. Biophys.* **100**, 360.
Daly, J. M., Knoche, H.-W., and Wiese, M. V. (1967). *Plant Physiol.* **42**, 1633.
Dauvillier, P., de Haas, G. H., van Deenen, L. M., and Raulin, J. (1964). *C. R. Acad. Sci.* **259**, 4865.
Davidson, F. M., and Long, C. (1958). *Biochem. J.* **69**, 458.
Davies, J. T., and Rideal, E. K. (1963). "Interfacial Phenomena," 2nd ed. Academic Press, New York.
Davies, M. E. (1954). *J. Gen. Microbiol.* **11**, 37.
Dawson, R. M. C. (1956). *Biochem. J.* **64**, 192.
Dawson, R. M. C. (1959a). *Biochim. Biophys. Acta* **33**, 68.
Dawson, R. M. C. (1959b). *Nature (London)* **183**, 1822.
Dawson, R. M. C. (1963a). *Biochem. J.* **88**, 414.
Dawson, R. M. C. (1963b). *Biochim. Biophys. Acta* **70**, 697.
Dawson, R. M. C. (1964). *Metab. Physiol. Significance Lipids, Proc. Advan. Study Course, 1963* p. 179.
Dawson, R. M. C. (1965). *Colloq. Ges. Physiol. Chem.* **16**, 29.
Dawson, R. M. C. (1967). *Biochem. J.* **102**, 205.
Dawson, R. M. C. (1968). *In* "Biological Membranes" (D. Chapman, ed.), p. 203. Academic Press, New York.
Dawson, R. M. C. (1969). *In* "Methods in Enzymology" (J. M. Lowenstein, ed.), Vol. 14, p. 633. Academic Press, New York.
Dawson, R. M. C., and Hauser, H. (1967). *Biochim. Biophys. Acta* **137**, 518.
Dawson, R. M. C., and Hemington, N. (1967). *Biochem. J.* **102**, 76.

Fryer, T. F., Reiter, B., and Lawrence, R. C. (1967). *J. Dairy Sci.* **50,** 477.

Fukumoto, J., Iwai, M., and Tsujisaka, Y. (1963). *J. Gen. Appl. Microbiol.* **9,** 353.

Funatsu, M., Aizono, Y., Hayashi, K., Watanabe, M., and Eto, M. (1971). *Agr. Biol. Chem.* **35,** 734.

Fung, C. K., and Proulx, P. (1969). *Can. J. Biochem.* **47,** 371.

Gaffney, P. J., and Harper, W. J. (1965). *J. Dairy Sci.* **48,** 615.

Gaffney, P. J., Harper, W. J., and Gould, I. A. (1966). *J. Dairy Sci.* **49,** 921.

Gaffney, P. J., Harper, W. J., and Gould, I. A. (1968). *J. Dairy Sci.* **51,** 1161.

Gaines, G. L. (1966). "Insoluble Monolayers at Liquid-Gas Interfaces," Wiley, New York.

Gallai-Hatchard, J., and Thompson, R. H. (1965). *Biochim. Biophys. Acta* **98,** 128.

Galliard, T. (1970). *Phytochemistry* **9,** 1725.

Galliard, T. (1971a). *Biochem. J.* **121,** 379.

Galliard, T. (1971b). *Eur. J. Biochem.* **21,** 90.

Galton, D. J. (1971). *In* "The Human Adipose Cell," p. 108. Appleton-Century-Crofts, New York.

Gander, G. W., and Jensen, R. G. (1960). *J. Dairy Sci.* **43,** 1762.

Gander G. W., Jensen, R. G., and Sampugna, J. (1961). *J. Dairy Sci.* **44,** 1980.

Ganesan, D. and Bradford, R. H. (1971). *Biochem. Biophys. Res. Commun.* **43,** 544.

Garfinkel, A. S., and Schotz, M. C. (1973). *Biochim. Biophys. Acta* **306,** 128.

Garner, C. W., and Smith, L. C. (1970a). *Arch. Biochem. Biophys.* **140,** 503.

Garner, C. W., and Smith, L. C. (1970b). *Biochem. Biophys. Res. Commun.* **39,** 672.

Garner, C. W., and Smith, L. C. (1972). *J. Biol. Chem.* **247,** 561.

Gatt, S. (1963). *J. Biol. Chem.* **238,** PC3131.

Gatt, S. (1966). *J. Biol. Chem.* **241,** 3724.

Gatt, S. (1968). *Biochim. Biophys. Acta* **159,** 304.

Gatt, S., and Barenholz, Y. (1969a). *In* "Methods in Enzymology" (J. M. Lowenstein, ed.), Vol. 14, p. 167. Academic Press, New York.

Gatt, S., and Barenholz, Y. (1969b). *In* "Methods in Enzymology" (J. M. Lowenstein, ed.), Vol. 14, p. 144.

Gatt, S., and Barenholz, Y. (1973). *Annu. Rev. Biochem.* **42,** 61.

Gatt, S., Barenholz, Y., and Roitman, A. (1966). *Biochem. Biophys. Res. Commun.* **24,** 169.

Gatt, S., Barenholz, Y., Borkovski-Kubiler, I., and Leibovitz-Ben Gershon, Z. (1972). *Advan. Exp. Med. Biol.* **19,** 237.

Gatt, S., Herzl, A., and Barenholz, Y. (1973). *FEBS Lett.* **30,** 281.

Gelotte, B. (1964). *Acta Chem. Scand.* **18,** 1282.

Gidez, L. I. (1968). *J. Lipid Res.* **7,** 794.

Gidez, L. I. (1973). *J. Lipid Res.* **14,** 169.

Gillespie, W. H., and Alder, V. G. (1952). *J. Pathol. Microbiol.* **64,** 187.

Goerke, J., de Gier, J., and Bonsen, P. P. M. (1971). *Biochim. Biophys. Acta* **248,** 245.

Goller, H. J., and Sgoutas, D. S. (1970). *Biochemistry* **9,** 4801.

Goller, H. J., Sgoutas, D. S., Ismail, D. A., and Gunstone, F. D. (1970). *Biochemistry* **9,** 3072.

Gollub, S., Feldman, D., Schechter, D. C., Kaplan, F. E., and Merange, D. R. (1953). *Proc. Soc. Exp. Biol. Med.* **83,** 858.

Good, N. E., Winget, G. D., Winter, W., Connoly, T. N., Izawa, S., and Singh, R. R. M. (1966). *Biochemistry* **5**, 467.

Gorin, E., and Shafrir, E. (1964). *Biochim. Biophys. Acta* **84**, 24.

Gorman, R. R., Tepperman, H. M., and Tepperman, J. (1973). *J. Lipid Res.* **14**, 279.

Gottfried, E. L., and Rapport, M. M. (1962). *J. Biol. Chem.* **237**, 329.

Graham, E. R. B., McKenzie, H. A., and Murphy, W. H. (1970). in "Milk Proteins: Chemistry and Molecular Biology" (H. A. McKenzie, ed.). *In* Vol. 1, p. 219. Academic Press, New York.

Green, J. R. (1890). *Proc. Roy. Soc., London* **48**, 370.

Greten, H. (1972). *Klin. Wochenschr.* **50**, 39.

Greten, H., Levy, R. I., and Fredrickson, D. S. (1968). *Biochim. Biophys. Acta* **164**, 185.

Greten, H., Levy, R. I., and Fredrickson, D. S. (1969). *J. Lipid Res.* **10**, 326.

Greten, H., Levy, R. I., Fales, H., and Fredrickson, D. S. (1970). *Biochim. Biophys. Acta* **210**, 39.

Greten, H., Walter, B., and Brown, W. V. (1972). *FEBS Lett.* **27**, 306.

Grosskopf, J. F. W. (1965). *Onderstepoort J. Vet. Res.* **32**, 153.

Groves, M. L. (1971a). *In* "Milk Proteins: Chemistry and Molecular Biology" (H. A. McKenzie, ed.), Vol. 2, p. 367. Academic Press, New York.

Groves, M. L. (1971b). *In* "Milk Proteins: Chemistry and Molecular Biology" (H. A. McKenzie, ed.), Vol. 2, p. 399. Academic Press, New York.

Guder, W., Weiss, L., and Wieland, O. (1969). *Biochim. Biophys. Acta* **187**, 173.

Guilbault, G. G., and Kramer, D. N. (1966). *Anal. Biochem.* **14**, 28.

Gunstone, F. D., Ismail, I. A., and Lie Ken Yie, M. (1967). *Chem. Phys. Lipids* **1**, 376.

Gurr, M. I., Blades, J., and Appleby, R. S. (1972). *Eur. J. Biochem.* **29**, 362.

Habermann, E. (1972). *Science* **177**, 314.

Habermann E., and Hardt, K. L. (1972). *Anal. Biochem.* **50**, 163.

Habermann, E., and Reiz, K. G. (1965). *Biochem. Z.* **343**, 192.

Hachimori, Y., Wells, M. A., and Hanahan, D. J. (1971). *Biochemistry* **11**, 4084.

Hahn, P. F. (1943). *Science* **98**, 19.

Hair, M. L., ed. (1971). "The Chemistry of Biosurfaces," Vol. I. Dekker, New York.

Hajra, A. K., and Agranoff, B. W. (1969). *In* "Methods in Enzymology" (J. M. Lowenstein, ed.), Vol. 14, p. 185. Academic Press, New York.

Hammett, L. P. (1935). *Chem. Rev.* **17**, 125.

Hamosh, M., and Scow, R. O. (1973). *J. Clin. Invest.* **52**, 88.

Hanahan, D. J. (1952). *J. Biol. Chem.* **195**, 199.

Hanahan, D. J. (1954a). *J. Biol. Chem.* **207**, 879.

Hanahan, D. J. (1954b). *J. Biol. Chem.* **211**, 313.

Hanahan, D. J. (1957). *Progr. Chem. Fats, Other Lipids* **4**, 141.

Hanahan, D. J. (1960). "Lipide Chemistry," Wiley, New York.

Hanahan, D. J. (1971). *In* "The Enzymes" (P. D. Boyer, ed.), 3rd ed., Vol. 5, p. 71. Academic Press, New York.

Hanahan, D. J., and Brockerhoff, H. (1960). *Arch. Biochem. Biophys.* **91**, 326.

Hanahan, D. J., and Chaikoff, I. L. (1947a). *J. Biol. Chem.* **168**, 233.

Hanahan, D. J., and Chaikoff, I. L. (1947b). *J. Biol. Chem.* **169**, 699.

Hanahan, D. J., and Chaikoff, I. L. (1948). *J. Biol. Chem.* **172**, 191.

Hanahan, D. J., and Vercamer, R. (1954). *J. Amer. Chem. Soc.* **76**, 1804.

Hanahan, D. J., Brockerhoff, H., and Barron, E. J. (1960). *J. Biol. Chem.* **235**, 1917.

Harper, W. J., Schwartz, D. P., and El-Hagarawy, I. S. (1956). *J. Dairy Sci.* **39**, 45.

Hartsuck, J. A., and Lipscomb, W. N. (1971). *In* "The Enzymes" (P. D. Boyer, ed.), 3rd ed., Vol. 3 p. 1. Academic Press, New York.

Hassing, G. S. (1971). *Biochim. Biophys. Acta* **242**, 381.

Hatch, F. T. (1965). *Nature (London)* **206**, 777.

Havel, R. J., Shore, V. G., Shore, B., and Bier, D. M. (1970). *Circ. Res.* **27**, 595.

Havel, R. J., Fielding, C. J., Olivecrona, T., Shore, V. G., Fielding, P. E., and Egelrud, T. (1973). *Biochemistry* **12**, 1828.

Haverkate, F., and van Deenen, L. L. M. (1964). *Biochim. Biophys. Acta* **84**, 106.

Hayaishi, O. (1955). *In* "Methods in Enzymology" Vol. 1, p. 660. (S. P. Colowick and N. O. Kaplan, eds.), Academic Press, New York.

Hayaishi, O., and Kornberg, A. (1954). *J. Biol. Chem.* **206**, 647.

Hayase, K., and Tappel, A. L. (1970). *J. Biol. Chem.* **245**, 169.

Heard, D. H., and Seaman, G. V. F. (1960). *J. Gen. Physiol.* **43**, 635.

Heimermann, W. H., Holman, R. T., Gordon, D. T., Kowalyshyn, D. E., and Jensen, R. G. (1973). *Lipids* **8**, 45.

Heller, M., and Arad, R. (1970). *Biochim. Biophys. Acta* **210**, 276.

Heller, M., and Shapiro, B. (1966). *Biochem. J.* **98**, 763.

Heller, M., Aladjem, E., and Shapiro, B. (1968). *Bull. Soc. Chim. Biol.* **50**, 1395.

Heller, R. A., and Steinberg, D. (1972). *Biochim. Biophys. Acta* **270**, 65.

Helmsing, P. J. (1967). *Biochim. Biophys. Acta* **144**, 470.

Helmsing, P. J. (1969). *Biochim. Biophys. Acta* **178**, 519.

Henderson, C. (1971). *J. Gen. Microbiol.* **65**, 81.

Hernandez, H. H., and Chaikoff, I. L. (1957). *J. Biol. Chem.* **228**, 447.

Heymann, E., Krisch, K., and Pahlich, E. (1970). *Hoppe-Seyler's Z. Physiol. Chem.* **351**, 931.

Hill, E. E., and Lands, W. E. M. (1970). *In* "Lipid Metabolism" (S. J. Wakil, ed.), p. 185. Academic Press, New York.

Hine, J. (1956). "Physical Organic Chemistry." McGraw-Hill, New York.

Hirs, C. H. W., and Timasheff, S. N., eds. (1972). "Methods in Enzymology," Vol. 25, Part B. Academic Press, New York.

Hirschmann, H. (1960). *J. Biol. Chem.* **235**, 2762.

Hoertnagl, H., Winkler, H., and Hoertnagl, H. (1969). *Eur. J. Biochem.* **10**, 243.

Hofmann, A. F. (1963). *Biochem. J.* **89**, 57.

Hofmann, A. F., and Borgström, B. (1964). *J. Clin. Invest.* **43**, 247.

Hofmann, A. F., and Small, D. M. (1967). *Annu. Rev. Med.* **18**, 333.

Hokin, L. E. (1967). *In* "Handbook of Physiology" (Amer. Physiol. Soc., J. Field, ed.), Sect. 6, Vol. II, p. 935. Williams & Wilkins, Baltimore, Maryland.

Hokin, L. E., and Hokin, M. R. (1961). *Nature (London)* **189**, 836.

Hokin, L. E., Hokin, M. R., and Mathison, D. (1963). *Biochim. Biophys. Acta* **67**, 418.

Hollenberg, C. H., Raben, M. S., and Astwood, E. B. (1961). *Endocrinology* **68**, 589.

Holman, R. T. (1966). *Prog. Chem. Fats Other Lipids* **9**, 1.

Hori, T., Sugita, M., and Itasaka, O. (1969). *Shiga Daigaku Kyoiku Gakubu Kiyo, Shizenkagaku* **19**, 39.

Howard, C. F., and Portman, O. W. (1966). *Biochim. Biophys. Acta* **125**, 623.

Hughes, A. (1935). *Biochem. J.* **29**, 437.

Huttunen, J. K., and Steinberg, D. (1971). *Biochim. Biophys. Acta* **239**, 410.

Huttunen, J. K., Ellingboe, J., Pittman, R. C., and Steinberg, D. (1970a). *Biochim. Biophys. Acta* **218**, 333.

Huttunen, J. K., Ellingboe, J., Pittman, R. C., and Steinberg, D. (1970b). *Fed. Proc., Fed. Amer. Soc. Exp. Biol.* **29**, 267.

Huttunen, J. K., Ellingboe, J., Pittman, R. C., and Steinberg, D. (1970c). *Clin. Res.* **18**, 140.

Huttunen, J. K., Aquino, A. A., and Steinberg, D. (1970d). *Biochim. Biophys. Acta* **224**, 295.

Huttunen, J. K., Steinberg, D., and Mayer, S. E. (1970e). *Biochem. Biophys. Res. Commun.* **41**, 1350.

Hyun, J., Kothari, H., Hern, E., Mortenson, J., Treadwell, C. R., and Vahouny, G. V. (1969). *J. Biol. Chem.* **244**, 1937.

Hyun, J., Steinberg, M., Treadwell, C. R., and Vahouny, G. V. (1971). *Biochem. Biophys. Res. Commun.* **44**, 819.

Hyun, J., Treadwell, C. R., and Vahouny, G. V. (1972). *Arch. Biochem. Biophys.* **152**, 233.

Hyun, S. A., Vahouny, G., and Treadwell, C. R. (1964). *Arch. Biochem. Biophys.* **104**, 139.

Ibrahim, S. A. (1967). *Biochim. Biophys. Acta* **137**, 413.

Ibrahim, S. A. (1970). *Toxicon* **8**, 221.

Ibrahim, S. A., Sanders, H., and Thompson, R. H. S. (1964). *Biochem. J.* **93**, 588.

Ichishima, E., Yoshida, F., and Motai, H. (1968). *Biochim. Biophys. Acta* **154**, 586.

Ikezawa, H., Yamamoto, A., and Murata, R. (1964). *J. Biochem. (Tokyo)* **56**, 480.

Illingworth, D. R., and Glover, J. (1969). *Biochim. J.* **115**, 16P.

Infante, R., Koumanov, K., and Polonovski, J. (1968). *Biochim. Biophys. Acta* **164**, 436.

Ispolatovskaya, M. V. (1970). *Biokhimiya* **35**, 434.

Ispolatovskaya, M. V. (1971). *Vop. Med. Khim.* **17**, 137.

IUPAC-IUB Commission on Biochemical Nomenclature. (1967). *Eur. J. Biochem.* **2**, 127.

Iverius, P. H. (1971). *Biochem. J.* **124**, 677.

Iverius, P. H., Lindahl, U., Egelrud, T., and Olivecrona, T. (1972). *J. Biol. Chem.* **247**, 6610.

Jacks, T. J., and Kircher, H. W. (1967). *Anal. Biochem.* **21**, 279.

Jackson, R. L., and Hirs, C. H. (1970). *J. Biol. Chem.* **245**, 624.

Jaillard, J., Sezille, G., Fruchart, J. C., and Schenpeneel, P. (1972). *Clin. Chim. Acta* **38**, 277.

James, L. K., and Augenstein, L. G. (1966). *Advan. Enzymol. Relat. Areas Mol. Biol.* **28**, 1.

Jansen, H., and Hülsmann W. C. (1973). *Biochim. Biophys. Acta* **296**, 241.

Janssen, L. H. M., de Bruin, S. H., and de Haas, G. H. (1972). *Eur. J. Biochem.* **28**, 156.

Jenness, R. (1970). *In* "Milk Proteins: Chemistry and Molecular Biology" (H. A. McKenzie, ed.), Vol. 1, p. 17. Academic Press, New York.

Jensen, R. G. (1964). *J. Dairy Sci.* **47**, 210.

Jensen, R. G. (1971). *Progr. Chem. Fats Other Lipids* **11**, 347.

Jensen, R. G. (1972). *Top. Lipid Chem.* **3**, 1.

Jensen, R. G. (1973). *Lipids* **8**.

Jensen, R. G., Duthie, P. M., Gander, G. W., and Morgan, M. E. (1960). *J. Dairy Sci.* **43**, 96.

Jensen, R. G., Gander, G. W., Sampugna, J., and Forster, T. L. (1961a). *J. Dairy Sci.* **44**, 943.

Jensen, R. G., Sampugna, J., Parry, R. M., Shahani, K. M., and Chandan, R. C. (1961b). *J. Dairy Sci.* **45**, 1527.

Jensen, R. G., Sampugna, J., Parry, R. M., Shahani, K. M., and Chandan, R. C. (1962). *J. Dairy Sci.* **45**, 1527.

Jensen, R. G., Sampugna, J., Parry, R. M., and Shahani, K. M. (1963). *J. Dairy Sci.* **46**, 907.

Jensen, R. G., Sampugna, J., and Pereira, R. L. (1964a). *Biochim. Biophys. Acta* **84**, 481.

Jensen, R. G., Sampugna, J., and Pereira, R. L. (1964b). *J. Dairy Sci.* **47**, 727.

Jensen, R. G., Sampugna, J., Pereira, R. L., Chandan, R. C., and Shahani, K. M. (1964c). *J. Dairy Sci.* **47**, 1012.

Jensen, R. G., Sampugna, J., Quinn, J. G., Carpenter, D. L., and Marks, T. A. (1965). *J. Amer. Oil Chem. Soc.* **42**, 1029.

Jensen, R. G., Sampugna, J., and Quinn, J. G. (1966a). *Lipids* **1**, 294.

Jensen, R. G., Marks, T. A., Sampugna, J., Quinn, J. G., and Carpenter, D. L. (1966b). *Lipids* **1**, 451.

Jensen, R. G., Pitas, R. E., Quinn, J. G., and Sampugna, J. (1970). *Lipids* **5**, 580.

Jensen, R. G., Gordon, D. T., Heimermann, W. H., and Holman, R. T. (1972). *Lipids* **7**, 738.

Jensen, R. G., Gordon, D. T., and Scholfield, E. R. (1973). *Lipids* **8**, 323.

Jentsch, J. (1972). *Naturwiss. Rundsch.* **25**, 68.

Jentsch, J., and Dielenberg, D. (1972). *Justus Liebigs Ann. Chem.* **757**, 187.

Johnson, C. E., and Bonventre, P. F. (1967). *J. Bacteriol.* **94**, 306.

Johnston, J. M., and Bearden, J. H. (1965). *Biochim. Biophys. Acta* **56**, 365.

Johnston, J. M., Ras, G. A., Lowe, P. A., and Schwarz, B. E. (1967). *Lipids* **2**, 14.

Jubelin, J., and Boyer, J. (1972). *Eur. J. Clin. Invest.* **2**, 417.

Julien, R., Canioni, P., Rathelol, J., Sarda, L., and Plummer, T. H., Jr. (1972). *Biochim. Biophys. Acta* **280**, 215.

Jungalwala, F. B., Freinkel, N., and Dawson, R. M. C. (1971). *Biochem. J.* **123**, 19.

Kalle, G. P., Gadkari, S. V., and Deshpande, S. Y. (1972). *Indian J. Biochem. Biophys.* **9**, 171.

Kaneko, H., and Hara, I. (1967). *Bull. Soc. Chim. Biol.* **49**, 59.

Kanfer, J. N., and Brady, R. O. (1969). *In* "Methods in Enzymology" Vol. 14, p. 131. (J. M. Lowenstein, ed.), Academic Press, New York.

Kanfer, J. N., Young, O. M., Shapiro, D., and Brady, R. O. (1966). *J. Biol. Chem.* **241**, 1081.

Kaplan, A. (1970). *Anal. Biochem.* **33**, 218.

Kariya, M., and Kaplan, A. (1973). *J. Lipid Res.* **14**, 243.

Karnovsky, M. L., and Wolff, D. (1960). *Biochem. Lipids Proc. Int. Conf. Biochem. Probl. Lipids 5th, 1958,* p. 53. Pergamon, Oxford.

Kason, C. M., Pavamani, I. V. P., and Nakai, S. (1972). *J. Dairy Sci.* **55,** 1420.

Kates, M. (1953). *Nature (London)* **172,** 814.

Kates, M. (1955). *Can. J. Biochem. Physiol.* **33,** 575.

Kates, M. (1956). *Can. J. Biochem. Physiol.* **34,** 967.

Kates, M. (1957). *Can. J. Biochem. Physiol.* **35,** 127.

Kates, M. (1960). *In* "Lipid Metabolism" (K. Bloch, ed.), p. 165. Wiley, New York.

Kates, M., and Gorham, P. R. (1957). *Can. J. Biochem. Physiol.* **35,** 119.

Kates, M., and Sastry, P. S. (1969). *In* "Methods in Enzymology" (J. M. Lowenstein, ed.), Vol. 14, p. 197. Academic Press, New York.

Kates, M., Madeley, J. R., and Beare, J. L. (1965). *Biochim. Biophys. Acta* **106,** 630.

Katocs, A. S., Jr., Calvert, D. N., and Lech, J. J. (1971). *Biochim. Biophys. Acta* **229,** 608.

Katocs, A. S., Jr., Gnewuch, T., Lech, J. J., and Calvert, D. N. (1972). *Biochim. Biophys. Acta* **270,** 209.

Katz, S. (1957). *J. Appl. Physiol.* **10,** 519.

Kawasaki, N., and Saito, K. (1973). *Biochim. Biophys. Acta* **296,** 426.

Kawauchi, S., Iwanaga, S., Samejima, Y., and Suzuki, T. (1971a). *Biochim. Biophys. Acta* **236,** 142.

Kawauchi, S., Samejima, Y., Iwanaga, S., and Suzuki, T. (1971b). *J. Biochem. (Tokyo)* **69,** 433.

Kelley, T. F. (1968). *J. Lipid Res.* **9,** 799.

Kellum, R. E., and Strangfeld, K. (1969). *J. Invest. Dermatol.* **52,** 255.

Kellum, R. E., Strangfeld, K., and Ray, L. F. (1970). *Arch. Dermatol.* **101,** 41.

Kelly, R. A., and Newman, H. A. (1971). *Biochim. Biophys. Acta* **231,** 558.

Kemp, P., Hübscher, G., and Hawthorne, J. N. (1959). *Biochim. Biophys. Acta* **31,** 585.

Kennedy, E. P. (1961). *Fed. Proc., Fed. Amer. Soc. Exp. Biol.* **20,** 934.

Kent, C., and Lennerz, W. J. (1972). *Proc. Nat. Acad. Sci. U.S.* **69,** 2793.

Keough, K. M., and Thompson, W. (1970). *J. Neurochem.* **17,** 1.

Keough, K. M., and Thompson, W. (1972). *Biochim. Biophys. Acta* **270,** 324.

Khan, M. A. Q., and Hodgson, E. (1967). *Comp. Biochem. Physiol.* **23,** 899.

Khoo, J. C., Fong, W. C., and Steinberg, D. (1972a). *Biochem. Biophys. Res. Commun.* **49,** 407.

Khoo, J. C., Jarett, L., Mayer, S. E., and Steinberg, D. (1972b). *J. Biol. Chem.* **247,** 4812.

Kleiman, J. H., and Lands, W. E. M. (1969). *Biochim. Biophys. Acta* **187,** 477.

Kleiman, R., Earle, F. R., Tallent, W. H., and Wolff, I. A. (1970). *Lipids* **5,** 513.

Klein, F., and Mandel, P. (1972). *Biochimie* **54,** 371.

Klein, W. (1938). *Hoppe-Seyler's Z. Physiol. Chem.* **254,** 1.

Klibansky, C., London, Y., Frenkel, A., and de Vries, A. (1968). *Biochim. Biophys. Acta* **150,** 15.

Knoche, H. W., and Horner, T. L. (1970). *Plant Physiol.* **46,** 401.

Kocholaty, W. (1966a). *Toxicon* **3,** 175.

Kocholaty, W. (1966b). *Toxicon* **4,** 1.

Kocholaty, W., and Ashley, B. D. (1966). *Toxicon* **3,** 187.

Kohn, J. (1959). *Nature (London)* **183,** 1055.

Kondo, K. (1910a). *Biochem. Z.* **27,** 427.

Kondo, K. (1910b). *Biochem. Z.* **36**, 243.
Korhonen, H., and Antila, M. (1972). *Fette, Seifen, Anstrichm.* **24**, 399.
Korn, E. D. (1954). *Science* **120**, 399.
Korn, E. D. (1955a). *J. Biol. Chem.* **215**, 1.
Korn, E. D. (1955b). *J. Biol. Chem.* **215**, 15.
Korn, E. D. (1959). *Methods Biochem. Anal.* **7**, 145.
Korn, E. D. (1961). *J. Biol. Chem.* **236**, 1638.
Korn, E. D. (1962). *J. Lipid Res.* **3**, 246.
Korn, E. D., and Quigley, T. W., Jr. (1955). *Biochim. Biophys. Acta* **18**, 143.
Korn, E. D., and Quigley, T. W., Jr. (1957). *J. Biol. Chem.* **226**, 833.
Kornalik, F. (1964). *Proc. Int. Pharmacol. Meet, 2nd, 1963* Vol. 9, p. 81.
Korzenovsky, M., Diller, E. R., Marshall, A. C., and Auda, B. M. (1960a). *Biochem. J.* **76**, 238.
Korzenovsky, M., Walters, C. P., Harvey, O. A., and Diller, E. R., (1960b). *Proc. Soc. Exp. Biol. Med.* **105**, 303.
Kothari, M. V., Bonner, M. J., and Miller, B. F. (1970). *Biochim. Biophys. Acta* **202**, 325.
Kothari, H. V., Miller, B. F., and Krithchevsky, D. (1973). *Biochim. Biophys. Acta* **296**, 446.
Kozhukhov, V. I., Molotkovskii, Y. G., and Bergel'son, L. D. (1969). *Biokhimiya* **34**, 1236.
Kramer, S. P., Aronson, L. D., Rosenfeld, M. G., Sulkin, M. D., Chang, A., and Seligman, A. M. (1963). *Arch. Biochem. Biophys.* **102**, 1.
Krauss, R. M., Windmueller, H. G., Levy, R. I., and Fredrickson, D. S. (1973). *J. Lipid Res.* **14**, 286.
Krewson, C. F., and Scott, W. F. (1964). *J. Amer. Oil Chem. Soc.* **41**, 422.
Krewson, C. F., Ard, J. S., and Riemenschneider, R. W. (1962). *J. Amer. Oil Chem. Soc.* **39**, 334.
Kroll, J., Franzke, C., and Genz, S. (1973). *Pharmazie* **28**, 263.
Krysan, J. L., and Guss, P. L. (1971). *Biochim. Biophys. Acta* **239**, 349.
Kumar, S. S., Millay, R. H., and Bieber, L. L. (1970). *Biochemistry* **9**, 754.
Kupiecki, F. P. (1966). *J. Lipid Res.* **7**, 230.
Kupiecki, F. P. (1971). *Progr. Biochem. Pharmacol.* **6**, 274.
Kurioka, S. (1968). *J. Biochem. (Tokyo)* **63**, 678.
Kurioka, S., and Liu, P. V. (1967). *Appl. Microbiol.* **15**, 551.
Kurup, P. A. (1965). *Naturwissenchaften* **52**, 478.
Kurup, P. A. (1966). *Indian J. Biochem.* **3**, 164.
Kyes, P. (1903). *Berlin. Klin. Wochenschr.* **40**, 956.
Kyriakides, E. C., Paul, B., and Balint, J. A. (1972). *J. Lab. Clin. Med.* **80**, 810.
Laboureur, P., and Labrousse, M. (1966). *Bull. Soc. Chim. Biol.* **48**, 747.
Laboureur, P., and Labrousse, M. (1968). *Bull. Soc. Chim. Biol.* **50**, 2179.
Lagocki, J. W., Boyd, N. D., Law, J. H., and Kézdy, F. J. (1970). *J. Amer. Chem. Soc.* **92**, 2923.
Lagocki, J. W., Law, J. H., and Kézdy, F. J. (1973). *J. Biol. Chem.* **248**, 580.
Lake, B. D., and Patrick, A. D. (1970). *J. Pediat.* **76**, 262.
Landes, D. R. (1972). *Nutr. Rep. Int.* **5**, 421.
Lands, W. E. M. (1958). *J. Biol. Chem.* **231**, 883.
Lands, W. E. M. (1960). *J. Biol. Chem.* **231**, 2233.
Lands, W. E. M. (1965). *Annu. Rev. Biochem.* **34**, 313.

Lands, W. E. M., and Hart, P. (1965). *Biochim. Biophys. Acta* **98**, 532.

Lapetina, E. G., and Michell, R. H. (1973). *Biochem. J.* **131**, 433.

LaRosa, J. C., Levy, R. I., Windmueller, H. G., and Fredrickson, D. S. (1970a). *J. Clin. Invest.* **49**, 55a.

LeRosa, J. C., Levy, R. I., Herbert, P., Lux, S. E., and Fredrickson, D. S. (1970b). *Biochem. Biophys. Res. Commun.* **41**, 57.

LaRosa, J. C., Levy, R. I., Windmueller, H. G., and Fredrickson, D. S. (1972). *J. Lipid Res.* **13**, 356.

Laurell, S. (1968). *Biochim. Biophys. Acta* **152**, 80.

Lawrence, R. C. (1967). *Dairy Sci. Abstr.* **29**, 1.

Lawrence, R. C., Fryer, T. F., and Reiter, B. (1967a). *J. Gen. Microbiol.* **48**, 401.

Lawrence, R. C., Fryer, T. F., and Reiter, B. (1967b). *Nature (London)* **213**, 1264.

Lech, J. J., and Calvert, D. N. (1968). *Can. J. Biochem.* **46**, 707.

Leger, C. (1972). *Ann. Biol. Anim., Biochim., Biophys.* **12**, 341.

Leger, C., and Bauchart, D. (1972). *C. R. Acad. Sci.* **275**, 2419.

Leger, C., Bergot, P., Flanzy, J., and Francois, A. C. (1970). *C. R. Acad. Sci., Ser. D* **270**, 2813.

Lehmann, V. (1971). *Acta Pathol. Microbiol. Scand., Sect. B* **79**, 372.

Lehmann, V. (1972). *Acta Pathol. Microbiol. Scand., Sect. B* **80**, 827.

Leibovitz, Z., and Gatt, S. (1968). *Biochim. Biophys. Acta* **164**, 439.

Leibovitz-Ben Gershon, Z., Kobiler, I., and Gatt, S. (1972). *J. Biol. Chem.* **247**, 6840.

Lengle, E., and Geyer, R. P. (1973). *Biochim. Biophys. Acta* **296**, 411.

Lewis, G. M., and Macfarlane, M. G. (1953). *Biochem. J.* **54**, 138.

Lewis, J., Day, A., and de la Lande, I. S. (1968). *Toxicon* **6**, 109.

Lineweaver, H., and Burk, D. (1934). *J. Amer. Chem. Soc.* **56**, 658.

Litchfield, C. (1972). "Analysis of Triglycerides." Academic Press, New York.

Lloveras, J., and Douste-Blazy, L. (1968a). *Bull. Soc. Chim. Biol.* **50**, 1487.

Lloveras, J., and Douste-Blazy, L. (1968b). *Bull. Soc. Chim. Biol.* **50**, 1493.

Lloveras, J., Douste-Blazy, L., and Valdiguié, P. (1963). *C. R. Acad. Sci.* **256**, 1861.

Löffler, G., and Weiss, L. (1970). *In* "Adipose Tissue; Regulation and Function" (B. Jeanrenand and D. Hepp, eds.), p. 32. Academic Press, New York.

Long, C., and Penny, J. F. (1957). *Biochem. J.* **65**, 382.

Long, C., Odavic, R., and Sargent, E. J. (1962). *Biochem. J.* **85**, 33.

Lu, J. Y., and Liska, B. J. (1969a). *Appl. Microbiol.* **18**, 104.

Lu, J. Y., and Liska, B. J. (1969b). *Appl. Microbiol.* **18**, 108.

Lucy, J. A. (1968). *In* "Biological Membranes" (D. Chapman, ed.), p. 71. Academic Press, New York.

Luddy, F. E., Barford, R. A., Herb, S. F., Magidman, P., and Riemenschneider, R. W. (1964). *J. Amer. Oil Chem. Soc.* **41**, 693.

Luddy, F. E., Menna, A. J., and Calhoun, R. R. (1969). *J. Amer. Oil Chem. Soc.* **46**, 506.

Lüdecke, K. (1905). Dissertation, Munich.

Luhtala, A., and Antila, M. (1968). *Suomen Kemistilchti B* **41**, 386.

Lynn, W. S., Jr., and Perryman, N. C. (1960). *J. Biol. Chem.* **235**, 1912.

Lysenko, O. (1972). *Folia Microbiol. (Prague)* **17**, 221.

McBride, O. W., and Korn, E. D. (1963). *J. Lipid Res.* **4**, 17.

McBride, O. W., and Korn, E. D. (1964). *J. Lipid Res.* **5**, 459.

McCaman, R. E., Smith, M., and Cook, K. (1965). *J. Biol. Chem.* **240**, 3513.

Macchia, V., and Pastan, I. (1967). *J. Biol. Chem.* **242**, 1864.

Macfarlane, M. G. (1948). *Biochem. J.* **42**, 587.

Macfarlane, M. G., and Knight, B. C. (1941). *Biochem. J.* **35**, 884.

Machovich, R., Csillag, J., and Naray, A. (1970). *FEBS Lett.* **9**, 119.

Mackinlay, A. G., and Wake, R. G. (1971). *In* "Milk Proteins: Chemistry and Molecular Biology" (H. A. McKenzie, ed.), Vol. 2, p. 175. Academic Press, New York.

McLaren, A. D., and Packer, L. (1970). *Advan. Enzymol. Relat. Areas Mol. Biol.* **33**, 245.

McPherson, J. C., Askins, R. E., and Pope, J. L. (1962). *Proc. Soc. Exp. Biol. Med.* **110**, 744.

Magee, W. L., and Thompson, R. H. S. (1960). *Biochem. J.* **77**, 526.

Magee, W. L., Gallai-Hatchard, J., Sanders, H., and Thompson, R. H. S. (1962). *Biochem. J.* **82**, 17.

Mahadevan, S., and Tappel, H. L. (1968). *J. Biol. Chem.* **243**, 2849.

Mahadevan, S., Killard, C. J., and Tappel, H. L. (1969). *Anal. Biochem.* **27**, 387.

Mahler, H. R., and Cordes, E. H. (1971). "Biological Chemistry." Harper, New York.

Malcolm, B. R. (1968). *Proc. Roy. Soc., Ser. A* **305**, 363.

Mani, V. V. S., and Lakshminarayana, G. (1970). *Biochim. Biophys. Acta* **202**, 547.

Mann, J. T., III, and Tove, S. B. (1966). *J. Biol. Chem.* **241**, 3595.

Marchis-Mouren, G., Sarda, L., and Desnuelle, P. (1959). *Arch. Biochem. Biophys.* **83**, 309.

Marinetti, G. V. (1964). *Biochim. Biophys. Acta* **84**, 55.

Marinetti, G. V. (1965). *Biochim. Biophys. Acta* **98**, 554.

Marks, T. A., Quinn, J. G., Sampugna, J., and Jensen, R. G., (1968). *Lipids* **3**, 143.

Maroux, S., Puigserver, S., Dlouha, V., Denuelle, P, de Haas, G. H., Slotboom, A. J., Bonsen, P. P. M., Nieuwenhuizen, W., and van Deenen, L. L. M. (1969). *Biochim. Biophys. Acta* **188**, 351.

Marples, E. A., and Thompson, R. H. S. (1960) *J. Biol. Chem.* **74**, 123.

Marples, R. R., Downing, D. T., and Kligman, A. M. (1971). *J. Invest. Dermatol.* **56**, 127.

Marsh, D. G., and George, J. M. (1969). *J. Biol. Chem.* **244**, 1381.

Marsh, W. H., and Fitzgerald, P. J. (1972). *J. Lipid Res.* **13**, 284.

Martin, H. F., and Peers, F. G. (1953). *Biochem. J.* **55**, 523.

Massion, G. G., and Seligson, D. (1967). *Amer. J. Clin. Pathol.* **48**, 307.

Mattson, F. H., and Beck, L. W. (1955). *J. Biol. Chem.* **214**, 115.

Mattson, F. H., and Beck, L. W. (1956). *J. Biol. Chem.* **219**, 735.

Mattson, F. H., and Volpenhein, R. A. (1962). *J. Lipid Res.* **3**, 281.

Mattson, F. H., and Volpenhein, R. A. (1966a). *J. Amer. Oil Chem. Soc.* **43**, 286.

Mattson, F. H., and Volpenhein, R. A. (1966b). *J. Lipid Res.* **7**, 536.

Mattson, F. H., and Volpenhein, R. A. (1968). *J. Lipid Res.* **9**, 79.

Mattson, F. H., and Volpenhein, R. A. (1969). *J. Lipid Res.* **10**, 271.

Mattson, F. H., and Volpenhein, R. A. (1972a). *J. Lipid Res.* **13**, 256.

Mattson, F. H., and Volpenhein, R. A. (1972b). *J. Lipid Res.* **13**, 325.

Mattson, F. H., and Volpenhein, R. A. (1972c). *J. Lipid Res.* **13**, 777.

Mattson, F. H., Volpenhein, R. A., and Benjamin, L. (1970). *J. Biol. Chem.* **245,** 5335.

Maylié, M. F., Charles, M., Sarda, L., and Desnuelle, P. (1969). *Biochim. Biophys. Acta* **178,** 196.

Maylié, M. F., Charles, M., Gache, C., and Desnuelle, P. (1971). *Biochim. Biophys. Acta* **229,** 286.

Maylié, M. F., Charles, M., and Desnuelle, P. (1972). *Biochim. Biophys. Acta* **270,** 162.

Maylié, M. F., Charles, M., Astker, M., and Desnuelle, P. (1973). *Biochem. Biophys. Res. Commun.* **52,** 291.

Mebs, D. (1970). *Int. J. Biochem.* **1,** 335.

Meinertz, H. (1971). *Progr. Biochem. Pharmacol.* **6,** 317.

Melius, P., and Doster, M. S. (1970). *Anal. Biochem.* **37,** 395.

Melius, P., and Simmons, W. S. (1965). *Biochim. Biophys. Acta* **105,** 600.

Mellors, A., and Tappel, A. L. (1967). *J. Lipid Res.* **8,** 479.

Mencher, J. R., and Alford, J. A. (1967). *J. Gen. Microbiol.* **48,** 317.

Michell, R. H., and Coleman, R. (1971). *Biochem. J.* **124,** 49P.

Michell, R. H., and Lapetina, E. G. (1972). *Nature (London), New Biol.* **240,** 258.

Miller, A. L., and Smith, L. C. (1973). *J. Biol. Chem.* **248,** 3359.

Miller, I. R., and Ruysschaert, J. M. (1971). *J. Colloid Interface Sci.* **35,** 340.

Misiorowski, R. L., and Wells, M. A. (1971). Referred to in Wells (1971c).

Mohamed, A. H., Kamel, A., and Ayobe, M. H. (1969). *Toxicon* **6,** 293.

Molnar, D. M. (1962). *J. Bacteriol.* **84,** 147.

Montgomery, M. W., and Forster, T. L. J. (1961). *Dairy Sci.* **44,** 721.

Morel, L., and Terroine, E. (1909). *C. R. Acad. Sci.* **149,** 236.

Morgan, R. G. H., and Hoffman, N. E. (1971). *Biochim. Biophys. Acta* **248,** 143.

Morgan, R. G. H., Barrowman, J., Filipek-Wender, H., and Borgström, B. (1968). *Biochim. Biophys. Acta* **167,** 355.

Morgan, R. G. H., Barrowman, J., and Borgström B. (1969). *Biochim. Biophys. Acta* **175,** 65.

Morin, R. J. (1972). *Lipids* **7,** 795.

Morin, R. J. (1973). *Biochim. Biophys. Acta* **296,** 203.

Morley, N., and Kuksis, A. (1972). *J. Biol. Chem.* **247,** 6389.

Moskowitz, M., Deverell, M. W., and McKinney, R. (1956). *Science* **123,** 1077.

Motai, H., Ichishima, E., and Yoshida, F. (1966). *Nature (London)* **210,** 308.

Mueller, G. (1972). *Clin. Chim. Acta* **39,** 89.

Mueller, J. H. (1915). *J. Biol. Chem.* **22,** 1.

Mueller, J. H. (1916). *J. Biol. Chem.* **27,** 463.

Mulder, E., van den Berg, J. W. O., and van Deenen, L. L. M. (1965). *Biochim. Biophys. Acta* **106,** 118.

Muller, L., and Alaupovic, P. (1970). *FEBS Lett.* **10,** 117.

Munjal, D., and Elliott, W. B. (1971). *Toxicon* **9,** 403.

Munjal, D., and Elliott, W. B. (1972). *Toxicon* **10,** 367.

Murthy, S. K., and Ganguly, J. (1962). *Biochem. J.* **83,** 460.

Murthy, S. K., Mahadevan, S., Sastry, P. S., and Ganguly, J. (1961). *Nature (London)* **189,** 482.

Nachbaur, J., and Vignais, P. M. (1968). *Biochem. Biophys. Res. Commun.* **33,** 315.

Nakai, S., Perrin, J. J., and Wright, V. (1970). *J. Dairy Sci.* **53**, 537.

Nashif, S. H., and Nelson, F. E. (1953a). *J. Dairy Sci.* **36**, 459.

Nashif, S. H., and Nelson, F. E. (1953b). *J. Dairy Sci.* **36**, 471.

Nashif, S. H., and Nelson, F. E. (1953c). *J. Dairy Sci.* **36**, 481.

Nedswedski, S. W. (1936). *Hoppe-Sayler's Z. Physiol. Chem.* **239**, 165.

Nedswedski, S. W. (1937). *Biokhimiya* **2**, 758.

Negishi, T., Ito, S., and Fujino, Y. (1971). *Nippon Nogei Kagaku Kaishi* **45**, 426.

Neumann, W., and Hagermann, E. (1965). *Naunyn-Schmiedebergs Arch. Exp. Pathol. Pharmakol.* **222**, 367.

Neumann, W., and Habermann, E. (1956). *In* "Venoms," Publ. No. 44, p. 171. Amer. Ass. Advan. Sci., Washington, D.C.

Newkirk, J. D., and Waite, M. (1971). *Biochim. Biophys. Acta* **225**, 224.

Nieuwenhuizen, W., and de Haas, G. H. (1972). *FEBS Abstr., 8th, 1972* p. 434.

Nieuwenhuizen, W., Kunze, H., and de Haas, G. H. (1974). *In* "Methods in Enzymology," Vol. 32. Academic Press, New York (in press).

Nikkila, E. A. (1958). *Biochim. Biophys. Acta* **27**, 612.

Nilsson, A. (1969). *Biochim. Biophys. Acta* **176**, 339.

Nilsson-Ehle, P., Belpage, P., and Borgström, B. (1971). *Biochim. Biophys. Acta* **248**, 114.

Noguchi, S. (1944). *J. Biochem. (Tokyo)* **36**, 113.

Noma, A., and Borgström, B. (1971). *Biochim. Biophys. Acta* **227**, 106.

Ogston, A. G. (1948). *Nature (London)* **196**, 963.

Ohsaka, A., and Sugahara, T. (1968). *J. Biochem. (Tokyo)* **64**, 335.

Ohta, M., Hasegawa, H., and Ohna, K. (1972). *Biochim. Biophys. Acta* **280**, 552.

Oi, S., and Satomura, Y. (1963a). *Agr. Biol. Chem.* **27**, 397.

Oi, S., and Satomura, Y. (1963b). *Agr. Biol. Chem.* **27**, 405.

Okuyama, H., and Nojima, S. (1965). *J. Biochem. (Tokyo)* **57**, 529.

Okuyama, H., and Nojima, S. (1969). *Biochim. Biophys. Acta* **176**, 120.

Okuyama, H., Lands, W. E. M., Gunstone, F. D., and Barve, J. A. (1972). *Biochemistry* **11**, 4392.

O'Leary, W. M. (1967). *In* "The Chemistry and Metabolism of Microbiol Lipids," p. 70. World Publ. Co., Cleveland, Ohio.

O'Leary, W. M., and Weld, J. T. (1964). *J. Bacteriol.* **88**, 1356.

Olive, J., and Dervichian, D. (1968). *Bull. Soc. Chim. Biol.* **50**, 1409.

Olive, J., and Dervichian, D. G. (1971). *Biochimie* **53**, 207.

Olivecrona, T., and Lindahl, U. (1969). *Acta Chem. Scand.* **23**, 3587.

Olivecrona, T., Egelrud, T., Iverius, P. H., and Lindahl, U. (1971). *Biochem. Biophys. Res. Commun.* **43**, 524.

Olney, C. E., Jensen, R. G., Sampugna J., and Quinn, J. G. (1968). *Lipids* **3**, 498.

Olson, A. C., and Alaupovic, P. (1966). *Biochim. Biophys. Acta* **125**, 185.

Ono, Y., and Nojima, S. (1969). *Biochim. Biophys. Acta* **176**, 111.

Ono, Y., and White, D. C. (1970a). *J. Bacteriol.* **103**, 111.

Ono, Y., and White, D. C. (1970b). *J. Bacteriol.* **104**, 712.

Oosterbaan, R. A., and Cohen, J. A. (1964). *In* "Structure and Activity of Enzymes" (T. W. Goodwin, J. I. Harris, and B. S. Hartley, eds.), p. 87. Academic Press, New York.

Ory, R. L. (1969). *Lipids* **4**, 177.

Ory, R. L., and Altschul, A. M. (1962). *Biochem. Biophys. Res. Commun.* **5**, 375.

Ory, R. L., and Altschul, A. M. (1964). *Biochem. Biophys. Res. Commun.* **17,** 12.

Ory, R. L., and St. Angelo, A. J. (1971). *Lipids* **6,** 54.

Ory, R. L., St. Angelo, A. J., and Altschul, A. M. (1960). *J. Lipid Res.* **1,** 208.

Ory, R. L., St. Angelo, A. J., and Altschul, A. M. (1962). *J. Lipid Res.* **3,** 99.

Ory, R. L., Barker, R. H., and Boudreaux, G. J. (1964). *Biochemistry* **3,** 2013.

Ory, R. L., St. Angelo, A. J., DeGruy, I. V., and Altschul, A. A. (1967a). *Can. J. Biochem.* **45,** 1445.

Ory, R. L., Kircher, H. W., and Altschul, A. A. (1967b). *Biochim. Biophys. Acta* **147,** 200.

Ory, R. L., Yatsu, L. Y., and Kircher, H. W. (1968). *Arch. Biochem. Biophys.* **123,** 255.

Ory, R. L., Kiser, J., and Pradel, P. A. (1969). *Lipids* **4,** 261.

Osmond, D. H., Holub, B. J., and Ross, L. J. (1973). *Can. J. Biochem.* **51,** 855.

Ota, Y., and Yamada, K. (1966a). *Agr. Biol. Chem.* **30,** 351.

Ota, Y., and Yamada, K. (1966b). *Agr. Biol. Chem.* **30,** 1030.

Ota, Y., and Yamada, K. (1967). *Agr. Biol. Chem.* **31,** 809.

Ota, Y., Suzuki, M., and Yamada, K. (1968a). *Agr. Biol. Chem.* **32,** 390.

Ota, Y., Miyairi, S., and Yamada, K. (1968b). *Agr. Biol. Chem.* **32,** 1476.

Ota, Y., Nakamiya, T., and Yamada, K. (1970). *Agr. Biol. Chem.* **34,** 1368.

Oterholm, A., Ordal, Z. J., and Witter, L. D. (1970a). *J. Dairy Sci.* **53,** 592.

Oterholm, A., Ordal, Z. J., and Witter, L. D. (1970b). *Appl. Microbiol.* **20,** 16.

Otnaess, A. B., Prydz, H., Bjoerklid, E., and Berre, A. (1972). *Eur. J. Biochem.* **27,** 238.

Otterby, D. E., Ramsay, H. A., and Wise, G. H. (1964). *J. Dairy Sci.* **47,** 997.

Ottolenghi, A. C. (1963). *Anal. Biochem.* **5,** 37.

Ottolenghi, A. C. (1965). *Biochim. Biophys. Acta* **106,** 510.

Ottolenghi, A. C. (1969). *In* "Methods in Enzymology" (J. M. Lowenstein, ed.), Vol. 14, p. 188. Academic Press, New York.

Ottolenghi, A. (1973). *Lipids* **8,** 415.

Pancholy, S. K., and Lynd, J. Q. (1972). *Phytochemistry* **11,** 643.

Parry, R. M., Jr., Chandan, R. C., and Shahani, K. M. (1966). *J. Dairy Sci.* **49,** 356.

Pastan, I., Macchia, V., and Katzen, R. (1968). *J. Biol. Chem.* **243,** 3750.

Patel, C. V., Goldberg, H. S., and Blenden, D. C. (1964). *J. Bacteriol.* **88,** 877.

Patel, C. V., Fox, P. F., and Tarassuk, N. P. (1968). *J. Dairy Sci.* **51,** 1879.

Patelski, J., Waligora, Z., and Szulc, S. (1971). *Enzyme* **12,** 299.

Patriarca, P., Beckerdite, S., and Elsbach, P. (1972a). *Biochim. Biophys. Acta* **260,** 593.

Patriarca, P., Beckerdite, S., Pettis, P., and Elsbach, P. (1927b). *Biochim. Biophys. Acta* **280,** 45.

Patrick, A. D., and Lake, B. D. (1969). *Nature (London)* **222,** 1067.

Patten, R. L., and Hollenberg, C. H. (1969). *J. Lipid Res.* **10,** 374.

Patton, J. S., and Quinn, J. G. (1972). *J. Amer. Oil Chem. Soc.* **49,** 308A.

Patton, S. (1973a). *J. Amer. Oil Chem. Soc.* **50,** 158.

Patton, S. (1973b). *J. Amer. Oil Chem. Soc.* **50,** 193.

Paysant, M., and Polonovski, J. (1966). *C. R. Acad. Sci., Ser. D* **263,** 1419.

Paysant, M., Delbauffe, D., Wald, R., and Polonovski, J. (1967). *Bull. Soc. Chim. Biol.* **49,** 169.

Paysant, M., Bitran, M., and Polonovski, J. (1969a). *C. R. Acad. Sci.* **269,** 93.

Paysant, M., Bitran, M., Etienne, J., and Polonovski, J. (1969b). *Bull. Soc. Chim. Biol.* **51**, 863.

Paysant, M., Bitran, M., Wald, R., and Polonovski, J. (1970). *Bull. Soc. Chim. Biol.* **52**, 1257.

Pieterson, W. A., Volwerk, J. J., and de Haas, G. H. (1974a). *Biochemistry* **13**, 1439.

Pieterson, W. A., Vidal, J. C., Volwerk, J. J., and de Haas, G. H. (1974b). *Biochemistry* **13**, 1455.

Pitas, R. E., and Jensen, R. G. (1970). *J. Dairy Sci.* **53**, 1083.

Pitas, R. E., Sampugna, J., and Jensen, R. G. (1967). *J. Dairy Sci.* **50**, 1332.

Pittman, R. C., Golanty, E., and Steinberg, D. (1972). *Biochim. Biophys. Acta* **270**, 81.

Plummer, T. H. (1970). Cited by Verger (1970).

Pokorny, J., Zwain, H., and Janicek, G. (1965). *Nahrung* **9**, 382.

Polonovski, J., Paysant, M., and Bure, J. (1969). *Exposés Annu. Biochim. Méd.* **29**, 267.

Pope, J. L., McPherson, J. C., and Tidwell, H. C. (1966). *J. Biol. Chem.* **241**, 2306.

Posner, I., and Morales, A. (1972). *J. Biol. Chem.* **247**, 2255.

Pritchard, E. T., and Nichol, N. E. (1964). *Biochim. Biophys. Acta* **84**, 781.

Proudlock, J. W., and Day, A. S. (1972). *Biochim. Biophys. Acta* **260**, 716.

Proulx, P., and Fung, C. K. (1969). *Can. J. Biochem.* **47**, 1125.

Proulx, P., and van Deenen, L. L. M. (1967). *Biochim. Biophys. Acta* **144**, 171.

Puhvel, S. M., and Reisner, R. M. (1972). *Arch. Dermatol.* **106**, 45.

Quarles, R. H., and Dawson, R. M. C. (1969a). *Biochem. J.* **112**, 787.

Quarles, R. H., and Dawson, R. M. C. (1969b). *Biochem. J.* **112**, 795.

Quarles, R. H., and Dawson, R. M. C. (1969c). *Biochem. J.* **113**, 697.

Quigley, T. W. C., Roe, C. E., and Dallansch, M. J. (1958). *Fed. Proc., Fed. Amer. Soc. Exp. Biol.* **17**, 292.

Quinn, J. G., Sampugna, J., and Jensen, R. G. (1967). *J. Amer. Oil. Chem. Soc.* **44**, 439.

Quinn, P. J. (1973). *Biochem. J.* **133**, 273.

Rachford, B. K. (1891). *J. Physiol. (London)* **12**, 72.

Ragab, H., Beck, C., Dillard, C., and Tappel, A. L. (1967). *Biochim. Biophys. Acta* **148**, 501.

Rahman, Y. E., and Verhagen, J. (1970). *Biochem. Biophys. Res. Commun.* **38**, 670.

Rahman, Y. E., Verhagen, J., and van der Wiel, D. F. M. (1969). *Biochem. Biophys. Res. Commun.* **38**, 670.

Ramachandran, S., Yip, Y. K., and Wagle, S. R. (1970). *Eur. J. Biochem.* **12**, 201.

Rao, R. H., and Subrahmanyam, D. (1969a). *Life Sci.* **8**, 447.

Rao, R. H., and Subrahmanyam, D. (1969b). *J. Lipid Res.* **10**, 636.

Rao, R. H., and Subrahmanyam, D. (1970). *Arch. Biochem. Biophys.* **140**, 443.

Raybin, D. M., Bertsch, L. L., and Kornberg, A. (1972). *Biochemistry* **11**, 1754.

Reichl, D. (1972). *Biochem. J.* **128**, 79.

Reid, T. W., and Wilson, I. B. (1971). *In* "The Enzymes" (P. D. Boyer, ed.), 3rd ed., Vol. 4, p. 373. Academic Press, New York.

Reisner, R. M., Silver, S. Z., Puhvel, M., and Sternberg, T. H., (1968). *J. Invest. Dermatol.* **51**, 190.

Renshaw, E. C., and San Clemente, C. L. (1967). *Develop. Ind. Microbiol.* **8**, 206.

Ribeiro, L. P. (1971). *Biochimie* **53**, 865.

Richardson, G. H., Nelson, J. H., and Farnham, M. G. (1971). *J. Dairy Sci.* **54**, 643.

Richter, R. L., and Randolph, H. E. (1971). *J. Dairy Sci.* **54,** 1275.

Riddle, M. C., and Glomset, J. A. (1973). *Fed. Proc., Fed. Amer. Soc. Exp. Biol.* **32,** 363 (abstr.).

Rimon, A., and Shapiro, B. (1959). *Biochem. J.* **71,** 620.

Rizack, M. A. (1961). *J. Biol. Chem.* **236,** 657.

Robertson, A. F., and Lands, W. E. M. (1962). *Biochemistry* **1,** 804.

Robertson, J. A., Harper, W. J., and Gould, I. A. (1966). *J. Dairy Sci.* **49,** 1384.

Robinson, D. S. (1963). *Advan. Lipid Res.* **1,** 134.

Robinson, D. S. (1965). *In* "Handbook of Physiology" (Amer. Physiol. Soc., J. Field, ed.), Sect. 5, p. 297. Williams & Wilkins, Baltimore, Maryland.

Robinson, D. S. (1970). *Compr. Biochem.* **18,** 51.

Robinson, D. S., and French, J. E. (1960). *Pharmacol. Rev.* **12,** 241.

Rodbell, M. (1970). *Advan. Biochem. Psychopharmacol.* **3,** 185.

Rodnight, R. (1956). *Biochem. J.* **63,** 223.

Roholt, O. A., and Schlamowitz, M. (1958). *Arch. Biochem. Biophys.* **77,** 510.

Roholt, O. A., and Schlamowitz, M. (1961). *Arch. Biochem. Biophys.* **94,** 364.

Rona, P., and Lasnitzki, A. (1924). *Biochem. Z.* **152,** 504.

Rona, P., and Michaelis, L. (1911). *Biochem. Z.* **31,** 345.

Rosenberg, E. W. (1969). *Annu. Rev. Med.* **20,** 201.

Rosendal, K., and Bulow, P. (1965). *J. Gen. Microbiol.* **41,** 349.

Rosenthal, A. F., and Geyer, R. P. (1962). *Arch. Biochem. Biophys.* **96,** 240.

Rosenthal, A. F., and Han, S. C. H. (1970). *Biochim. Biophys. Acta* **218,** 213.

Rosenthal, A. F., and Pousada, M. (1968). *Biochim. Biophys. Acta* **164,** 226.

Rosenthal, A. F., Chodsky, S. V., and Han, S. C. H. (1969). *Biochim. Biophys. Acta* **187,** 385.

Rossi, C. R., Sartorelli, L., Tato, L., Baretta, L., and Siliprandi, N. (1965). *Biochim. Biophys. Acta* **98,** 207.

Rottem, S., and Razin, S. (1964). *J. Gen. Microbiol.* **37,** 123.

Ruebsamen, K., Breithaupt, H., and Habermann, E. (1971). *Naunyn-Schmiedebergs Arch. Pharmakol. Exp. Pathol.* **270,** 274.

St. Clair, R. W., Clarkson, T. B., and Lofland, H. B. (1972). *Circulation Res.* **31,** 664.

Saito, K., and Hanahan, D. J. (1962). *Biochemistry* **1,** 521.

Saito, K., and Mukoyama, K. (1968). *Biochim. Biophys. Acta* **164,** 596.

Saito, K., and Sato, K. (1968a). *Biochim. Biophys. Acta* **151,** 706.

Saito, K., and Sato, K. (1968b). *J. Biochem. (Tokyo)* **64,** 293.

Saito, K., Okada, Y., and Kawasaki, N. (1972). *J. Biochem. (Tokyo)* **72,** 213.

Saito, Z. (1963). *Jap. J. Zootech. Sci.* **34,** 94.

Saito, Z., and Igarashi, Y. (1971). *Bull. Fac. Agr., Hirosaki Univ.* **17,** 126.

Sakhibov, D. N., Sohokin, V. M., and Yukel'son, L. Y. (1970). *Biokhimiya* **35,** 13.

Salach, J. I., Turini, P., Hauber, J., Seng, R., Tisdale, H. D., and Singer, T. P. (1968). *Biochem. Biophys. Res. Commun.* **33,** 936.

Salach, J. I., Turini, P., Seng, R., Hauber, J., and Singer, T. P. (1971a). *J. Biol. Chem.* **246,** 331.

Salach, J. I., Seng, R., Tisdale, H. D., and Singer, T. P. (1971b). *J. Biol. Chem.* **246,** 340.

Samejima, Y., Iwanaga, S., Suzuki, T., and Kawauchi, S. (1970). *Biochim. Biophys. Acta* **221,** 417.

Sampugna, J., and Jensen, R. G. (1968). *Lipids* **3,** 519.

Sampugna, J., Quinn, J. G., Pitas, R. E., Carpenter, D. L., and Jensen, R. G. (1967). *Lipids* **2**, 397.

San Clemente, C. L., and Vadehra, D. V. (1967). *Appl. Microbiol.* **15**, 110.

Sarda, L., and Desnuelle, P. (1958). *Biochim. Biophys. Acta* **30**, 513.

Sarda, L., Marchis-Mouren, G., and Desnuelle, P. (1957). *Biochim. Biophys. Acta* **24**, 425.

Sarda, L., Ailhaud, G., and Desnuelle, P. (1960). *Biochim. Biophys. Acta* **37**, 570.

Sarda, L., Maylié, M. F., Roger, J., and Desnuelle, P. (1964). *Biochim. Biophys. Acta* **89**, 183.

Sarzala, M. G. (1969). *Bull. Acad. Pol. Sci., Ser. Sci. Biol.* **17**, 285.

Sastry, P. S., and Kates, M. (1964). *Biochemistry* **3**, 1280.

Sastry, P. S., and Kates, M. (1969). *In* "Methods in Enzymology" (J. M. Lowenstein, ed.), Vol. 14, p. 204. Academic Press, New York.

Savary, P. (1971). *Biochim. Biophys. Acta* **248**, 149.

Savary, P. (1972). *Biochim. Biophys. Acta* **270**, 463.

Savary, P., and Desnuelle, P. (1956). *Biochim. Biophys. Acta* **21**, 349.

Savary, P., Flanzy, J., and Desnuelle, P. (1958). *Bull. Soc. Chim. Biol.* **40**, 637.

Scandella, C. J., and Kornberg, A. (1971). *Biochemistry* **10**, 4447.

Scanu, A. M. (1966). *Science* **153**, 640.

Scanu, A. M. (1967). *J. Biol. Chem.* **242**, 711.

Scanu, A. M., and Page, I. H. (1959). *J. Exp. Med.* **109**, 239.

Scanu, A. M., and Wisdom, C. (1972). *Annu. Rev. Biochem.* **41**, 704.

Scanu, A. M., van Deenen, L. L. M., and de Haas, G. H. (1969). *Biochim. Biophys. Acta.* **18**, 471.

Scherphof, G. L., and van Deenen, L. L. M. (1965). *Biochim. Biophys. Acta* **98**, 204.

Scherphof, G. L., Waite, M., and van Deenen, L. L. M. (1966). *Biochim. Biophys. Acta* **125**, 406.

Schmidt, G., Bessman, M. J., and Thannhauser, S. J. (1957). *Biochim. Biophys. Acta* **23**, 127.

Schmidt-Nielsen, K. (1946). *Acta Physiol. Scand.* **12**, Suppl. 37, 1.

Schnatz, D. J. (1964). *Biochim. Biophys. Acta* **116**, 243.

Schnatz, D. J. (1966). *Biochim. Biophys. Acta* **166**, 243.

Schnatz, D. J., and Cortner, J. A. (1967). *J. Biol. Chem.* **242**, 3850.

Schnatz, D. J., and Cummiskey, T. C. (1969). *Life Sci.* **8**, 1273.

Schneider, P. B., and Kennedy, E. P. (1967). *J. Lipid Res.* **8**, 202.

Schoefl, G. I., and French, J. E. (1968). *Proc. Roy. Soc., Ser. B* **169**, 153.

Schønheyder, F., and Volqvartz, K. (1944a). *Enzymologia* **11**, 178.

Schønheyder, F., and Volqvartz, K. (1944b). *Acta Physiol. Scand.* **7**, 376.

Schønheyder, F., and Volqvartz, K. (1945). *Acta Physiol. Scand.* **9**, 57.

Schønheyder, F., and Volqvartz, K. (1952). *Biochim. Biophys. Acta* **8**, 407.

Schønheyder, F., and Volqvartz, K. (1954). *Biochim. Biophys. Acta* **15**, 288.

Schoor, W. P., and Melius, P. (1969). *Biochim. Biophys. Acta* **187**, 186.

Schoor, W. P., and Melius, P. (1970). *Biochim. Biophys. Acta* **212**, 173.

Schotz, M. C., and Garfinkel A. S. (1972a). *J. Lipid Res.* **13**, 824.

Schotz, M. C., and Garfinkel, A. S. (1972b). *Biochim. Biophys. Acta* **270**, 472.

Schotz, M. C., Garfinkel, A. S., Huebotter, R. J., and Stewart, J. E. (1970). *J. Lipid Res.* **11**, 68.

Schultz, J. H. (1912). *Biochem. Z.* **42**, 255.

Schumaker, V. N., and Adams, G. H. (1969). *Annu. Rev. Biochem.* **38**, 113.

Schwartz, J. P., and Jungas, R. L. (1971). *J. Lipid Res.* **12,** 553.

Scow, R. O., Hamosh, M., Blanchette-Mackie, E. J., and Evans, A. J. (1972). *Lipids* **7,** 497.

Sedgwick, B., and Hübscher, G. (1965). *Biochim. Biophys. Acta* **106,** 63.

Sedgwick, B., and Hübscher, G. (1967). *Biochim. Biophys. Acta* **144,** 397.

Seitz, E. W. (1973). *Lipids* **8,** in press.

Sémériva, M., and Dufour, C. (1972). *Biochim. Biophys. Acta* **260,** 393.

Sémériva, M., Benzonana, G., and Desnuelle, P. (1967a). *Biochim. Biophys. Acta* **144,** 703.

Sémériva, M., Benzonana, G., and Desnuelle, P. (1967b). *Bull. Soc. Chim. Biol.* **49,** 71.

Sémériva, M., Benzonana, G., and Desnuelle, P. (1969). *Biochim. Biophys. Acta* **191,** 598.

Sémériva, M., Dufour, C., and Desnuelle, P. (1971). *Biochemistry* **10,** 2143.

Sémériva, M., Dufour, C., and Desnuelle, P. (1972). *FEBS Abstr., 8th, 1972* p. 433.

Senior, J. R., and Isselbacher, K. J. (1963) *J. Clin. Invest.* **42,** 187.

Sgoutas, D. (1968). *Biochim. Biophys. Acta* **164,** 317.

Shafrir, E., and Biale, Y. (1970). *Eur. J. Clin. Invest.* **1,** 19.

Shah, D. B., and Wilson, J. B. (1963). *J. Bacteriol.* **85,** 516.

Shah, D. B., and Wilson, J. B. (1965). *J. Bacteriol.* **89,** 949.

Shah, D. O., and Schulman, J. H. (1967). *J. Colloid Interface Sci.* **25,** 107.

Shah, S. N., Lossow, W. J., and Chaikoff, I. L. (1965). *J. Lipid Res.* **6,** 228.

Shahani, K. M. (1966). *J. Dairy Sci.* **49,** 907.

Shahani, K. M., and Chandan, R. C. (1962). *J. Dairy Sci.* **45,** 1178.

Shahani, K. M., Harper, W. J., Jensen, R. G., Parry, R. M., and Zittle, C. A. (1973). *J. Dairy Sci.* **56,** 531.

Shapiro, B. (1953). *Biochem. J.* **53,** 663.

Shastry, B. S., and Rao, M. R. (1971). *Indian J. Biochem. Biophys.* **8,** 327.

Shemanova, G. F., Vlasova, E. V., and Tsvetkov, V. S. (1965). *Biochemistry (USSR)* **30,** 634.

Shemanova, G. F., Vlasova, E. V., Tsvetkov, V. S., Logunov, A. I., and Levin, F. B. (1968). *Biochemistry (USSR)* **33,** 110.

Shibko, S., and Tappel, A. L. (1964). *Arch. Biochem. Biophys.* **106,** 259.

Shiloah, J., Klibansky, C., de Vries, A., and Berger, A. (1973). *J. Lipid Res.* **14,** 267.

Shipe, W. F., Jr. (1951). *Arch. Biochem.* **30,** 165.

Shipolini, R. A., Callewaert, G. L., Cottrell, R. C., Doonan, S., Vernon, C. A., and Banks, B. E. C. (1971a). *Eur. J. Biochem.* **20,** 459.

Shipolini, R. A., Callewaert, G. L., Cottrell, R. C., and Vernon, C. A. (1971b). *FEBS Lett.* **17,** 39.

Shore, B. (1955). *Proc. Soc. Exp. Biol. Med.* **88,** 73.

Shore, B., and Shore, V. (1961). *Amer. J. Physiol.* **201,** 915.

Shore, B., and Shore, V. (1969). *Biochemistry* **8,** 4510.

Shyamala, G., Lossow, W. J., and Chaikoff, I. L. (1965). *Proc. Soc. Exp. Biol. Med.* **118,** 138.

Siewert, K. L., and Otterby, D. E. (1968). *J. Dairy Sci.* **51,** 1305.

Singer, T. P. (1948). *J. Biol. Chem.* **174,** 11.

Singer, T. P., and Hofstee, B. H. J. (1948a). *Arch. Biochem.* **18,** 229.

Singer, T. P., and Hofstee, B. H. J. (1948b). *Arch. Biochem.* **18**, 245.

Skean, J. D., and Overcast, W. W. (1961). *J. Dairy Sci.* **44**, 823.

Skipski, V. P. (1972). *In* "Blood Lipids and Lipoproteins: Quantitation, Composition, and Metabolism" (G. J. Nelson, ed.), p. 471. Wiley (Interscience), New York.

Slein, M. W., and Logan, G. J., Jr. (1965). *J. Bacteriol.* **90**, 69.

Sloan, H. R. (1972). *In* "Methods in Enzymology," Vol. 28, Part B, p. 874. Academic Press, New York.

Sloan, H. R., and Fredrickson, D. S. (1972a). *In* "The Metabolic Basis of Inherited Disease" (J. B. Stanbury, J. B. Wyngaarden, and D. S. Fredrickson, eds.), 3rd ed., p. 808. McGraw-Hill, New York.

Sloan, H. R., and Fredrickson, D. S. (1972b). *J. Clin. Invest.* **51**, 1923.

Sloane-Stanley, G. H. (1953). *Biochem. J.* **53**, 513.

Slotboom, A. J., de Haas, G. H., Bonsen, P. P. M., Burbach-Westerhuis, G. J., and van Deenen, L. L. M. (1970a). *Chem. Phys. Lipids* **4**, 15.

Slotboom, A. J., de Haas, G. H., Burbach-Westerhuis, G. J., and van Deenen, L. L. M. (1970b). *Chem. Phys. Lipids* **4**, 30.

Slotta, K. H. (1960). *In* "The Enzymes", (P. D. Boyer, H. Lardy, and K. Myrback, eds.), 2nd rev. ed., Vol. 4, p. 551. Academic Press, New York.

Small, D. M. (1971). *In* "The Bile Acids" (P. P. Nair and D. Kritchevsky, eds.), Vol. I, p. 249. Plenum, New York.

Smith, A. D., and Winkler, H. (1968). *Biochem. J.* **108**, 867.

Smith, A. D., Gul, S., and Thompson, R. H. S. (1972). *Biochim. Biophys. Acta* **289**, 147.

Smith, J. B., and Silver, M. J. (1973). *Biochem. J.* **131**, 615.

Smith, L. C. (1972). *J. Lipid Res.* **13**, 769.

Smith, L. D., and Gardner, M. V. (1950). *Arch. Biochem.* **25**, 54.

Smith, R. H. (1954). *Biochem. J.* **56**, 240.

Smith, S. W., Weiss, S. B., and Kennedy, E. P. (1957). *J. Biol. Chem.* **228**, 915.

Soderling, T. R., Corbin, J. D., and Park, C. R. (1973). *J. Biol. Chem.* **248**, 1822.

Somkuti, G. A., and Babel, F. J. (1968). *Appl. Microbiol.* **16**, 617.

Somkuti, G. A., Babel, F. J., and Somkuti, A. C. (1969). *Appl. Microbiol.* **17**, 606.

Souček, A., and Souckova, A. (1966). *J. Hyg., Epidemiol., Microbiol., Immunol.* **10**, 125.

Souček, A., Michalec, C., and Souckova, A. (1971). *Biochim. Biophys. Acta* **227**, 116.

Sperry, W. M., and Webb, M. (1950). *J. Biol. Chem.* **187**, 97.

Stanacev, N. Z., and Stuhne-Sekalec, L. (1970). *Biochim. Biophys. Acta* **210**, 350.

Stanacev, N. Z., Stuhne-Sekalec, L., and Domazet, Z. (1973). *Can. J. Biochem.* **51**, 743.

Stauffer, C. E., and Glass, R. L. (1966). *Cereal Chem.* **43**, 644.

Steinberg, D. (1972). *In* "Pharmacological Control of Lipid Metabolism" (W. L. Holmes, R. Paoletti, and D. Kritchevsky, eds.), p. 77. Plenum, New York.

Steinberg, D., and Vaughan, M. (1965). *In* "Handbook of Physiology" (Amer. Physiol. Soc., J. Field, ed.), Sect. 5, p. 335. Williams & Wilkins, Baltimore, Maryland.

Stevenson, E. (1969). *J. Insect Physiol.* **15**, 1537.

Stevenson, E. (1972). *J. Invest. Physiol.* **18**, 1751.

Stillway, L. W., and Lane, C. E. (1971). *Toxicon* **9**, 193.

Stoffel, W., and Greten, H. (1967). *Hoppe-Seyler's Z. Physiol. Chem.* **348**, 1145.

Stoffel, W., and Trabert, U. (1969). *Hoppe-Seyler's Z. Physiol. Chem.* **350,** 836.
Stokke, K. T. (1972a). *Biochim. Biophys. Acta* **270,** 156.
Stokke, K. T. (1972b). *Biochim. Biophys. Acta* **280,** 329.
Strand, O., Vaughan, M., and Steinberg, D. (1964). *J. Lipid Res.* **5,** 554.
Strickland, K. P., Subrahmanyan, D., Pritchard, E. T., Thompson, N., and Rossiter, R. J. (1963). *Biochem. J.* **87,** 128.
Struijk, C. B., Houtsmuller, U. M. T., Jansen, H., and Hulsman, W. C. (1973). *Biochim. Biophys. Acta* **296,** 253.
Subbaiah, P. V., and Ganguly, J. (1970). *Biochem. J.* **118,** 233.
Swell, L., and Treadwell, C. R. (1955). *J. Biol. Chem.* **212,** 141.
Swell, L., Field, H., Jr., and Treadwell, C. R. (1953). *Proc. Soc. Exp. Biol. Med.* **84,** 417.
Swell, L., Field, H., Jr., and Treadwell, C. R. (1954). *Proc. Soc. Exp. Biol. Med.* **87,** 216.
Tallent, W. H., and Kleiman, R. (1968). *J. Lipid Res.* **9,** 146.
Tallent, W. H., Kleiman, R., and Cope, D. G. (1966). *J. Lipid Res.* **7,** 531.
Tarassuk, N. P., and Frankel, E. M. (1957). *J. Dairy Sci.* **40,** 418.
Tarassuk, N. P., and Yaguchi, M. (1959). *J. Dairy Sci.* **42,** 864.
Tarassuk, N. P., Nickerson, T. A., and Yaguchi, M. (1964). *Nature (London)* **201,** 298.
Tattrie, N. H. (1959). *J. Lipid Res.* **1,** 60.
Tattrie, N. H., and Cyr, R. (1963). *Biochim. Biophys. Acta* **70,** 693.
Tattrie, N. H., Bailey, R. A., and Kates, M. (1958). *Arch. Biochem. Biophys.* **78,** 319.
Tavener, R. J. A., and Laidman, D. L. (1972). *Phytochemistry* **11,** 989.
Teale, J. D., Davies, T., and Hall, D. A. (1972). *Biochem. Biophys. Res. Commun.* **47,** 234.
Tejasen, P., and Ottolenghi, A. (1970). *Toxicon* **8,** 225.
Thanki, R. J., Patel, K. C., and Patel, R. D. (1970). *Lipids* **5,** 519.
Thiele, F. H. (1913). *Biochem. J.* **7,** 275.
Thomas, A. E., III, Scharoun, J. E., and Ralston, H. (1965). *J. Amer. Oil Chem. Soc.* **42,** 789.
Thomas, E. L., Nicolsen, A. J., and Olson, J. C., Jr. (1955). *Amer. Milk Rev.* **17,** 50.
Thompson, G. A., Jr. (1969). *J. Protozool.* **16,** 397.
Thompson, W. (1967). *Can. J. Biochem.* **45,** 853.
Thompson, W., and Dawson, R. M. C. (1964a). *Biochem. J.* **91,** 233.
Thompson, W., and Dawson, R. M. C. (1964b). *Biochem. J.* **91,** 237.
Tidwell, H. C., and Johnston, J. M. (1960). *Arch. Biochem. Biophys.* **89,** 79.
Tietz, N. W., and Fiereck, E. A. (1972). *Standard Methods Clin. Chem.* **7,** 19.
Tinker, D. O., and Pinteric, L. (1971). *Biochemistry* **10,** 860.
Tomizuka, N., Ota, Y., and Yamada, K. (1966a). *Agr. Biol. Chem.* **30,** 576.
Tomizuka, N., Ota, Y., and Yamada, K. (1966b). *Agr. Biol. Chem.* **30,** 1090.
Tookey, H. L., and Balls, A. K. (1956). *J. Biol. Chem.* **218,** 213.
Tornqvist, H., Krabisch, R., and Belfrage, P. (1972). *J. Lipid Res.* **13,** 424.
Tou, J. S., Hurst, M. W., Baricos, W. H., and Huggins, C. G. (1973). *Arch. Biochem. Biophys.* **154,** 593.
Treadwell, C. R., and Vahouny, G. V. (1968). *In* "Handbook of Physiology", (Amer. Physiol. Soc., J. Field, ed.), Sect. 6, Vol. III, p. 1407. Williams & Wilkins, Baltimore, Maryland.

Troller, J. A., and Bozeman, M. A. (1970). *Appl. Microbiol.* **20**, 480.

True, L. C., Dooley, S. M., and Mickle, J. B. (1969). *J. Dairy Sci.* **52**, 2046.

Tsai, S. C., and Vaughan, M. (1972). *J. Biol. Chem.* **247**, 6253.

Tsai, S. C., Belfrage, P., and Vaughan, M. (1970). *J. Lipid Res.* **11**, 466.

Tseng, T.-C., and Bateman, D. F. (1969). *Phytopathology* **59**, 359.

Tu, A. T., Passey, R. B., and Toom, P. M. (1970). *Arch. Biochem. Biophys.* **140**, 96.

Tume, R. K., and Day, A. Y. (1970). *RES, J. Reticuloendothel. Soc.* **7**, 338.

Tzur, R., and Shapiro, B. (1972). *Biochim. Biophys. Acta* **280**, 290.

Tsujisaka, Y., Iwai, M., and Tominaga, Y. (1973). *Agr. Biol. Chem.*, **37**, 1457.

Uthe, J. F., and Magee, W. L. (1971). *Can. J. Biochem.* **49**, 776.

Uwatoko-Setoguchi, Y. (1970). *Acta Med. Univ. Kagoshima.* **12**, 73.

Uwatoko-Setoguchi, Y., and Ohbo, F. (1969). *Acta Med. Univ. Kagoshima.* **11**, 139.

Vadehra, D. V., and Harmon, L. G. (1965). *Appl. Microbiol.* **13**, 335.

Vadehra, D. V., and Harmon, L. G. (1967a). *Appl. Microbiol.* **15**, 292.

Vadehra, D. V., and Harmon, L. G. (1967b). *Appl. Microbiol.* **15**, 480.

Vahouny, G. V., and Treadwell, C. R. (1964). *Proc. Soc. Exp. Biol. Med.* **116**, 496.

Vahouny, G. V., and Treadwell, C. R. (1968). *Methods Biochem. Anal.* **16**, 219.

Vahouny, G. V., Borja, C. R., and Treadwell, C. R. (1964a). *Arch. Biochem. Biophys.* **106**, 440.

Vahouny, G. V., Weersing, S., and Treadwell, C. R. (1964b). *Arch. Biochem. Biophys.* **107**, 7.

Vahouny, G. V., Weersing, S., and Treadwell, C. R. (1964c). *Biochem. Biophys. Res. Commun.* **15**, 224.

Vahouny, G. V., Weersing, S., and Treadwell, C. R. (1965). *Biochim. Biophys. Acta* **98**, 607.

Vahouny, G. V., Kothari, H., and Treadwell, C. R. (1967). *Arch. Biochem. Biophys.* **121**, 242.

van Deenen, L. L. M. (1964). *Metab. Physiol. Significance Lipids, Proc. Advan. Study Course, 1963* p. 155.

van Deenen, L. L. M., and de Haas, G. H. (1963). *Biochim. Biophys. Acta* **70**, 538.

van Deenen, L. L. M., and de Haas, G. H. (1966). *Annu. Rev. Biochem.* **35**, 157.

van Deenen, L. L. M., de Haas, G. H., Heemskerk, C. H. T., and Meduski, J. (1961). *Biochem. Biophys. Res. Commun.* **4**, 183.

van Deenen, L. L. M., de Haas, G. H., and Heemskerk, C. H. T. (1963). *Biochim. Biophys. Acta* **67**, 295.

van den Bosch, H., and van Deenen, L. L. M. (1964). *Biochim. Biophys. Acta* **84**, 234.

van den Bosch, H., and van Deenen, L. L. M. (1965). *Biochim. Biophys. Acta* **106**, 326.

van den Bosch, H., Postema, N. M., de Haas, G. H., and van Deenen, L. L. M. (1965). *Biochim. Biophys. Acta* **98**, 657.

van den Bosch, H., van der Elzen, H. M., and van Deenen, L. L. M. (1967). *Lipids* **2**, 279.

van den Bosch, H., Aarsman, A. J., Slotboom, A. J., and van Deenen, L. L. M. (1968). *Biochim. Biophys. Acta* **164**, 215.

van den Bosch, H., Aarsman, A. J., de Jong, J. G. N., and van Deenen, L. L. M., (1973). *Biochim. Biophys. Acta* **296**, 94.

van den Bosch. H., van Golde, L. M. G., and van Deenen, L. L. M. (1972). *In* "Reviews of Physiology, Biochemistry and Experimental Pharmacology," p. 61. Springer-Verlag, Berlin and New York.

Vandermeers, A., and Christophe, J. (1968). *Biochim. Biophys. Acta* **154,** 11.

Van Golde, L. M. G., McElhaney, R. N., and van Deenen, L. L. M. (1971a). *Biochim. Biophys. Acta* **231,** 245.

Van Golde, L. M. G., Fleischer, B., and Fleischer, S. (1971b). *Biochim. Biophys. Acta* **249,** 318.

Van Heyningen, W. E. (1941). *Biochem. J.* **35,** 1246.

Vaskovsky, V. E., Gorovoi, P. G., and Suppes, Z. S. (1972). *Int. J. Biochem.* **3,** 647.

Vaughan, M., and Steinberg, D. (1965). *In* "Handbook of Physiology" (Amer. Physiol. Soc., J. Field, ed.), Sect. 5, p. 239. Williams & Wilkins, Baltimore, Maryland.

Vaughan, M., Berger, J. E., and Steinberg, D. (1964). *J. Biol. Chem.* **239,** 401.

Vaughan, M., Steinberg, D., Lieberman, F., and Stanley, S. (1965). *Life Sci.* **4,** 1077.

Vavrinkova, H., and Mosinger, B. (1965). *Physiol. Bohemoslov.* **14,** 46.

Verger, R. (1970). Thesis, Faculté des Sciences de Marseille.

Verger, R., and de Haas, G. H. (1973). *Chem. Phys. Lipids* **10,** 127.

Verger, R., de Haas, G. H., Sarda, L., and Desnuelle, P. (1969). *Biochim. Biophys. Acta* **188,** 272.

Verger, R., Sarda, L., and Desnuelle, P. (1971). *Biochim. Biophys. Acta* **242,** 580.

Verger, R., Mieras, M. C. E., and de Haas, G. H. (1973). *J. Biol. Chem.* **248,** 4023.

Verzar, F., and Laszt, L. (1934). *Biochem. Z.* **270,** 24.

Victoria, E. J., van Golde, L. M. G., Hostetler, K. Y., Scherphof, G. L., and van Deenen, L L. M. (1971). *Biochim. Biophys. Acta* **239,** 443.

Vidal, J. C., and Stoppani, A. O. M. (1966). *Rev. Soc. Argent. Biol.* **42,** 138.

Vidal, J. C., and Stoppani, A. O. M. (1970). *Experientia* **26,** 831.

Vidal, J. C., and Stoppani, A. O. M. (1971a). *Arch. Biochem. Biophys.* **145,** 543.

Vidal, J. C., and Stoppani, A. O. M. (1971b). *Arch. Biochem. Biophys.* **147,** 66.

Vidal, J. C., Cattaneo, P., and Stoppani, A. O. M. (1972a). *Arch. Biochem. Biophys.* **151,** 168.

Vidal, J. C., Molina, H., and Stoppani, A. O. M. (1972b). *Acta Physiol. Lat. Amer.* **22,** 91.

Vignais, P. M., and Nachbaur, J. (1968). *Biochem. Biophys. Res. Commun.* **33,** 307.

Vignais, P. M., Nachbaur, J., Vignais, P. V., and Andre, J. (1968). *Mitochondria: Struct. Funct., Fed. Eur. Biochem. Soc., Meet., 5th, 1968* p. 43.

Vignais, P. M., Nachbaur, J., Huet, J., and Vignais, P. V. (1970). *Biochem. J.* **116,** 42P.

Vogel, W. C., and Bierman, E. L. (1965). *Fed. Proc., Fed. Amer. Soc. Exp. Biol.* **24,** 439. (abstr.).

Vogel, W. C., and Bierman, E. L. (1967). *J. Lipid Res.* **8,** 46.

Vogel, W. C. and Zieve, L. (1963). *Clin. Chem.* **9,** 168.

Vogel, W. C., Ryan, W. G., Koppel, J. L., and Olivin, J. H. (1965). *J. Lipid Res.* **6,** 335.

Vogel, W. C., Brunzell, J. D., and Bierman, E. L. (1971). *Lipids* **11**, 805.

Volwerk, J. J., Pieterson, W. A., and de Haas, G. H. (1974). *Biochemistry* **13**, 1446.

Vyvoda, O. S., and Rowe, C. E. (1973). *Biochem. J.* **132**, 233.

Wahlström, A. (1971). *Toxicon* **9**, 45.

Waite, M. (1973). *Fed. Proc., Fed. Amer. Soc. Exp. Biol.* **32**, 561 (abstr.).

Waite, M., and Sisson, P. (1971). *Biochemistry* **10**, 2377.

Waite, M., and van Deenen, L. L. M. (1967). *Biochim. Biophys. Acta* **137**, 498.

Waite, M., Scherphof, G. L., Boshouwers, F. G., and van Deenen, L. L. M. (1969). *J. Lipid Res.* **10**, 411.

Wallach, D. P. (1968). *J. Lipid Res.* **9**, 200.

Wallach, D. P., Ko, H., and Marshall, N. B. (1962). *Biochim. Biophys. Acta* **59**, 690.

Wang, M. C., and Meng, H. C. (1973). *Fed. Proc., Fed. Amer. Soc. Exp. Biol.* **32**, 363 (abstr.).

Weaber, K., Freedman, R., and Eudy, W. W. (1971). *Appl. Microbiol.* **21**, 639.

Webster, G. R. (1970). *Biochem. J.* **117**, 10P.

Webster, G. R., and Cooper, M. F. (1968). *J. Neurochem.* **15**, 795.

Weglicki, W. B., Waite, M., Sisson, P., and Shohet, S. B. (1971). *Biochim. Biophys. Acta* **231**, 512.

Weinreb, N. J., Brady, R. O., and Tappel, A. L. (1968). *Biochim. Biophys. Acta* **159**, 141.

Weiss, H., Spiegel, H. E., and Titus E. (1959). *Nature (London)* **183**, 1393.

Weiss, S. B., Smith, S. W., and Kennedy, E. P. (1956). *Nature (London)* **178**, 594.

Weld, J. T., and O'Leary, W. M. (1963). *Nature (London)* **199**, 510.

Wells, M. A. (1971a). *Biochemistry* **10**, 4074.

Wells, M. A. (1971b). *Biochemistry* **10**, 4078.

Wells, M. A. (1971c). *Biochim. Biophys. Acta* **248**, 80.

Wells, M. A. (1972). *Biochemistry* **11**, 1030.

Wells, M. A. (1973). *Biochemistry* **12**, 1086.

Wells, M. A., and Hanahan, D. J. (1969). *Biochemistry* **8**, 414.

Whayne, T. F., Jr., and Felts, J. M. (1970). *Circ. Res.* **27**, 941.

Whayne, T. F., Jr., and Felts, J. M. (1972). *Clin. Biochem.* **5**, 109.

White, D. A., Pounder, D. J., and Hawthorne, J. N. (1971). *Biochim. Biophys. Acta* **242**, 99.

Wilcox, J. C., Nelson, W. O., and Wood, W. A. (1955). *J. Dairy Sci.* **38**, 775.

Wilgram, G. F., and Kennedy, E. P. (1963). *J. Biol. Chem.* **238**, 2615.

Wills, E. D. (1954). *Biochem. J.* **57**, 109.

Wills, E. D. (1955). *Biochem. J.* **60**, 529.

Wills, E. D. (1960). *Biochim. Biophys. Acta* **40**, 481.

Wills, E. D. (1961b). *In* "The Enzymes of Lipid Metabolism" (P. Desnuelle, ed.), p. 13. Pergamon, Oxford.

Wills, E. D. (1965). *Advan. Lipid Res.* **3**, 197.

Willstätter, R., and Memmen, F. (1928). *Hoppe-Seyler's Z. Physiol. Chem.* **133**, 229.

Willstätter, R., Waldschmitz-Leitz, E., and Memmen, F. (1923). *Hoppe-Seyler's Z. Physiol. Chem.* **125**, 93.

Wilson, D. E., Flowers, C. M., and Reading, J. C. (1973). *J. Lipid Res.* **14**, 124.

Winkler, H., Smith, A. D., Dubois, F., and van den Bosch, H. (1967). *Biochem. J.* **105**, 38C.

Wittcoff, H. (1951). "The Phosphatides." Van Nostrand-Reinhold, Princeton, New Jersey.

Wittich, K. A., and Schmidt, H. (1969). *Enzymol. Biol. Clin.* **10,** 477.

Woelk, H., and Debuch, H. (1971). *Hoppe-Seyler's Z. Physiol. Chem.* **352,** 1275.

Woelk, H., Furniss, H., and Debuch, H. (1972). *Hoppe-Seyler's Z. Physiol. Chem.* **353,** 1111.

Wright, W. R., and Tove, S. B. (1967). *Biochim. Biophys. Acta* **137,** 54.

Wu, T-W., and Tinker, D. O. (1969). *Biochemistry* **8,** 1558.

Yagi, T., and Benson, A. A. (1962). *Biochim. Biophys. Acta* **57,** 601.

Yaguchi, M., Tarassuk, N. P., and Abe, N. (1964). *J. Dairy Sci.* **47,** 1167.

Yamada, M. (1957). *Sci. Pap. Coll. Gen. Educ., Univ. Tokyo* **7,** 97.

Yamamoto, M., and Drummond, G. I. (1967). *Amer. J. Physiol.* **213,** 1365.

Yang, S.-F. (1969). *In* "Methods in Enzymology" (J. M. Lovenstein, ed.), Vol. 14, p. 208. Academic Press, New York.

Yang, S.-F., Freer, S., and Benson, A. A. (1967). *J. Biol. Chem.* **242,** 477.

Yavin, E., and Gatt, S. (1969). *Biochemistry* **8,** 1692.

Yurkowski, M., and Brockerhoff, H. (1965). *J. Fish. Res. Bd. Can.* **22,** 643.

Zeller, E. A. (1952). *Fed. Proc., Fed. Amer. Soc. Exp. Biol.* **11,** 316.

Zieve, F. J. (1973). *Anal. Biochem.* **51,** 436.

Zieve, F. J., and Zieve, L. (1972). *Biochem. Biophys. Res. Commun.* **47,** 1480.

Zografi, G., Verger, R., and de Haas, G. H. (1971). *Chem. Phys. Lipids* **7,** 185.

Zwaal, R. F., Roelofsen, B., Comfurius, P., and van Deenen, L. L. M. (1971). *Biochim. Biophys. Acta* **233,** 474.

Subject Index

A

Achromobacter lipolyticum, lipase of, 165
ACTH, activation of tissue lipases by, 108
Activation
  of lipases, 111
  of lipoprotein lipase, 98, 103, 105
  of lysophospholipase, 261
  of phospholipases-3, 276
  of phospholipases-4, 285
Acyl isomerase
  in plant lipase, 108, 132
  in *Vernonia anthelmintica* lipase, 135
Acyl migration, 27, 28
Adipose tissue, 108–114, 114–115
  as source of lipase phosphatase, 113
  as source of lipoprotein lipase, 99
  as source of monoglyceride lipases, 171
Adrenals, as source of cholesterol esterase, 189
Albumin, effect of pancreatic lipase, 82
Alcohol esters, primary, hydrolysis with pancreatic lipase, 58
Amino acid composition
  of *Candida cylindracea* lipase, 153
  of pancreatic cholesterol esterases, in rat, 182
  of phospholipase-2, 211, 212, 214
  of rat pancreatic lipase, 43
  of *Rhizopus arrhizus* lipase, 147
Anaerovibrio lipolytica, lipase of, 166
Anemone, lipases in, 93
Aorta tissue, as source of lipoprotein lipase, 99

Apolipoproteins, defined, 96
Arterial tissue
  as source of cholesterol esterase, 191
  as source of sphingomyelinase, 280
Aspergillus niger, lipase of,
Assay
  of cholesterol esterase, 177–178
  of lipoprotein lipase, 97–99
  of lysophospholipase, 257–258
  of milk lipase, 120
  of phospholipase-1, 246–248
  of phospholipase-2, 199–201
  of phospholipase-3, 270–271
  of phospholipase-4, 283–284
  of sphingomyelinase, 281
  of substrates for, 25–27
  of tissue lipases, 109

B

Bacterial lipases, 157–168
  Achromobacter lipolyticum, 165
  Anaerovibrio lipolytica, 166
  Corynebacterium acnes, 164–165
  Leptospirae, 157–158
  Mycoplasma, 166
  Propionibacterium shermanii, 165
  Pseudomonads, 158–161
  Staphylococci, 161–164
Bile salts
  effect on pancreatic lipolysis, 77–82
Blabenus cranifer,
  lipase of, 94
Brain
  as source of cholesterol esterase, 190
  as source of monoglyceride lipases, in pig, 173

325

A 4
B 5
C 6
D 7
E 8
F 9
G 0
H 1
I 2
J 3